Communications
in Computer and Information Science 129

Joaquim Filipe   Ana Fred   Bernadette Sharp (Eds.)

Agents and Artificial Intelligence

Second International Conference, ICAART 2010
Valencia, Spain, January 22-24, 2010
Revised Selected Papers

Volume Editors

Joaquim Filipe
INSTICC and IPS
Estefanilha, Setúbal, Portugal
E-mail: joaquim.filipe@estsetubal.ips.pt

Ana Fred
IST - Technical University of Lisbon
Lisbon, Portugal
E-mail: afred@lx.it.pt

Bernadette Sharp
Staffordshire University
Beaconside, Stafford, UK
E-mail: b.sharp@staffs.ac.uk

ISSN 1865-0929
e-ISSN 1865-0937
ISBN 978-3-642-19889-2
e-ISBN 978-3-642-19890-8
DOI 10.1007/978-3-642-19890-8
Springer Heidelberg Dordrecht London New York

Library of Congress Control Number: 2011923680

CR Subject Classification (1998): I.2, I.2.6, H.3-5, C.2, F.1, J.1

Typesetting: Camera-ready by author, data conversion by Scientific Publishing Services, Chennai, India

Printed on acid-free paper

Springer is part of Springer Science+Business Media (www.springer.com)

Preface

The present book includes a set of selected papers from the Second International Conference on Agents and Artificial Intelligence (ICAART 2010), held in Valencia, Spain, during January 22–24, 2010. The conference was organized in two simultaneous tracks: Artificial Intelligence and Agents. The book is based on the same structure.

ICAART 2010 received 364 paper submissions, from 58 different countries in all continents. From these, after a blind review process, only 31 were accepted as full papers, of which 17 were selected for inclusion in this book, based on the classifications provided by the Program Committee. The selected papers reflect the interdisciplinary nature of the conference. The diversity of topics is an important feature of this conference, enabling an overall perception of several important scientific and technological trends. These high-quality standards will be maintained and reinforced at ICAART 2011, to be held in Rome, Italy, and in future editions of this conference.

Furthermore, ICAART 2010 included five plenary keynote lectures given by Yves Demazeau (Laboratoire d'Informatique de Grenoble), Tim Finin (University of Maryland), Vicent J. Botti i Navarro (Universidad Politécnica de Valencia), Peter D. Karp (Bioinformatics Research Group within the Artificial Intelligence Center at SRI International) and Amilcar Cardoso (University of Coimbra). We would like to express our appreciation to all of them and in particular to those who took the time to contribute with a paper to this book.

On behalf of the conference Organizing Committee, we would like to thank all participants. First of all the authors, whose quality work is the essence of the conference and the members of the Program Committee, who helped us with their expertise and diligence in reviewing the papers. As we all know, producing a conference requires the effort of many individuals. We wish to thank also all the members of our Organizing Committee, whose work and commitment were invaluable.

July 2010

Joaquim Filipe
Ana Fred
Bernadette Sharp

Conference Committee

Conference Co-chairs

Joaquim Filipe — Polytechnic Institute of Setúbal / INSTICC, Portugal
Ana Fred — Technical University of Lisbon / IT, Portugal

Program Chair

Bernadette Sharp — Staffordshire University, UK

Organizing Committee

Patrícia Alves — INSTICC, Portugal
Sérgio Brissos — INSTICC, Portugal
Helder Coelhas — INSTICC, Portugal
Andreia Costa — INSTICC, Portugal
Bruno Encarnação — INSTICC, Portugal
Bárbara Lima — INSTICC, Portugal
Liliana Medina — INSTICC, Portugal
Elton Mendes — INSTICC, Portugal
Carla Mota — INSTICC, Portugal
Vitor Pedrosa — INSTICC, Portugal
José Varela — INSTICC, Portugal
Pedro Varela — INSTICC, Portugal

Program Committee

Stéphane Airiau, The Netherlands
Natasha Alechina, UK
Sander Bohte, The Netherlands
Luigi Ceccaroni, Spain
Berndt Farwer, UK
Nicoletta Fornara, Switzerland
Cees Witteveen, The Netherlands
Thomas Ågotnes, Norway
Rudolf Albrecht, Austria
Cesar Analide, Portugal
Andreas Andreou, Cyprus
Ioannis N. Athanasiadis, Switzerland
Antonio Bahamonde, Spain
Matteo Baldoni, Italy
Bikramjit Banerjee, USA
Ana Lúcia C. Bazzan, Brazil
Orlando Belo, Portugal
Carole Bernon, France
Daniel Berrar, UK
Ateet Bhalla, India
Olivier Boissier, France
Tibor Bosse, The Netherlands
Luis Botelho, Portugal
Danielle Boulanger, France

Paolo Bresciani, Belgium
Egon L. van den Broek, The Netherlands
Silvia Calegari, Italy
Valérie Camps, France
Jose Jesus Castro-schez, Spain
Radovan Cervenka, Slovak Republic
Georgios Chalkiadakis, UK
Shyi-Ming Chen, Taiwan
Carlos Iván Chesñevar, Argentina
Adam Cheyer, USA
Paolo Ciancarini, Italy
Diane Cook, USA
Paulo Cortez, Portugal
Jacob W. Crandall, UAE
Yves Demazeau, France
Andreas Dengel, Germany
Gaël Dias, Portugal
Ian Dickinson, UK
Edith Elkind, Singapore
Stefano Ferilli, Italy
Antonio Fernández-caballero, Spain
Edilson Ferneda, Brazil
Dariusz Frejlichowski, Poland
Naoki Fukuta, Japan
Wai-Keung Fung, Canada
Arnulfo Alanis Garza, Mexico
Chiara Ghidini, Italy
Joseph Giampapa, USA
Marie-Pierre Gleizes, France
Dominic Greenwood, Switzerland
Sven Groppe, Germany
Zahia Guessoum, France
Renata Guizzardi, Brazil
Andreas Herzig, France
Wladyslaw Homenda, Poland
Wei-Chiang Hong, Taiwan
Mark Hoogendoorn, The Netherlands
Eyke Hüllermeier, Germany
Carlos Iglesias, Spain
François Jacquenet, France
Yanguo Jing, UK
W. Lewis Johnson, USA
Catholijn Jonker, The Netherlands
Lalana Kagal, USA
Matthias Klusch, Germany
Joanna Kolodziej, Poland
Sébastien Konieczny, France
Igor Kotenko, Russian Federation
Stan Kurkovsky, USA
Nuno Lau, Portugal
Ho-fung Leung, Hong Kong
Baoding Liu, China
Chao-Lin Liu, Taiwan
Stephane Loiseau, France
Gabriel Pereira Lopes, Portugal
Hongen Lu, Australia
Bernd Ludwig, Germany
José Machado, Portugal
Nadia Magnenat-thalmann, Switzerland
Pierre Maret, France
Tokuro Matsuo, Japan
Nicola Di Mauro, Italy
Amnon Meisels, Israel
Benito Mendoza, USA
Daniel Merkle, Denmark
Bernd Meyer, Australia
Simon Miles, UK
Pavlos Moraitis, France
Thierry Moyaux, France
Tracy Mullen, USA
Radhakrishnan Nagarajan, USA
Tomoharu Nakashima, Japan
Paulo Novais, Portugal
Charlie L. Ortiz, USA
Jeng-Shyang Pan, Taiwan
Oscar Pastor, Spain
Krzysztof Patan, Poland
Catherine Pelachaud, France
Wojciech Penczek, Poland
Dana Petcu, Romania
Steve Phelps, UK
Rui Prada, Portugal
Rong Qu, UK
T. Ramayah, Malaysia
Martin Rehák, Czech Republic
Luís Paulo Reis, Portugal
Daniel Rodriguez, Spain
Pilar Rodriguez, Spain

Rosaldo Rossetti, Portugal
Alessandro Saffiotti, Sweden
Manuel Filipe Santos, Portugal
Jurek Sasiadek, Canada
Camilla Schwind, France
Candace L. Sidner, USA
Peer Olaf Siebers, UK
Viviane Silva, Brazil
Adam Slowik, Poland
Alexander Smirnov, Russian Federation
Safeeullah Soomro, Saudi Arabia
Kostas Stathis, UK
Kathleen Steinhofel, UK
Thomas Stützle, Belgium
Chun-Yi Su, Canada
Shiliang Sun, China
Ryszard Tadeusiewicz, Poland
Paola Turci, Italy
Franco Turini, Italy
Paulo Urbano, Portugal
Ioannis Vlahavas, Greece
Wutnipong Warakraisawad, Thailand
Bozena Wozna-Szczesniak, Poland
Seiji Yamada, Japan
Xin-She Yang, UK
Huiyu Zhou, UK

Auxiliary Reviewers

Juan José Del Coz, Spain
Jorge Diez, Spain
Alessandro Farinelli, Italy
Ya Gao, China
Sergio Alejandro Gómez, Argentina
Tal Grinshpoun, Israel
Rongqing Huang, China
You Ji, China
Carson K. Leung, Canada
Jinbo Li, China
Oscar Luaces, Spain
Pedro Moreira, Portugal
Maxime Morge, France
Matthijs Pontier, The Netherlands
José Ramón Quevedo, Spain
Daniel Castro Silva, Portugal
Frans Van der Sluis, The Netherlands
Jung-woo Sohn, USA
Wenting Tu, China
Sylvain Videau, France
Perukrishnen Vytelingum, UK
Zhijie Xu, China

Invited Speakers

Yves Demazeau — Laboratoire d'Informatique de Grenoble, France
Tim Finin — University of Maryland, USA
Vicent J. Botti i Navarro — Universidad Politécnica de Valencia, Spain
Peter D. Karp — Bioinformatics Research Group, Artificial Intelligence Center, SRI International, USA
Amilcar Cardoso — University of Coimbra, Portugal

Table of Contents

Invited Paper

Part I: Artificial Intelligence

Part II: Agents

Invited Papers

Creating and Exploiting a Hybrid Knowledge Base for Linked Data

Zareen Syed and Tim Finin

University of Maryland, Baltimore County
Baltimore MD 21250, U.S.A.
zarsyed1@umbc.edu, finin@cs.umbc.edu

Abstract. Twenty years ago Tim Berners-Lee proposed a distributed hypertext system based on standard Internet protocols. The Web that resulted fundamentally changed the ways we share information and services, both on the public Internet and within organizations. That original proposal contained the seeds of another effort that has not yet fully blossomed: a Semantic Web designed to enable computer programs to share and understand structured and semi-structured information easily. We will review the evolution of the idea and technologies to realize a Web of Data and describe how we are exploiting them to enhance information retrieval and information extraction. A key resource in our work is Wikitology, a hybrid knowledge base of structured and unstructured information extracted from Wikipedia.

Keywords: Semantic web, Wikipedia, Information extraction, Knowledge base, Linked data.

1 Introduction

Twenty-five years ago Doug Lenat started a project to develop Cyc [1] as a broad knowledge base filled with the common sense knowledge and information needed to support a new generation of intelligent systems. The project was visionary and ambitious, requiring broad ontological scope as well as detailed encyclopedia information about the world. While the research and development around Cyc has contributed much to our understanding of building complex, large scale knowledge representation and reasoning systems relatively few applications have been built that exploit it.

Not long after the Cyc project began, Tim Berners-Lee proposed a distributed hypertext system based on standard Internet protocols [2]. The Web that resulted fundamentally changed the ways we share information and services, both on the public Internet and within organizations. One success story is Wikipedia, the familiar Web-based, collaborative encyclopedia comprising millions of articles in dozens of languages. The original Web proposal contained the seeds of another effort that is beginning to gain traction: a Semantic Web designed to enable computer programs to share and understand structured and semi-structured information easily as a Web of Data.

Resources like Wikipedia and the Semantic Web's Linked Open Data [3] are now being integrated to provide experimental knowledge bases containing both general purpose knowledge as well as a host of specific facts about significant people, places, organizations, events and many other entities of interest. The results are finding immediate

J. Filipe, A. Fred, and B. Sharp (Eds.): ICAART 2010, CCIS 129, pp. 3–21, 2011.

applications in many areas, including improving information retrieval, text mining, and information extraction. As a motivating example, consider the problem of processing a mention of entity in the text of a newspaper article such as Michael Jordan. There are many people named Michael Jordan and while one, the former basketball star, is currently famous, many of the others appear in newspaper articles as well. Current information extraction systems are good at identifying mentions of named entities and even recognizing the set of them (e.g., Mr. Jordan, he, Jordan) as co-referent within a document. A more challenging problem is to predict when named entities in different documents co-refer or to link mentions to entities in a knowledge base.

An ultimate goal for many text mining applications is to map the discovered entities and the information learned about them from the documents to an instance in a knowledge base. If that can be done reliably it is easy to imagine building systems that can read documents like newspaper articles and build and populate a knowledge base of basic facts about the named entities found in them: people, organizations, places, and even products and services.

We have been exploring the use of Web-derived knowledge bases through the development of Wikitology [4] - a hybrid knowledge base of structured and unstructured information extracted from Wikipedia augmented by RDF data from DBpedia and other Linked Open Data resources. Wikitology is not unique in using Wikipedia to form the backbone of a knowledge base, see [5] and [6] for examples, however, it is unique in incorporating and integrating structured, semi-structured and unstructured information accessible through a single query interface. The core idea exploited by most approaches is to use references to Wikipedia articles and categories as terms in an ontology. For example, the reference to the Wikipedia page on weapons of mass destruction can be used to represent the WMD concept and the page on Alan Turing that individual person.

These basic Wikipedia pages are further augmented by category pages (e.g., biological weapons) representing concepts associated with articles and other categories. Finally, the Wikipedia pages are rich with other data that has semantic impact, including links to and from other Wikipedia articles, links to disambiguation pages, redirection links, in and out-links from the external Web, popularity values computed by search engines, and history pages indicating when and how often a page has been edited. Wikipedia's infobox structure flesh out the nascent ontology with links corresponding to properties or attributes.

There are many advantages in deriving a knowledge base from Wikipedia. The current English Wikipedia has over three million articles and 200 thousand categories, resulting in a concept space developed by consensus and exhibiting broad coverage. The ontology concepts are kept current and maintained by a diverse collection of people who volunteer their effort. Evaluations have shown that the quality of the content is high [7]. The intended meaning of the pages, as concepts, is self evident to humans, who can read the text and examine the images on the pages. Using text similarity algorithms and other techniques, it is easy to associate a set of Wikipedia pages with a short or long piece of text. Finally, Wikipedia exists in many languages with links between articles that are intended to denote the same thing, providing opportunities to use it in applications involving multiple languages.

In the remainder of this paper we review the evolution of our hybrid Wikitology system and some of the applications used to evaluate its utility. We conclude with a brief section summarizing our approach, sketching our ongoing work and speculating on the relationship between a knowledge base and an information extraction system.

2 Wikitology

World knowledge may be available in different forms such as relational databases, triple stores, link graphs, meta-data and free text. Human minds are capable of understanding and reasoning over knowledge represented in different ways and are influenced by different social, contextual and environmental factors. By following a similar model, we can integrate a variety of knowledge sources in a novel way to produce a single hybrid knowledge base enabling applications to better access and exploit knowledge hidden in different forms.

Some applications may require querying knowledge in the form of a relational database or triple store, whereas, others might need to process free text or benefit from exploiting knowledge in multiple forms using several complex algorithms at the same time. For example, in order to exploit the knowledge available in free text and knowledge available in a triple store, a possible approach is to convert one form into another, such as using natural language processing techniques to extract triples from free text to populate a knowledge base. This will enable the applications to query the knowledge available in free text and the triple store using SPARQL-like [8] queries.

Populating a knowledge base with information extracted from text is still an open problem and active research is taking place in this direction. Another approach is to augment the knowledge available in the form of free text with triples from the knowledge base, enabling the applications to access the knowledge by submitting free text queries to an information retrieval index. However, in this case we will lose much of the information that is available through the highly structured triple representation and other benefits such as reasoning over the knowledge. We approach this problem in a different way and favor an approach that does not depend on converting one form of data into another and benefits from the hybrid nature of the data that is available in different forms.

Wikipedia proves to be an invaluable resource for generating a hybrid knowledge base due to the availability and interlinking of structured, semi-structured and unstructured encyclopedic information. However, Wikipedia is designed in a way that facilitates human understanding and contribution by providing interlinking of articles and categories for better browsing and search of information, making the content easily understandable to humans but requiring intelligent approaches for being exploited by applications directly.

Wikipedia has structured knowledge available in the form of database tables with metadata, inter-article links, category hierarchy, article-category links and infoboxes whereas, unstructured knowledge in the form of free text of articles. Infoboxes are the most structured form of information and are composed of a set of subject-attribute-value triples that summarize the key features of the concept or subject of the article. Resources like DBpedia [9] and Freebase [10] have harvested this structured data and have made it available as triples for semantic querying. While infoboxes are a readily available source

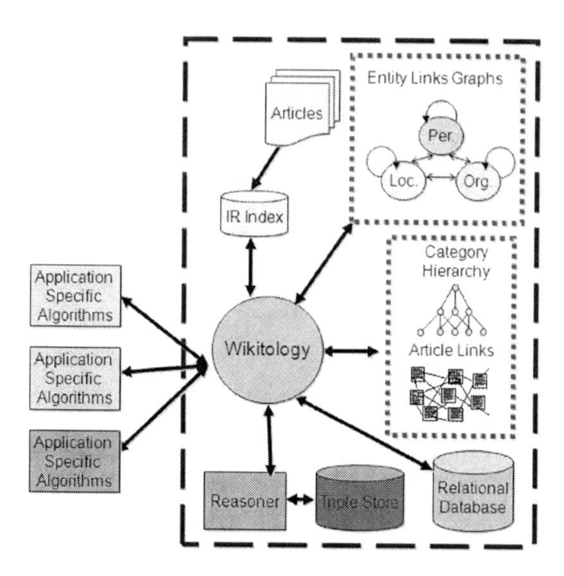

Fig. 1. Wikitology is a hybrid knowledge based storing information in structured and unstructured forms and reasoning over it using a variety of techniques

of structured data, the free text of the article contains much more information about the concept. Exploiting both the structured, semi-structured and unstructured information in a useful way will expose greater knowledge hidden in different forms to applications.

Since it might not always be practical to convert knowledge available in one form to another, we favor keeping the knowledge in the form it is available in the real world however, in order to integrate it we need to provide meaningful links between data available in different forms. Our approach is different from the Linked Open Data [11] community as it is targeted towards linking triples whereas in our approach we are interested in linking data available in different forms either free text, semi structured data, relational tables, graphs or triples and providing an integrated interface for applications needing access to the knowledge base. An overview of our system is given in Figure 1.

Our Wikitology knowledge base includes data structures and algorithms that enable applications to query knowledge available in different forms whether an IR index, graphs (category links, page links and entity links), relational database and a triple store. Different applications can use application specific algorithms to exploit the Wikitology knowledge base in different ways.

2.1 The Hybrid Knowledge Base

One of the challenges in developing a hybrid knowledge base is the selection of appropriate data structures to represent or link data available in different forms and provide a query interface giving an integrated view. For Wikipedia, most of the knowledge is available in the form of natural language text. Approaches for indexing and querying IR indices are more efficient and scalable as compared to triple stores. Therefore, an information retrieval (IR) index is our basic information substrate. To integrate it with

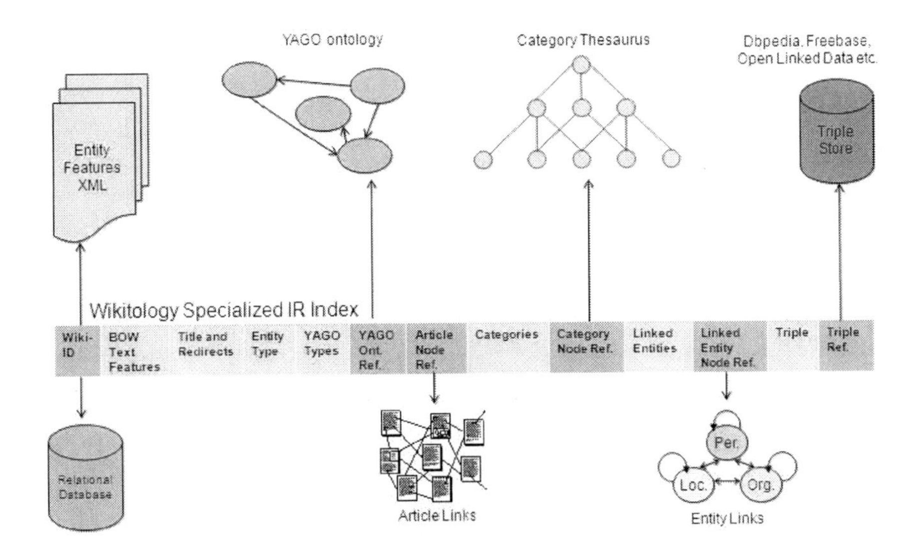

Fig. 2. Wikitology uses a specialized information retrieval index comprising text, instance fields and reference fields with references to data available in other data structures

other forms of knowledge, we enhance the IR index with fields containing instance data taken from other data structures and links to related information available in other data structures such as graphs or triples in an RDF triple store.

Using a specialized IR index enables applications to query the knowledge base using either simple free text queries or complex queries over multiple index fields. The presence of reference fields (having references to related instances in different data structures) enables the applications to exploit and harvest related information available in other forms by running data structure specific algorithms as well. An overview of our specialized IR index is given in Figure 2 with instance fields and reference fields.

The first field is the *Wiki-Id* field which has the ID associated with each Wikipedia article or concept in Wikipedia, this ID field helps in referencing related data in Wikipedia MySQL tables as well as other tables generated having relevant information to the concept, for example, we generated a table with disambiguation entries for a concept which can be referenced using the concept ID. The *Wiki-Id* field can also be used to refer to Entity Features XML documents. These documents allow us to represent and incorporate computed features that we do not want to add as separate index fields but would like to have available to applications.

The second field is the *BOW Text features* field which contains the bag of words text features extracted from Wikipedia articles.

The *Title and Redirects* field contains the titles of concepts and redirects to those concepts in Wikipedia. The Entity Type field currently contains entity types such as Person, Location or Organization for Wikipedia concepts denoting entities. We used the Freebase resource [10] to label Wikipedia articles on entities as persons, locations and organizations. We plan to add other entity types such as products, diseases etc. in the future.

We imported structured data in RDF from the Yago ontology [12]. The structured data was encoded in an RDFa-like format in the *YAGO Type* field for the Wikipedia page. This enables one to query the Wikitology knowledge base using both text (e.g., an entity document) and structured constraints (e.g., rdfs:type = yago:President). We also have a reference field *YAGO Ont. Ref.* to reference the particular node in the YAGO ontology in case the applications need to process information in the ontology.

The *Article Node Ref.* field contains a link to the article or concept node in the article links graph derived from Wikipedia inter-article links. This reference enables applications to refer to particular concept nodes and run graph algorithms on Wikipedia article links graph.

The *Categories* field contains the list of associated categories with the Wikipedia article and the *Category Node Ref.* field contains references to the specific nodes representing the category in the Category Thesaurus.

The *Linked Entities* field lists the Persons, Locations and Organizations linked with this concept encoded in an RDFa-like format (e.g., LinkedPerson = Michelle_Obama, LinkedLocation = Chicago). This enables the applications to use information about associated entities for tasks like named entity disambiguation. We derived the linked entities information using the inter-article links in Wikipedia. We extracted the links between articles that were on persons, locations and organizations that we labeled earlier using the Freebase resource. We generated person, location and organization graphs with links between entities of each type along with a general graph having interlinks between these individual graphs. We plan to use these graphs for entity linking and named entity disambiguation tasks in the future.

The *Triple* field currently contains the triples or properties present in Infoboxes in Wikipedia articles and the *Triple Ref.* field contains references to respective triples in the DBpedia triple store. We can also add related triples from other sources in this field.

The specialized IR index enables applications to query the knowledge base using either simple free text queries or complex queries over multiple fields with or without structured constraints and the presence of reference fields enables the applications to exploit and harvest related information available in other forms by running data structure specific algorithms.

Initial work done in this direction resulted in the development of our Wikitology 1.0 system which was a blend of the statistical and ontological approach for predicting concepts represented in documents [13]. Our algorithm first queries the specialized index and then exploits the page links and category links graphs using spreading activation algorithm for predicting document concepts.

The results of our first system were quite encouraging which motivated us to enhance the original system and incorporate knowledge from other knowledge resources and develop Wikitology 2.0 system which was employed and evaluated for cross-document entity co-reference resolution task [14].

Developing the novel hybrid knowledge base and evaluating it using different real world applications will allow us to improve the knowledge base in order to better serve the needs of variety of applications especially in the Information Extraction domain.

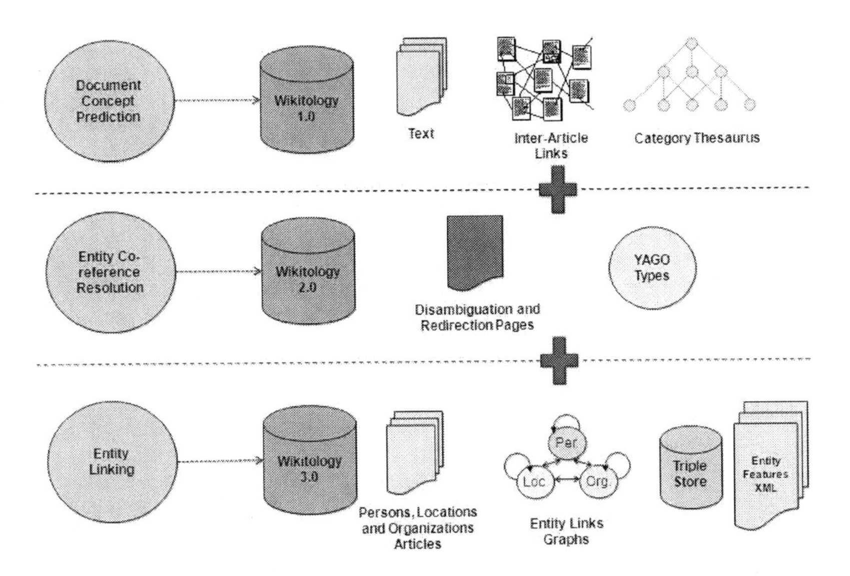

Fig. 3. The Wikitology system has evolved over the course of our research. As we used the system in different applications we enhanced if by adding data structures and knowledge sources and the reasoning mechanisms needed to support them.

We are currently working on our Wikitology 3.0 system which is targeted towards entity linking task. A comparison of the enhancements done in different versions of Wikitology and the applications that exploited our system is given in Figure 3.

2.2 Enriching the Knowledge Base

Another important challenge regarding a knowledge base is developing well defined approaches of enriching the knowledge base with new information so that it can be kept current when new information becomes available. The three tasks i.e., document concept prediction, cross document co-reference resolution and entity linking not only provide a way to evaluate our knowledge base but also directly contribute to enriching the knowledge base itself.

For example, to add new documents into the knowledge base we can use our Wikitology 1.0 system which predicts document concepts and links documents to appropriate articles and categories within Wikipedia. The same algorithm can suggest new categories in Wikipedia defined as a union of articles within Wikipedia.

Cross document co-reference resolution can also contribute to enriching the knowledge base, for example, knowing which documents are related to the same entity can provide additional information about that entity and the knowledge base could be updated and kept current by keeping references and links between articles that are related to a particular entity. For example, the knowledge base could be enriched by adding news articles and cross document entity co-reference resolution can help in linking articles talking about the same entities.

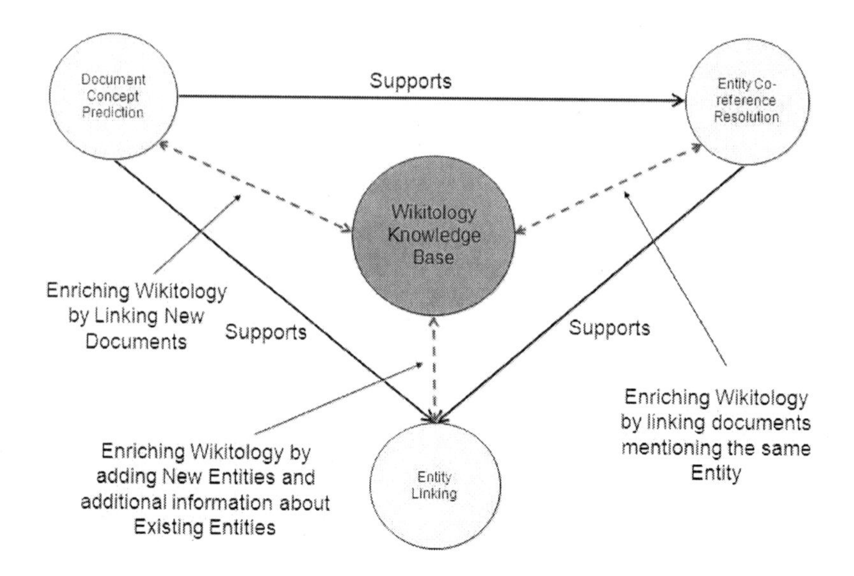

Fig. 4. Document concept prediction, cross document entity co-reference resolution, and entity linking are three tasks support each other and enrich Wikitology

Another way to enrich the knowledge base is to directly link the entities in the articles to relevant entities in the knowledge base and in case the entity doesn't exist in the knowledge base then the entity could be added, thereby enriching the knowledge base with new entities. These three applications also support each other. We used an enhanced system for document concept prediction to support cross document entity co-reference resolution in our Wikitology 2.0 system. Both document concept prediction and cross document co-reference resolution systems can also directly support entity linking task (Figure 4). We are currently working on our Wikitology 3.0 system which is targeted towards entity linking task and plan to incorporate document concept prediction and co-reference resolution for the entity linking task.

The next section gives a brief review of our earlier work on the Wikitology 1.0 and 2.0 systems. We then discuss our ongoing work on Wikitology 3.0 system and how we are using it to address the entity linking problem.

3 Wikitology 1.0

The first version of our Wikitology hybrid knowledge base, Wikitology 1.0 [13] used structured information in the form of a loose ontology and unstructured information in the form of free text in Wikipedia articles. Wikipedia articles served as specialized concepts whereas, the Wikipedia categories served as generalized concepts with inter-article, article-category and inter-category links representing relations between concepts. These concepts interlinked with each other exposed knowledge in the form of loose concept ontology. The article text served as a way to map free text to terms in the ontology i.e. article titles and categories. We developed algorithms to select, rank and aggregate concepts using the hybrid knowledge base.

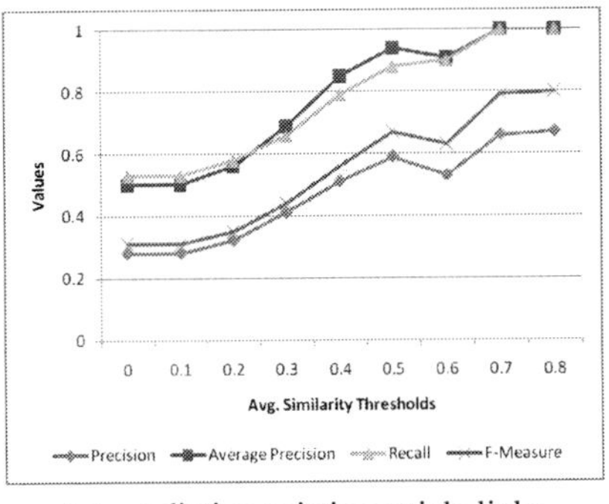

(a) predicting existing article links

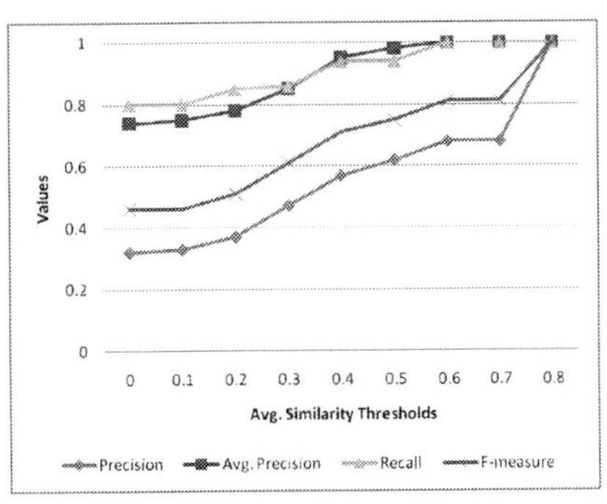

(b) predicting existing categories

Fig. 5. Using a high similarity threshold, Wikitology can predict article links and categories of existing Wikipedia articles with good results

We used the initial version of Wikitology [13] to predict individual document topics as well as concepts common to a set of documents. Algorithms were implemented and evaluated to aggregate and refine results, including using spreading activation to select the most appropriate terms. Spreading activation is a technique that has been widely adopted for associative retrieval [15]. In associative retrieval the idea is that it is possible to retrieve relevant documents if they are associated with other documents that have been considered relevant by the user. In Wikipedia the links between articles show association between concepts of articles and hence can be used as such for finding related concepts to a given concept. The algorithm starts with a set of activated nodes and in each iteration the activation of nodes is spread to associated nodes. The spread of activation may be directed by addition of different constraints like distance constraints, fan out constraints, path constraints, threshold etc. These parameters are mostly domain specific.

Our Wikitology 1.0 system took as input a document or set of documents, retrieved top N similar Wikipedia articles from the Wikitology index and used them as seed nodes for spreading activation on the page links graph to predict specialized concepts related to the input document(s). The categories associated with the top N similar articles were used as seed nodes for spreading activation on the category links graph for predicting generalized concepts for the input document(s). We used different activation functions for spreading activation on the page links graph and the category links graph.

We evaluated the system by predicting the categories and article links of existing Wikipedia articles and compared them with the ground truth. We then computed measures for precision, average precisions, recall and f-measure. As Figures 5(a) and 5(b) show, a higher value for average similarity between the test documents and the retrieved Wikipedia articles produces much better the prediction scores. While the Wikipedia category graph can be used to predict generalized concepts, the article links graph helped by predicting more specific concepts and concepts not in the category hierarchy. Our experiments show that it is possible to suggest new category concepts identified as a union of pages from the page link graph. Such predicted concepts could also be used to define new categories or sub-categories within Wikipedia.

⟨DOC⟩ ⟨DOCNO⟩ABC19980430.1830.0091.LDC2000T44-E2 ⟨DOCNO⟩ ⟨TEXT⟩
LONGEST MENTION: Webb Hubbell
TYPE: PER
SUBTYPE: Individual
NAM: "Hubbell" "Hubbells" "Webb Hubbell" "Webb_Hubbell"
NAM: "Mr ." "friend" "income"
PRO: "he" "him" "his"
CONTEXT: abc's accountant again ago alleges alone arranged attorney avoid being betray came cat charges cheating circle clearly close concluded conspiracy cooperate counsel counsel's department disgrace ... today ultimately vernon washington webb webb_hubbell white whitewater wife years
⟨/TEXT⟩ ⟨/DOC⟩

Fig. 6. The input to Wikitology was a special EDOC created by the output of the information extraction system for each entity in each document. This example is for an entity with longest mention *Webb Hubbell*. Only some of the context words, i.e., words near any of the entity's mentions, are shown.

4 Wikitology 2.0

We constructed an enhanced version of the Wikitology system as a knowledge base of known individuals and organizations as well as general concepts for use in the ACE cross document co-reference task. This was used as a component of a system developed by the JHU Human Language Technology Center of Excellence [16]. For our ACE task we enhanced Wikitology in several ways and added a custom query front end to better support the cross-document co-reference resolution task. Starting with the original Wikitology, we imported structured data in RDF from DBpedia [9], Freebase [10] and Yago ontology [6]. Most of the data in DBpedia and Freebase were in fact derived from Wikipedia, but have been mapped onto various Ontologies and re-imported in structured form. The structured data was encoded in an RDFa-like format in a separate field in the Lucene index object [17] for the Wikipedia page. This allows one to query Wikitology using both text (e.g., an entity document) and structured constraints (e.g., rdfs:type=yago:Person).

We enriched the text associated with each article with titles of Wikipedia redirects. A Wikipedia redirect page is a pseudo page with a title that is an alternate name or misspelling for the article. We extracted type information for people and organizations from the Freebase system. This information was stored in a separate database and used by the ACE Wikitology query system. We extracted data from Wikipedia's disambiguation pages to identify Wikitology terms that might be easily confused, e.g., the many people named Michael Jordan that are in Wikipedia. This information was stored in a separate table and used in the Wikitology feature computation for a feature indicating that two document entities do not refer to the same individual.

We used special entity documents, or EDOCs, extracted from the output of the Serif [18] information extraction system's APF output [19] to find candidate matches in Wikitology. Each entity in a given document produced one EDOC that included the longest entity mention, all name mentions, all nominal mentions, all pronominal mentions, APF type and subtype and all words within fifteen tokens of each mention. Figure 4 shows an example for a mention of a person *Webb Hubbell*.

The EDOCs were processed by a custom query module for Wikitology that mapped the information in the EDOC into different components of Wikitology entries. The EDOCs name mention strings are compared to the text in Wikitology's title field, giving a slightly higher weight to the longest mention. The EDOC type information is mapped into the Wikitology type information terms imported from DBpedia which are expressed using the Yago ontology and matched against the RDF field of each Wikitology entry. Finally the name mention strings along with contextual text surrounding the mentions are matched against the text of the Wikitology entries. The Wikitology module returns two vectors: one for matches against article entries and the other against category articles. We produced twelve features based on Wikitology: seven that were intended to measure similarity of a pair of entities and five to measure their dissimilarity. To analyze and evaluate our approach and Wikitology knowledge base we constructed a training set and a test set from the EDOCs for which human judgments were available for the cross-document entity co-reference task. The evaluation results, shown in table 1 are positive and show that the Wikitology knowledge base can be used effectively for cross-document entity co-reference resolution task with high accuracy.

Table 1. Evaluation results for cross-document entity co-reference task from the 2008 Automatic Content Extraction conference using Wikitology features

Match	True positive rate	False positive rate	Precision	Recall	F-measure
Yes	0.72	0.001	0.966	0.722	0.826
No	0.99	0.278	0.990	0.999	0.994

5 Wikitology 3.0

We are currently developing version 3.0 of Wikitology [4] by focusing on enhancements originally targeted toward supporting the Knowledge Base Population (KBP) track [20] of the 2009 Text Analysis Conference. One part of the KBP track is to link entities found in text to those in an external knowledge base. In particular, the entity linking task [21] is defined as: given an entity mention string and an article with that entity mention, find the link to the right Wikipedia entity if one exists. In the next section we describe our approach in detail.

5.1 Approach for Entity Linking

We used Freebase resource [10] to label Wikipedia articles on entities as persons, locations and organizations. Freebase is an online, collaborative database of general information containing facts in different categories such as people, locations, books, movies, music, companies, science etc. While of Freebase's data is extracted from Wikipedia, it contains significant amounts of independent data and metadata. We used Freebase to identify Wikipedia articles that were on Persons (550,000), Locations (350,000) and Organizations (13,000). We updated the Wikitology index and included a field for Entity Type. We implemented different querying approaches for our specialized Wikitology index which are described below.

In approach one we query the index using the entity mentions against the titles and redirects field and then search within the returned results by querying the given entity document against the Wikipedia article contents field. For the second approach, we query the index using the entity mentions against the titles and redirects field and the given entity document against the Wikipedia article contents field. As a third approach, we query the entity mentions against the title and redirects fields and then search within the returned results by querying the entity mentions against the title and redirects fields and the given entity document against the Wikipedia article contents field. In the fourth approach, we query using the entity mentions against the title and redirects fields and then search within the returned results by querying the given entity document against the Wikipedia article contents field and against the Yago types field and the entity mentions against the title, redirects and contents field with a boost of four.

The different approaches return a list of ranked Wikipedia entities and the topmost is selected as the correct match. To detect if the test entity is a new entity and doesn't exist in our Wikitology system we learn a threshold and if the score of the top most entity is below the threshold we report a NIL match (i.e., not in the knowledge base) or a New Entity.

Heuristics Based on Entity Types. We developed a few heuristics based on the type of entity for persons and locations. In case of persons, we used approach four however, if a name initial was present in the entity mention we included that in the query by introducing the initial followed by a wild card in the query to also match any entity names that contain the full name rather than the initial, this helped in giving a higher score to Chris Broad for the entity mention C Broad as compared to Dan Broad.

For locations, we replaced all adjectival forms of place names with noun forms by using the list of adjectival place names in Wikipedia so words like Australian would get replaced by Australia in the entity mention strings. We also used another heuristic for matching locations. In Wikipedia place names are often followed by names of locations that would help in disambiguating them. For example, Baltimore is a name of several places in Wikipedia. The different Baltimores are distinguished by having another place name in the title such as Baltimore, Ohio or Baltimore, Indiana. To exploit this information present in the titles of Wikipedia pages, we extracted all the locations from the test entity article and if any of the locations appeared in the title of the top five matching locations, that Wikipedia entity was selected as the matching entity.

5.2 Experiments and Evaluation

In order to evaluate the entity linking task we used the Wikinews corpus from October 2009 dump [22], which consists of news articles that are linked manually by contributors to relevant Wikipedia articles. We extracted all the links to Wikipedia articles and the surface text associated with the links as entity mentions. We created a test set of 1000 articles and entity mentions for persons, locations and organizations each by randomly selecting articles which had links to persons, locations and organizations in Wikipedia.

Evaluation for Existing KB Entities. We conducted different experiments to evaluate the different approaches. For experiments number one to five we used separate

Wikitology indices for Persons, Locations and Organizations whereas for experiment six we used the full index with Persons, Locations and Organizations without any information about the entity type of the test entity being queried. The entity type of an entity can be predicted using the entity mention and the related article by locating the entity mention in the article and using any named entity recognition system to label the type of entity. Once the entity type is known, the entity could be queried against the respective index. In case it is not possible to know the entity type in advance, then the query could be directed to the full index as we do in experiment six.

Table 2 reports the accuracy obtained for Persons, Locations and Organizations using the different approaches. We observed that amongst the four different approaches, approach four in general gave the highest accuracy for all the three entity types i.e. 95.2% , 86.2 % and 86.1% for Persons, Locations and Organizations respectively. The specialized approaches for Persons and Locations further improved the accuracy from 95.2% to 96.9% and from 86.2% to 93.3% for Persons and Locations. The improvement in accuracy was seen more in case of Locations as compared to Persons. Using the fourth approach on the full index with all the three types of entities resulted in a slight drop in the accuracy for all the three entity types as compared to when approach four is used on individual indices for the entity types.

Table 2. Accuracy obtained for Entity Linking Task for entities that exist in Wikipedia using different approaches

Experiment	Approach	Person	Location	Organization
1	Approach 1	66.8%	51.6%	56.8%
2	Approach 2	94.2%	85.0%	84.7%
3	Approach 3	95.1%	85.8%	85.5%
4	Approach 4	95.2%	86.2%	86.1%
5	Specialized Approach	96.9%	93.3%	–
6	Approach 4 (Full Index)	94.9%	85.0%	82.8%

Evaluation for New KB Entities. In order to detect if the entities are not present in Wikitology we wanted to learn a threshold that we could use to distinguish between an existing entity in the knowledge base and a new entity. Our approach for Entity Linking for existing entities is totally unsupervised, however, to detect new entities we use a very small set of labeled examples. We constructed a list of scores for positive and negative entity matches using the following approach. We used approach four to query Wikitology and retrieved top ranking entities using our test data of 1000 entity mentions and entity articles for people, locations and organizations each. In case the top most match was not the right one, we labeled that score as a negative match. If the top most match was a correct match we label the score as a positive match and the score of the second ranking entity as a negative match, because in case the first entity was not in the Wikitology knowledge base it would have predicted the second entity as the top most match which would be an example of negative match.

We used the decision tree algorithm in Weka [23] to learn a score to split the positive and negative matches. We learned the threshold using only 5% of the data as the training set and then tested it using the 95% of the remaining scores. Table 3 shows the accuracy obtained for the positive and negative entity matches using the learned thresholds for people, locations and organizations dataset separately and then for the combined dataset.

The highest accuracy for predicting a positive and negative match using thresholds separately for each type of entity was for persons (88.1% accuracy) with positive matches being predicted more accurately (92.1%) as compared to negative matches (84.0%), followed by organizations (79.0% accuracy) in which case the accuracy for predicting a positive match (67.8%) was much lower than for predicting a negative match (92.2%). For locations (71.8%) the accuracy for predicting a positive match (83.4%) was higher than for predicting a negative match (59.8%). When a single threshold was used to detect positive and negative matches for the three type of entities, the overall accuracy was 79.9% with the accuracy of positive match being slightly higher than the accuracy for negative match prediction.

Table 3. Good accuracy was obtained for the positive and negative entity matches using the learned thresholds for people, locations and organizations dataset separately and for the combined dataset

Entity Type	Negative (match not in Knowledge Base)	Positive (match in Knowledge Base)	Combined Accuracy
Person	84.0 %	92.1%	88.1%
Location	59.8%	83.4%	71.8%
Organization	92.2%	67.8%	79.0%
All types	78.7%	81.1%	79.9%

5.3 Disambiguation Trees

We are developing an approach to disambiguate mentions that refer to a Wikipedia entity. Wikipedia has special, manually created disambiguation pages for sets of entities with identical or similar names. For example, the disambiguation page *Michael_Jackson_(Disambiguation)* lists 36 different entities to which the string Michael Jackson might refer. A short description is given for each, such as:

> "Michael Jackson (actor) (born 1970), Canadian actor, best known for his role as Trevor on Trailer Park Boys"

that identifies one or more facts that can help distinguish it from others in the set. Not all confusable entities have such disambiguation pages and an automated process for creating them could both contribute to Wikipedia and also support entity linking.

An initial prototype creates disambiguation pages for people. We modeled this problem as a multiclass classification problem where each person with a similar name is considered an individual class. We extract nouns and adjectives from the first sentence of the person articles and use them to construct a decision tree. Most commonly, the nodes that were selected to split on referred to either the persons nationality or profession.

We enhanced our approach by using a domain model [24] of nationalities and professions constructed from Wikipedia's list pages *list_of_nationalities* and *list_of_professions*. These were used to extract the nationality and profession of persons by selecting nouns and adjectives as features that appeared in the first sentence and were in one of the domain models. Using the nationality and profession as features, we constructed a decision tree using Weka [23] for different Persons having a confusable name. When we were not able to extract a profession or nationality of the entity from the first sentence, we gave that feature a value of zero. We refer to these decision trees that help in disambiguating entities as disambiguation trees.

We constructed several disambiguation trees for different sets of persons having the same name. Disambiguation trees constructed as a result of three of our experiments on persons having name Michael Jackson, Michael Jordan and George Harrison are shown in Tables 4, 5 and 6.

Table 4. This automatically generated disambiguation tree helps link the document mention *Michael Jackson* to the appropriate Wikitology entity using the extracted profession and nationality features

Profession = *musician*: Michael_Jackson
Profession = *0*
— Nationality = *english*: Michael_Jackson_(footballer)
— Nationality = *0*: Michael_Jackson_(Anglican_bishop)
Profession = *guitarist*: Michael_Gregory_(jazz_guitarist)
Profession = *sheriff*: Michael_A._Jackson_(sheriff)
Profession = *journalist*: Michael_Jackson_(journalist)
Profession = *player*: Michael_Jackson_(basketball)
Profession = *executive*: Michael_Jackson_(television_executive)
Profession = *writer*: Michael_Jackson_(writer)
Profession = *professor*: Michael_Jackson_(anthropologist)
Profession = *footballer*: Michael_Jackson_(rugby_league)
Profession = *scientist*: Michael_A._Jackson
Profession = *soldier*: Michael_Jackson_(American_Revolution)
Profession = *actor*
— Nationality = *english*: Michael_J._Jackson_(actor)
— Nationality = *canadian*: Michael_Jackson_(actor)

Out of 21 different people named Michael Jackson we were able to disambiguate 15 of them using just the profession and nationality features. For three of the remaining six we were unable to extract the profession and nationality features from the first sentence using our domain models either because the profession and nationality were not mentioned or our domain model did not contain a matching entry. We could generate a more comprehensive list of professions by adding new domain model entries using resources such as DBpedia [9], Freebase [10] and Yago ontology [6] or by extracting them from text using patterns, as described by Garera and Yarowsky [24] and McNamee et al. [21].

For one person we were not able to extract the profession feature and the nationality was not sufficient to distinguish it from others. For two of them the profession and the nationality were the same, e.g., *Michael_A._Jackson* and *Michael_C._Jackson* both are

British scientists and *Michael_Jackson_(basketball)* and *Michael_Jackson_(wide_recei ver)* are American professional athletes. The ambiguity can be reduced with a finer-grained professions hierarchy. This might help distinguish, for example, *Michael_A._ Jackson* the "Computer Scientist" and *Michael_C._Jackson* the "Systems Scientist". Similarly, *Michael_Jackson_(basket-ball)* is described as a Basketball Player whereas *Michael_Jackson_(wide_receiver)* is a Football Player.

Of six people named Michael Jordan, we were able to disambiguate five of them using the profession and nationality features. The one that could not be classified (*Mich ael_Jordan*) had the same profession and nationality as *Michael_ Hakim_ Jordan*. Both have the nationality American and profession as Player and at fine grained level both are Basketball Players. There is a need of additional features in order to disambiguate between them. Out of thirteen people named George Harrison we were able to disambiguate eight using the profession and nationality features. For five of them we were not able to extract the profession and nationality features from the first sentence.

Table 5. This automatically generated disambiguation tree helps distinguish and link *Michael Jordan* mentions

Profession = *politician*: Michael_Jordan_(Irish_politician)
Profession = *footballer*: Michael_Jordan_(footballer)
Profession = *player*: Michael-Hakim_Jordan
Profession = *researcher*: Michael_I._Jordan
Profession = *actor*: Michael_B._Jordan

Our disambiguation trees show that the profession and nationality features are very useful in disambiguating different entities with the same name in Wikipedia. We can extract the profession and nationality features from the first sentence in articles about people in most of the cases. From the profession and nationality attributes, profession is selected as the first node to split on by the decision tree algorithm in all of the disambiguation trees we discussed, which shows that the profession feature is more discriminatory and helpful in disambiguating people with the same name as compared to nationality of a person in Wikipedia. We are currently working on improving our approach by introducing a professions hierarchy in our disambiguation tree.

Table 6. A disambiguation tree was compiled for different persons named George Harrison in Wikipedia

Profession = *sailor*: George_H._Harrison
Profession = *swimmer*: George_Harrison_(swimmer)
Profession = *banker*: George_L._Harrison
Profession = *0*
— Nationality = *english*: George_Harrison_(civil_servant)
— Nationality = *irish*: George_Harrison_(Irish_Republican)
Profession = *guitarist*: George_Harrison
Profession = *president*: George_Harrison_(executive)
Profession = *editor*: Harrison_George

For Wikipedia articles about people, we were able to extract nationality and profession information from the first sentence in most of the cases. Since Wikipedia articles are encyclopedic in nature, their first sentence usually defines the entity and, if necessary, provides some disambiguating information. This is not likely to be the case in general for news articles and for entities mentioned in other Wikipedia articles that are not primarily about that entity. Extracting the profession and nationality of persons mentioned in non-encyclopedic articles is more challenging as the article might not directly state the profession and nationality of that person. We are developing approaches for extracting the profession and nationality of the person using other features. For example, places mentioned in the same article can offer evidence for nationality and verbs used with the person can suggest the profession.

6 Conclusions

A knowledge base incorporating information available in different forms can better meet the needs of real world applications than one focusing and exposing knowledge in a more restricted way such as through SQL, SPARQL or IR queries. Exploiting Wikipedia and related knowledge sources to develop a novel hybrid knowledge base brings advantages inherent to Wikipedia. Wikipedia provides a way to allow ordinary people to contribute knowledge as it is familiar and easy to use. This collaborative development process leads to a consensus model that is kept current and up-to-date and is also available in many languages. Incorporating these qualities in knowledge bases like Cyc [25] will be very expensive in terms of time, effort and cost.

Efforts like DBpedia, Freebase and Linked Open Data are focused on making knowledge available in structured forms. Our novel hybrid knowledge base can complement these valuable resources by integrating knowledge available in other forms and providing much more flexible access to knowledge. In the document concept prediction task, for example, we queried the specialized index available in Wikitology 1.0 and used the reference fields with links to relevant nodes in the page and category link graphs which were then analyzed by custom graph algorithms to predict document concepts. In the cross document entity co-reference resolution task, we distributed a complex query (EDOC) to different components in Wikitology 2.0 and then integrated the results to produce evidence for or against co-reference.

We have directly demonstrated through our earlier work that we can use world knowledge accessible through our Wikitology hybrid knowledge base system to go beyond the level of mere words and can predict the semantic concepts present in documents as well as resolve ambiguity in named entity recognition systems by mapping the entities mentioned in documents to unique entities in the real world. Our Wikitology knowledge base system can provide a way to access and utilize common-sense and background knowledge for solving real world problems.

Acknowledgements

This research was carried out with support from the Fulbright Foundation, Microsoft Research, NSF award IIS-0326460 and the Johns Hopkins University Human Language Technology Center of Excellence.

References

1. Lenat, D.B., Guha, R.V.: Building Large Knowledge-Based Systems; Representation and Inference in the Cyc Project. Addison-Wesley Longman Publishing Co., Inc., Boston (1989)
2. Berners-Lee, T.: Information management: A proposal. In: European Particle Physics Laboratory, CERN (1989) (unpublished report)
3. Bizer, C.: The emerging web of linked data. IEEE Intelligent Systems 24(5), 87–92 (2009)
4. Syed, Z., Finin, T.: Wikitology: A Wikipedia derived novel hybrid knowledge base. In: Grace Hopper Conference for Women in Computing (2009)
5. Wu, F., Weld, D.S.: Automatically refining the wikipedia infobox ontology. In: Proceeding of the 17th International Conference on World Wide Web, WWW 2008, pp. 635–644. ACM, New York (2008)
6. Suchanek, F.M., Kasneci, G., Weikum, G.: Yago: A large ontology from wikipedia and wordnet. Web Semant. 6(3), 203–217 (2008)
7. Hu, M., Lim, E.P., Sun, A., Lauw, H.W., Vuong, B.Q.: Measuring article quality in Wikipedia: models and evaluation. In: Proceedings of the Sixteenth ACM Conference on Conference on Information and Knowledge Management, CIKM 2007, pp. 243–252. ACM, New York (2007)
8. Prud'Hommeaux, E., Seaborne, A., et al.: SPARQL query language for RDF. W3C working draft 4 (2006)
9. Auer, S., Bizer, C., Kobilarov, G., Lehmann, J., Ives, Z.: Dbpedia: A nucleus for a web of open data. In: Aberer, K., Choi, K.-S., Noy, N., Allemang, D., Lee, K.-I., Nixon, L.J.B., Golbeck, J., Mika, P., Maynard, D., Mizoguchi, R., Schreiber, G., Cudré-Mauroux, P. (eds.) ASWC 2007 and ISWC 2007. LNCS, vol. 4825, pp. 722–735. Springer, Heidelberg (2007)
10. Bollacker, K., Evans, C., Paritosh, P., Sturge, T., Taylor, J.: Freebase: a collaboratively created graph database for structuring human knowledge. In: Proceedings of the 2008 ACM SIGMOD International Conference on Management of Data, SIGMOD 2008, pp. 1247–1250. ACM, New York (2008)
11. Bizer, C., Heath, T., Ayers, D., Raimond, Y.: Interlinking open data on the web. In: 4th European Semantic Web Conference (2007)
12. Suchanek, F.M., Kasneci, G., Weikum, G.: Yago: a core of semantic knowledge. In: Proceedings of the 16th International Conference on World Wide Web, p. 706. ACM, New York (2007)
13. Syed, Z., Finin, T., Joshi, A.: Wikipedia as an ontology for describing documents. In: Proceedings of the Second International Conference on Weblogs and Social Media. AAAI Press, Menlo Park (2008)
14. Finin, T., Syed, Z., Mayfield, J., McNamee, P., Piatko, C.: Using wikitology for cross-document entity coreference resolution. In: Proceedings of the AAAI Spring Symposium on Learning by Reading and Learning to Read. AAAI Press, Menlo Park (2009)
15. Crestani, F.: Application of spreading activation techniques in information retrieval. Artificial Intelligence Review 11(6), 453–482 (1997)
16. Mayfield, J., Alexander, D., Dorr, B., Eisner, J., Elsayed, T., Finin, T., Fink, C., Freedman, M., Garera, N., McNamee, P., et al.: Cross-Document Coreference Resolution: A Key Technology for Learning by Reading. In: AAAI 2009 Spring Symposium on Learning by Reading and Learning to Read (2009)
17. Hatcher, E., Gospodnetic, O.: Lucene in action. Manning Publications Co., Greenwich (2004)
18. Boschee, E., Weischedel, R., Zamanian, A.: Automatic Information Extraction. In: Proceedings of the 2005 International Conference on Intelligence Analysis, McLean, VA, pp. 2–4 (2005)

19. Doddington, G., Mitchell, A., Przybocki, M., Ramshaw, L., Strassel, S., Weischedel, R.: The automatic content extraction (ACE) program – tasks, data, and evaluation. In: Proceedings of the Language Resources and Evaluation Conference, pp. 837–840
20. McNamee, P., Dang, H.: Overview of the TAC 2009 knowledge base population track. In: Proceedings of the 2009 Text Analysis Conference, National Institute of Standards and Technology, Gaithersburg MD (2009)
21. McNamee, P., Dredze, M., Gerber, A., Garera, N., Finin, T., Mayfield, J., Piatko, C., Rao, D., Yarowsky, D., Dreyer, M.: HLTCOE approaches to knowledge base population at TAC 2009. In: Proceedings of the 2009 Text Analysis Conference, National Institute of Standards and Technology, Gaithersburg MD (2009)
22. Wikinews: Wikinews, the free news source, `http://en.wikinews.org/wiki` (accessed 2009)
23. Hall, M., Frank, E., Holmes, G., Pfahringer, B., Reutemann, P., Witten, I.H.: The weka data mining software: an update. SIGKDD Explorations 11(1), 10–18 (2009)
24. Garera, N., Yarowsky, D.: Structural, transitive and latent models for biographic fact extraction. In: Proceedings of the 12th Conference of the European Chapter of the Association for Computational Linguistics, EACL 2009, Morristown, NJ, USA, pp. 300–308. Association for Computational Linguistics (2009)
25. Lenat, D.B.: Cyc: a large-scale investment in knowledge infrastructure. ACM Commun. 38(11), 33–38 (1995)

Part I

Artificial Intelligence

Language Support to XML Data Mining: A Case Study

Andrea Romei and Franco Turini

Department of Computer Science, University of Pisa
Largo B. Pontecorvo, 3, Pisa 56127, Italy
{romei,turini}@di.unipi.it

Abstract. There are several reasons that justify the study of a powerful, expressive and efficient XML-based framework for intelligent data analysis. First of all, the proliferation of XML sources offer good opportunities to mine new data. Second, native XML databases appear to be a natural alternative to relational databases when the purpose is querying both data and the extracted models in an uniform manner. This work offers a new query language for XML Data Mining. In presenting the language, we show its versatility, expressiveness and efficiency by proposing a concise, yet comprehensive set of queries which cover the major aspects of the data mining. Queries are designed over the well-known `xmark` XML database, that is a scalable benchmark dataset modeling an Internet auction site.

Keywords: Data mining, Knowledge discovery, Inductive databases, XML, XQuery, Query language.

1 Introduction

Inductive Databases (IDBs) are general purpose databases in which both the data and the knowledge are represented, retrieved, and manipulated in an uniform way [1]. A critical aspect in IDBs is the choice of what kind of formalism is more suited to represent models, data sources, as well as the queries one might want to apply on them. A considerable number of different papers proposes to integrate a mining system with a relational DBMS. Relational databases are fine for storing data in a tabular form, but they are not well suited for representing large volumes of semi-structured data fields. However, data mining need not be necessarily supported by a relational DB.

The past few years have seen a growth in the adoption of the *eXtensible Markup Language* (XML). On the one hand, the flexible nature of XML makes it an ideal basis for defining arbitrary representation languages. One such example is the Predictive Modelling Markup Language (PMML) [2], an industry standard for the representation of mined models as XML documents. On the other hand, the increasing adoption of XML has also raised the challenge of mining large collections of semi-structured data, for example web pages, graphs, geographical information and so on. From the XML querying point of view, a relevant on-going effort of the W3C is the design of a standard query language for XML, called *XQuery* [3], which is drawing much research and for which a large number of implementations already exists.

J. Filipe, A. Fred, and B. Sharp (Eds.): ICAART 2010, CCIS 129, pp. 25–38, 2011.

The goal of our research is the design and implementation of a mining language in support of an IDB in which an XML native database is used as a storage for Knowledge Discovery in Databases (KDD) entities, while Data Mining (DM) tasks are expressed in an XQuery-like language, in the same way in which mining languages on relational databases are expressed in a SQL-like format. These efforts have permitted the definition of XQuake, a new query language for mining XML data [4,5]. Features of the language are the expressiveness and flexibility in specifying a variety of different mining tasks and a coherent formalism capable of dealing uniformly with raw data, induced knowledge and background knowledge.

As far as the frequent pattern mining task is concerned, a first group of experiments showing promising results has been presented in [4]. In this paper, we reinforce our claims by offering a set of queries where each query is intended to challenge a particular aspect of the data mining process, not only related to the extraction of frequent patterns. More important, such queries are designed over a scalable XML database, named `xmark`, frequently used by the database community as a real XML benchmark suite.

In short, after a brief overview of XQuake, we propose a concise, yet comprehensive, set of tasks which covers the major aspects of the data mining life cycle, ranging from pre-processing, via data mining to post-processing. We complement our research with results we obtained from running the set of queries on several scalable `xmark` documents. We shall assume that the reader is familiar with the fundamentals of XML, XQuery and the Data Mining as well.

2 XQuake

XQuake stands for XQUery-based Applications for Knowledge Extraction. Essentially, XQuery expressions are used to identify XML data as well as mining fields and metadata, to express constraints on the domain knowledge, to specify user preferences and the format of the XML output.

A mining query begins with a collection of XQuery functions and variable declarations followed by an XQuake operator. The syntax of each operator includes four basic statements: (i) *task and method specification*; (ii) *domain entities identification*; (iii) *exploitation of constraints and user preferences*; (iv) *output construction*. The outline of a generic operator is explained below.

2.1 Task and Method Specification

Each operator starts specifying the kind of KDD activity. Constructs for preprocessing, model extraction, knowledge filtering, model evaluation or model meta-reasoning are available. As an instance, the following XQuake fragment specifies a data sampling task:

prepare sampling `doc("my-out")` **using** `alg:my-sampling-alg(my-params ...).`

The `doc("my-out")` expression directs the result of the mining task to a specific native XML database for further processing or analysis. The `using` statement introduces the kind of mining or preprocessing algorithm used, together with atomic parameters.

2.2 Domain Entities Identification

Any KDD task may need to specify the set of relevant entities as input of the analysis. The syntax is an adaptation of the standard XQuery `FLWOR` syntax, in which the result of the evaluation of an expression is linked to a variable in `for` and `let` clauses. Below you can see a simple statement that can be used to locate input data.

```
for data $tuple in <XQuery expr>
let active field $field := <XQuery expr on $tuple>...
```

Input data is typically a sequence of XML nodes. The *<for>* expression above binds the variable `$tuple` to each item during the evaluation of the operator, while the *<let>* clause identifies an attribute of the data. The user specifies them through an XQuery condition that is typically processed before the mining task. The *<let>* clause defines a data attribute with name `$field`, whose values are obtained by means of the XQuery expression in the body and whose type is omitted. The keyword after the *<let>* refers to the role of such an attribute in the mining activity of interest. More specifically:

- *<active>* specifies that the field is used as an input to the analysis;
- *<predicted>* specifies that it is a prediction attribute;
- *<supplementary>* states that it holds additional descriptive information.

Mining fields in input to the mining task are required to be atomic, i.e. an instance of one of the built-in data types defined by XML schemas, such as string, integer, decimal and date. A richer set of types is included into the data model by extending the system type of XQuery to support discrete, ordering and cyclical data types. As an example, in the query fragment below, we are interested in the specification of an active mining attribute that indicates whether a person of the `xmark` document has a credit card (see Figure 2). An explicit boolean type is specified by the user for the field `$has-card`.

```
for data $person in doc("xmark")//person
let active field $has-card as xs:boolean := not(empty($person/payment)).
```

A special syntax is used to locate transaction data, that is a special type of record data where each record (transaction) involves a set of items[1]. The basic idea of XQuake is to bind a variable to each XML transaction and a new variable to each XML item of that transaction, as shown in the query fragment below, in which a special clause binds each XML transaction to the variable `auction` and each XML item to the variable `bidder` (see Figure 2). An additional attribute `name` captures the descriptive information of the items (i.e. the bidders).

```
for data $auction in doc("xmark")/site/open_auctions/open_auction
for item $bidder in $auction/bidder
let supplementary field $name := string($bidder/personref/@person).
```

From the mining models perspective, a similar syntax may be used to locate (parts of) a (new or extracted) mining model, represented via the international PMML standard [2].

[1] A typical example of transaction is the set of products purchased by a customer, whereas the individual products are the items.

2.3 Exploitation of Constraints and User Preferences

XQuake offers a special syntax to specify domain knowledge, i.e. the information provided by a domain expert about the domain to be mined, particularly useful for the definition of domain-based constraints. In contrast to active and predicted mining fields, a metadata field may include also non-atomic types, such as XML nodes or attributes. For example, below we assign the XML element `<profile>` to each bidder as a piece of metadata.

```
for data $auction in doc("xmark")/site/open_auctions/open_auction
for item $bidder in $auction/bidder
let metadata field $prof:= doc("xmark")//person[@id eq $bidder]/profile
```

The language also includes a *<having>* clause for complex querying on the domain knowledge as well as an easy and elegant way to express the user preferences. The key idea of XQuake is the integration of XQuery expressions inside an ad hoc custom statement that depends on the kind of target and on the kind of knowledge to be mined.

As an example, the *<mine itemsets>* operator, used for the extraction of frequent patterns, constrains the number of items of an itemset that satisfy a particular condition to comply with certain threshold. The *<having>* statement has the following format.

```
having at least < positive integer > item satisfies < XQuery predicate >
```

The operator *<at least>* (similar operators are *<at most>* and *<exactly>*) is true for all itemsets which have at least a specified number of items that satisfy the XQuery predicate. The latter one can be expressed on the variables previously defined that, in the example above, denote both the bidder's name (variable `$bidder`) and its profile (variable `$prof`). This paper reports several examples to show how the user can express personalized sophisticated constraints based on her/his domain knowledge.

2.4 Output Construction

XQuake offers a *<return>* clause with element and attribute constructors in order to produce an XML output, as it is usually done in XQuery. Again, the key idea is that such a clause can refer to all the variables defined in previous statements in order to customize the result.

However, we allow a customized output only for operators returning XML data (e.g. preprocessing operators), since we fix the output of the other kinds of operator to be a PMML model. From this perspective, in addition to ad-hoc constructs for the extraction of mining models, operators for pre-processing and for model application include (among the others):

- *<prepare discretization>* that addresses the problem of unsupervised discretization with the objective of producing k intervals of a numeric field according to a strategy.
- *<apply itemsets>* that provides the capability of connecting the extracted set of patterns with the data. Basically, it takes a PMML association model and a set of XML transactions as input. For each transaction, the operator yields the number of itemsets that are supported or violated by that transaction.

We refer the reader to [4] and [5] for the details about these operators and for an in-depth description of the system architecture of XQuake.

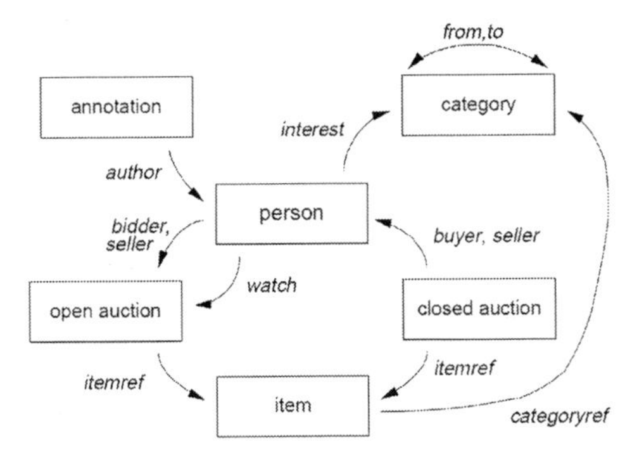

Fig. 1. References over the `xmark` database (from [6])

3 Case Study: Mining Internet Auction Data

This section discusses a concrete usage of XQuake to mine an artificial XML dataset modeling an Internet auction site.

3.1 Description of the Database

The aim of the `xmark` project[2] is to provide a benchmark suite that allows users and developers to gain insights into the characteristics of their XML repositories.

The structure of the document is modeled after a database as deployed by an Internet auction site. As stated in [6], the main entities are: `person`, `open auction`, `closed auction`, `item`, `category` and `annotation`. The relationships between such entities are expressed through references as depicted in Figure 1. The semantics of the main entities used in the experiments is as follows:

1. Items are the objects that are on for sale or that already have been sold. Each item carries a unique identifier and properties like payment (credit card, money order, etc.), a reference to the seller, a description, all encoded as elements. Each item is assigned a world region represented by the item's parent.
2. Open auctions are auctions in progress. Their properties are the privacy status, the bid history (i.e., increases or decreases over time) along with references to the bidders and the seller, the current bid, the time interval within which bids are accepted, the status of the transaction and a reference to the item being sold, among others.
3. Closed auctions are auctions that are finished. Their properties, among the others, are the seller (a reference to a person), the buyer (a reference to a person), a reference to the considered item, the price, the number of items sold, the date when the transaction was closed, the type of transaction.
4. Persons are characterized by name, email address, homepage URL, credit card number, profile of their interests, and the (possibly empty) set of open auctions they are interested in and get notifications about.
5. Categories feature a name and a description.

[2] XML Benchmark Project: `http://www.xml-benchmark.org/`

Table 1. The four `xmark` documents used in the experiments

Name	Tiny	Small	Standard	Large
Scaling Factor	0.1	1	5	10
Dimension	11 MB	111 MB	555 MB	1.164 MB
Table nodes	324.274	3.221.926	16.152.111	32.298.563
Tree height	13	13	13	13
`count(//person)`	2.550	25.500	127.500	255.000
`count(//closed_auction)`	975	9.750	48.750	97.500
`count(//open_auction)`	1.200	12.000	60.000	120.000
`count(//item)`	2.175	21.750	108.750	217.500
`count(//category)`	100	1.000	5.000	10.000

Fig. 2. Two XML fragments of the `xmark` database

The schema is regular on a per entity basis, and exceptions, such as that not every person has a homepage, are predictable.

Over the `xmark` project, an automatic generator program provides a scalable XML document database. Table 1 shows the main characteristics of the four databases used in the experiments (see Section 3.3). For each row, the table reports the name of the database, the scaling factor used to generate the document, the document size (in Megabytes), the number of XML nodes, the size of the XML tree and, finally, the number of elements in the main entities. Some snippets of the benchmark document are also reported in the Figure 2.

3.2 Description of the Scenario

This sub-section presents the queries of the benchmark. A brief motivation of their definition is also given. We are going to assume that the `xmark` analyst - in brief the user - is interested to DM techniques over the auctions registered into the database, to the aim of extracting high-level information about the navigational behaviour of the bidders.

As a first request, he/she aims at a mining model capable of classifying the profile of new persons that register to the site or that perform some relevant transaction over it. The first experiment tries to predict whether a person might be interested in visiting auctions belonging to a certain category (e.g. music, sport, books, etc.) according to her/his profile. The model extracted from this kind of analysis may serve for on-the-fly banners or page reorganization, thus giving a personalized view of the web site allowing to target banners, promotions or news. In order to accomplish these objectives, we combine some discretization techniques with a classification tree extraction.

[Q1] *Discretize the income (if any) of each* `person` *into three distinct bins having an equal width. In the result, append to each* `<person>` *tag a new XML element,* `<income-d>`, *containing both the original and the discretized value.*

```
prepare discretization doc("people-d") using alg:equal_width()
for data $person in doc("xmark")/site/people/person
let active field $income := xs:double($person/profile/@income)
having bin $income-d as ("low income", "med income", "rich income")
return <person-d> {
        ((for $i in $person/* return $i),
        <income-d income-num="{$income}">{$income-d}</income-d>)
  } </person-d>
```

[Q2] *On the basis of the result of the previous query, extract a classification tree able to determine whether a person may be interested to a particular category. Attributes of the task are: business information, education, a value indicating whether the person has a credit card or not, the discretized value of the income, the gender, a value indicating whether the number of open auctions he/she is interested in is greater than 5. The target is a boolean attribute that indicates whether such a person may be interested to "music" or "film" auctions*[3].

```
declare variable $music-cat :=
  for $i in doc("xmark")/site/categories/category
  where $i//text ftcontains "music" ftor "film"
                         occurs at least 3 times
  return $i;

mine tree doc("interested-tree") using alg:id3()
for data $person in
```

[3] This information is obtained by means of the XQuery Full Text, that is the upcoming W3C Recommendation for content-based querying. Specifically, we aim at determining whether some category of interest contains the words "music" or "film" in any order but occurring at least three times. For details see [7].

```
        doc("people-d")//person-discr[not(empty(profile/interest))]
let active field $business  := $person/profile/business/text()
let active field $education  := $person/profile/education/text()
let active field $has-credit-card  := string(not(empty($person/creditcard)))
let active field $income  := $person/income-d/text()
let active field $gender  := $person/profile/gender/text()
let active field $watch  := string(count($person/watches/watch) > 5)
let predicted field $has-a-music-interest  :=
        string(some $j in $person/profile/interest/@category
        satisfies some $k in $music-cat/@id satisfies $j eq $k)
```

The next goal is to extract a descriptive model to find frequent correlations among the bidders in all the open auctions. The output model contains patterns like {George, Stephen} supp=0.50, which states that George and Stephen appear together as bidders 50% of the the open auctions. Since at this stage the user has not idea on the parameters of the analysis, we suggest him/her to extract the largest number of frequent patterns from the bid history.

[Q_3] *Find frequent correlations among the bidders in all the open auctions. Try to use a minimum support as small as possible.*

```
mine itemsets doc("co-bidders") using alg:apriori(0.002)
for data $auction in doc("xmark")/site/open_auctions/open_auction
for item $bidder in $auction/bidder
let active field $name  := string($bidder/personref/@person)
```

Suppose now that by inspecting the result of the previous data mining extraction step, the user is interested in investigating all the pairs (seller, buyer) in the closed auctions history that comply with the set of extracted patterns. In particular, we suggest him/her to look, for each pair (seller, buyer), at the percentage of itemsets that comply with it. A high value of this variable may indicate that such persons tend also to bid together during the auctions.

[Q_4] *With reference to the itemsets extracted in the previous query, return as output the set of triplets, (b, s, c), where b (resp. s) is the buyer (resp. seller) identifier in all the closed auctions, and c is the percentage of the itemsets that comply with b and s, over the total number of itemsets.*

```
apply itemsets doc("supp-seller-buyer")
for association model $m in doc("co-bidders")
let metadata field $total  := xs:decimal($m/PMML//@numberOfItemsets)
for data $closed in doc("xmark")/site/closed_auctions/closed_auction
for item $sel-buy in $closed/(seller|buyer)/@person
let active field $item := string($sel-buy)
having supported $s,
return <closed_auction num-itemsets="$total"> {
        ($i/buyer, $i/seller, <perc>($s div $total)*100</perc>)
      } </closed_auction>
```

The query [Q_3] above generates a very large number of itemsets that are hardly human readable. In order to reduce the number of patters, we suggest the user to use some constraint expressed on the patterns themselves. Specifically, he/she is interested on three kinds of queries. In the first one, he/she decides to change the input data also including auctions that have been closed with a certain level of satisfaction. Also, he/she asks to filter out all the frequent 1-itemsets. In the second query he/she requires that at least one person in each itemset bought at least one item. Finally, in the third query he/she aims at extracting only the patters including at most two business men whose income is greater than the average income. In the latter one, a data filtering constraint also requires to compute the transactions whose bids fall into a certain time interval.

[Q_5] *Find frequent correlations among the bidders in open auctions and including also the seller and the buyer in each closed auction. The happiness of each transaction must be greater than 2. Look only for itemsets of length at least 2.*

```
mine itemsets doc("co-bidders-2") using alg:apriori(0.001)
for data $auction in (doc("xmark")/site/open_auctions/open_auction,
     doc("xmark")/site/closed_auctions/closed_auction)
     [.//happiness > 2]
for item $bid in if ($i/name() eq "open_auction")
                 then $i/bidder//@person
                 else $i/(seller|buyer)/@person
let active field $name := string($bid)
having at least 2 item satisfies true()
```

[Q_6] *Find frequent correlations among the bidders in all the open auctions. For each pattern, at least one person bought some item.*

```
mine itemsets doc("co-bidders-3") using alg:apriori(0.001)
for data $auction in doc("xmark")/site/open_auctions/open_auction
for item $bidder in $auction/bidder/personref/@person
let active field $name := string($bidder)
let metadata field $has-one-item :=
     (some $i in doc("xmark")/site/closed_auctions/closed_auction
      satisfies $i/buyer/@person eq $name)
having at least 1 item satisfies $has-one-item
```

[Q_7] *Find frequent correlations among the bidders in all the open auctions. Each auction must fall into a time interval of 50 days (i.e. the difference among the first bid and the last bid is less or equal than 50 days). Each pattern includes at most two business men whose income is greater than the averag e income.*

```
declare variable $mean :=
    fn:sum(doc("xmark")/site/people/person/profile/@income) div
    count(doc("xmark")/site/people/person);

declare function local:days-range($auction) {
  let $first := $auction/bidder[1]/date/text()
  let $last := $auction/bidder[last()]/date/text()
  return if (empty($first) or empty($last)) then 0
    else fn:days-from-duration($last - $first)
};
```

```
mine itemsets doc("co-bidders-4") using alg:apriori(0.01)
for data $a in doc("xmark")//open_auction[local:days-range(.)>50]
for item $bidder in $a/bidder/personref/@person
let active field $name := string($bidder)
let metadata field $p := doc("xmark")/site/people/person[@id eq $item]
having at most 2 item satisfies
  (($p//business eq "Yes") and ($p//income > $mean))
```

Finally, the user aims at extracting a semantically different kind of patterns. He/she is interested in frequent correlations among all the open auctions that young people are interested in and get notifications about. For example, the pattern {$\texttt{auction}_1$, $\texttt{auction}_2$, $\texttt{auction}_3$} `supp=0.03` means that, for 3% of the times, the three auctions were watched together. Such an information may indicate that persons that frequently tend to visit some open auction also tend to visit other kinds of auctions. The user also constrain the output selecting all the patterns in which a certain number of auctions have a difference among the current price and the initial price above/below certain thresholds.

[Q_8] *Find correlations expressing how frequently open auctions are watched together by under-30 people. In each pattern, at least 5 items increase their initial price of more than 30% w.r.t. the current price, but at most 10 items increase their initial price of less than 10%.*

```
mine itemsets doc("co-auctions") using alg:apriori(0.005)
for data$person in doc("xmark")/site/people/person[profile/age <= 30]
for item $auction in $i/watches/watch/@open_auction
let active field $item := string($auction)
let metadata field $inc :=
   let $a := doc("xmark")/site/open_auctions/open_auction
                                                    [@id eq $item]
   return $perc := (($a/current * 100) div $a/initial) - 100
having at least 5 item satisfies $inc > 30, at most 10 item satisfies $inc < 10
```

3.3 Experiments and Discussion

The query scenario discussed above has been designed by taking into account our real experience on data mining. It highlights the need of combining different tasks, such as classification with discretization (queries [Q_1] and [Q_2]), or to apply models to further data (query [Q_4]). More important, the queries defined over the frequent itemsets mining task (in particular queries [Q_6], [Q_7] and [Q_8]) show how more complex reasoning schemes can be formalized by combining context knowledge with mining. Finally, preprocessing and post-processing queries highlight a feature inherited by the XQuery world, that is the capability of constructing and personalizing the query result.

The experiments reported here are gathered from our work with XQuake. Since no similar systems exist in both the academic and industrial XML world (to the best of our knowledge), a comparison among our system with similar ones is very hard to achieve. So, the time evaluation is only meant as an impression on the performance of the architecture. The time were measured on a dual-core Athlon 4000+ running Windows Vista.

Table 2. Performance impression of XQuake (time is in seconds). †Query has been executed main memory. *Query has been executed with full-text indexes.

DB	Q_1	Q_2 †*	Q_3	Q_4	Q_5 †	Q_6 †	Q_7 †	Q_8
Tiny	≤1	≤1	≤1	≤1	≤1	≤1	≤1	≤1
Small	3	2	3	3	23	23	15	5
Standard	17	36	96	45	74	75	106	22
Large	35	213	85	67	322	158	131	63

We assigned the maximum memory possible to the Java Virtual Machine: 1.6Gbyte of memory. The benchmark documents are scaled at factors 0.1, 1, 5 and 10, as shown in Table 1. The execution time includes the time to serialize the XML output into a file, but it excludes the time needed to store such source into the database. Bulk load and full indexing for the largest dataset - more than 1GB - required about 140 seconds. Also, for the hardest mining queries (more precisely [Q_2], [Q_5], [Q_6] and [Q_7]), we used the main-memory option available in XQuake[4], to the aim of reducing the time for evaluating joins.

The running time (in seconds) is displayed in Table 2, while Figure 3 (left) depicts the scalability for four benchmark queries. As one could observe, queries [Q_2], [Q_5], [Q_6] and [Q_7] require some expensive join in order to select data for the mining step. Such joins affect the overall execution time. Consider for example the query [Q_7]. It requires a multi-join among some `<watch>` element of each person with the `<open_auction>` entity, in order to determine whether such a person is interested in or get notifications about that auction. Over such a query, XQuake spends about 83% of the total processing time in iterating over the data, and in evaluating the mining constraint. This time dominates the time of the mining algorithm to generate the required patterns. A similar consideration can be applied to the query [Q_2], in which more than the 90% of the time is spent in extracting the values of the classification field from the original data, while the remaining time is used to build the tree on such values. The graph of Figure 3 (right) shows that the time difference between main-memory and on-disk execution may be considerable whenever the query contains some expensive join. On the contrary, queries [Q_3] and [Q_4] highlight a different situation, in which the constraint evaluation is now trivial and only the mining engine of XQuake is stressed. In these cases, the mining time dominates the data preparation time.

Summing up, the performance of a mining query clearly depends both on the performance of the XQuery run-time support, and on the capability of the mining system to couple with the database. With a certain degree of generality, the total execution time T_{tot} is given as a sum of the iteration time T_i (i.e. the time to iterate over the data in order to prepare data structures), the mining time T_m (i.e. the time to extract the model from such data structures) and a certain overhead T_{over} due to the time spent in coupling the mining with the DBMS. The more such a coupling is well defined, the more the overhead time is reduced. While maintaining the schema of the algorithm in

[4] In the main-memory mode, both the table data and the text are kept in memory. This option will not create any file on the disk.

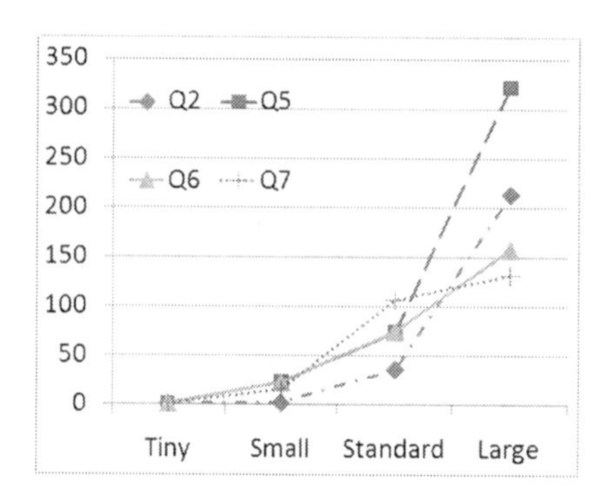

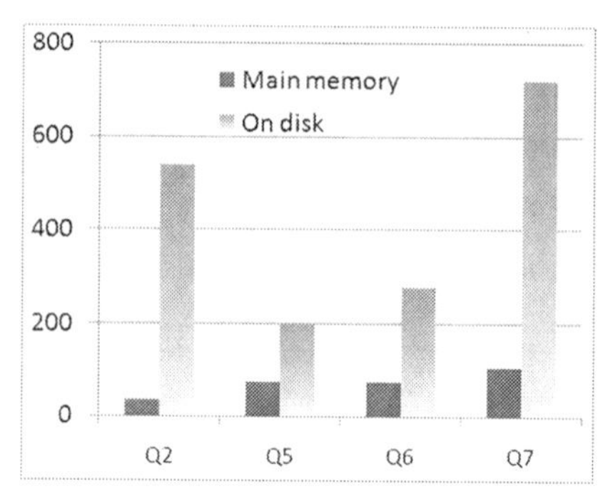

Fig. 3. Scalability w.r.t. the data size (left) and time difference on the standard `xmark` among main memory and on disk execution for the four hardest queries (right)

XQuery, the XQuake engine uses external Java functions to interact with the native XML database [4]. The encoding of the evaluation of XQuery expressions into more efficient Java structures is part of our T_{over}. These experiments show that this time is acceptable, since it is always dominated by T_i and/or T_m. However, other interesting ways of coping with efficiency need to be investigated as well. For example, since there already exist several open-sofware KDD systems, may be very interesting to leverage on their success and re-design XQuake as DM middleware loosely coupled with those systems. We expect our work to continue and evolve also in this direction in the near future.

4 Final Remarks

4.1 Related Work

An important issue in DM is how to make all the heterogeneous patterns, sources of data and other KDD objects coexist in a single framework. A solution considered in the last few years is the exploitation of XML as a flexible instrument for IDBs [8,9,10,11].

In [8] XML has been used as the basis for the definition of a semi-structured data model designed for KDD, called XDM. In this approach both data and mining models are stored in the same XML database. This allows the reuse of patterns by the inductive database management system. The perspective suggested by XDM is also taken in KDDML [9] and RapidMiner [10]. Essentially, the KDD process is modeled as an XML document and the description of an operator application is encoded by means of an XML element. Both KDDML and XDM integrate XQuery expressions into the mining process. For instance, XDM encodes XPath expressions into XML attributes to select sources for the mining, while KDDML uses an XQuery expression to evaluate a condition on a mining model. In our opinion, XQuake offers a deeper amalgamation with the XQuery language and consequently a better integration between DM and XML native databases.

Finally, the XMineRule operator [11] defines the basic concept of association rules for XML documents. Two are the main differences with respect to XQuake. From the physical point of view, XMineRule requires that the data are mapped to the relational model and it uses SQL-oriented algorithms to do the mining. Also the output rules are translated into an XML representation. As a consequence, the loosely-coupled architecture of XMineRule makes it difficult to use optimizations based on the pruning of the search space, since constraints can be evaluated only at pre- or post-processing time. From the semantics perspective, items have an XML-based hierarchical tree structure in which rules describe interesting relations among fragments of the XML source [12]. In contrast, in our approach, items are denoted by using simple structured data from the domains of basic data types, favouring both the implementation of efficient data structures and the design of powerful domain-specific optimizations, which are evaluated as deeper as possible in the extraction process. Domain knowledge is linked to items through XML metadata elements.

4.2 Conclusions and Future Work

In this work, we propose a new query language as a solution to the XML DM problem. In our view, an XML native database is used as a storage for KDD entities. Data mining tasks are expressed in an XQuery-like language. The syntax of the language is flexible enough to specify a variety of different mining tasks by means of user-defined functions in the statements. These ones provide the user with personalized sophisticated constraints, based, for example, on domain knowledge. The first empirical assessment, reported in [4,5] and finalized in this paper, exhibits promising results.

Summing up, our project aims at a completely general solution for data mining. Clearly, the generality is even more substantial in XML-based languages, since no general-purpose XML mining language has been yet proposed (at the best of our knowledge). Our on-going work includes the exploitation of ontologies to represent the metadata. For example, ontologies may represent enriched taxonomies, used to describe the application domain by means of data and object properties. As a consequence, they may provide enhanced possibilities to constrain the mining queries in a more expressive way. This opportunity is even more substantial in our project, since ontologies are typically represented via the Web Ontology Language (OWL) [13], de facto an XML-based language.

References

1. Imielinski, T., Mannila, H.: A database perspective on knowledge discovery. Comm. of the ACM 39(11), 58–64 (1996)
2. The Data Mining Group. The Predictive Model Markup Language (PMML), version 4.0, `http://www.dmg.org/v4-0/GeneralStructure.html`
3. W3C World Wide Web Consortium. XQuery 1.0: An XML Query Language (Recommendation January 23, 2007), `http://www.w3.org/TR/Query`
4. Romei, A., Turini, F.: XML data mining. Softw., Pract. Exper. 40(2), 101–130 (2010)
5. Romei, A., Turini, F.: XQuake - An XQuery-like Language for Mining XML Data. In: The Second International Conference on Agents and Artificial Intelligence (ICAART), Valencia, Spain, pp. 20–27 (2010)

6. Schmidt, A., Waas, F., Kersten, M.L., Carey, M.J., Manolescu, I., Busse, R.: XMark: A Benchmark for XML Data Management. In: The 28th International Conference on Very Large Data Bases (VLDB), Hong Kong, China, pp. 974–985 (2002)
7. W3C World Wide Web Consortium. XQuery and XPath Full Text 1.0 (Candidate Recommendation July 09, 2009), http://www.w3.org/TR/xpath-full-text-10/
8. Meo, R., Psaila, G.: An XML-Based Database for Knowledge Discovery. In: Grust, T., Höpfner, H., Illarramendi, A., Jablonski, S., Fischer, F., Müller, S., Patranjan, P.-L., Sattler, K.-U., Spiliopoulou, M., Wijsen, J. (eds.) EDBT 2006. LNCS, vol. 4254, pp. 814–828. Springer, Heidelberg (2006)
9. Romei, A., Ruggieri, S., Turini, F.: KDDML: a middleware language and system for knowledge discovery in databases. Data Knowl. Eng. 57(2), 179–220 (2006)
10. Euler, T., Klinkenberg, R., Mierswa, I., Scholz, M., Wurst, M.: YALE: rapid prototyping for complex data mining tasks. In: Twelfth ACM SIGKDD International Conference on Knowledge Discovery and Data Mining (KDD), Philadelphia, PA, USA, pp. 935–940 (2006)
11. Braga, D., Campi, A., Ceri, S., Klemettinen, M., Lanzi, P.: Discovering interesting information in XML data with association rules. In: The Eighteenth Annual ACM Symposium on Applied Computing (SAC), Melbourne, Florida, pp. 450–454 (2003)
12. Feng, L., Tharam, S.D.: Mining Interesting XML-Enabled Association Rules with Templates. In: Goethals, B., Siebes, A. (eds.) KDID 2004. LNCS, vol. 3377, pp. 66–88. Springer, Heidelberg (2005)
13. W3C World Wide Web Consortium. OWL Web Ontology Language (Recommendation February 10, 2004), http://www.w3.org/TR/owl-features

Improved Door Detection Fusing Camera and Laser Rangefinder Data with AdaBoosting

Jens Hensler, Michael Blaich, and Oliver Bittel

University of Applied Sciences
Brauneggerstr. 55, 78462 Konstanz, Germany
{jhensler,mblaich,bittel}@htwg-konstanz.de

Abstract. For many indoor robot applications it is crucial to recognize doors reliably. Doors connect rooms and hallways and therefore determine possible navigation paths. Moreover, doors are important landmarks for robot self-localization and map building. Existing algorithms for door detection are often limited to restricted environments. They do not consider the large intra-class variability of doors. In this paper we present an approach which combines a set of simple door detection classifiers. The classifiers are based either on visual information or on laser ranger data. Separately, these classifiers accomplish only a weak door detection rate. However, by combining them with the aid of AdaBoost Algorithm more than 82% of all doors with a false positive rate less than 3% are detected in static test data. Further improvement can easily be achieved by using different door perspectives from a moving robot. In a realtime mobile robot application we detect more than 90% of all doors with a very low false detection rate.

Keywords: Door detection, AdaBoost, Learning algorithm, Mobile robot.

1 Introduction

In an indoor environment doors constitute significant landmarks. They represent the entrance and exit points of rooms. Therefore, robust real-time door detection is an essential component for indoor robot applications (e.g. courier, observation or tour guide robots). To match maps from several mobile robots [1] used doors as unique landmarks for the map matching algorithm. In [2] and [3] doors connect rooms and corridors in an topological map. [4] present a cognitive map representation for mobile robots with high level features like doors. For all this applications a reliable door detection algorithm is necessary.

In the past, the problem of door detection has been studied several times. The approaches differ in the implemented sensor systems and the diversity of environments and doors, respectively. For example, in [5] and [6] only visual information was used. The approaches from [7] and [8] depends on a well structured environment. Others, like [9] apply an additional 2D laser range finder and thereby receive better results.

From these approaches we find that there are two major difficulties in autonomous door detection. Firstly, it is often impossible to cover the entire door in a single camera image. In our scenario, the robot camera is close to the ground so that the top of the door is often not captured by the robot's camera (see figure 1).

J. Filipe, A. Fred, and B. Sharp (Eds.): ICAART 2010, CCIS 129, pp. 39–48, 2011.

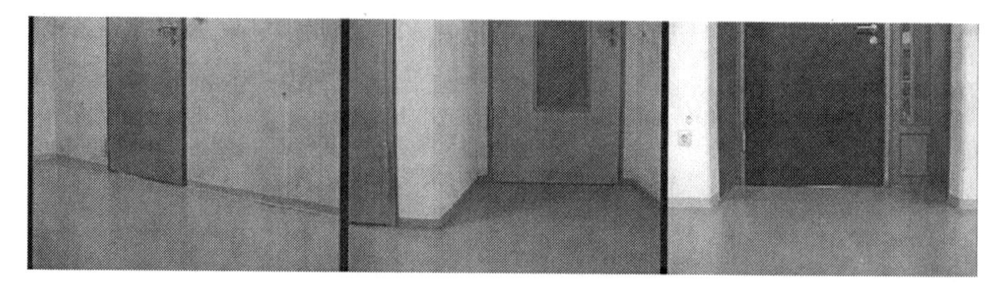

Fig. 1. Illustrates typical door images, taken by the robots camera. The top of the doors are cut off and also the diversity of doors are recognizable: The doors have different poses, colors, lighting situations as well as different features e.g. door gap or texture on the bottom.

The second difficulty is characterized by the large intra-class variability of doors (even for the same door types) in various environments. As shown in figure 1, doors can have different poses, lighting situations, reflections, as well as completely different features. The main features of a door are illustrated in figure 2. A door can be recognized e.g. by its color or texture with respect to the color or texture of the surrounding wall. Door gaps or door knobs are indicators of a door, too. Even if some of these features are detected in a single camera image, a robust algorithm should detect the door by using the remaining features.

In recent work [6], these two issues of door detection were solved by extracting several door features from the robots camera images and applying the AdaBoost algorithm [10]. The algorithm combines all weak features of a door candidate to receive a strong door classifier, which allows it to decide whether a door is found or not.

For our situation this approach is not sensitive enough. We could not reach the same high detection rate in our complex university environment with a similar system (see section 4). Therefore, we added a laser ranger. Further weak classifiers were used to improve the detection results. In the experimental result section we demonstrate the performance of our system on a large database of images from different environments and situations.

2 The AdaBoost Algorithm

The concept behind all boosting algorithms is to use multiple weak classifiers instead of one strong classifier and solve the decision problem by combining the results of the weak classifiers. Hereby, the weak classifiers are built up to solve binary decisions. The AdaBoost algorithm uses a training dataset to build up a strong classifier. For this purpose, the algorithm requires that each weak classifier reaches at least 50% success rate in the training process and the errors of the classifiers are independent. If this is given, the algorithm is able to improve the error rate by calculating optimal weight for each weak classifier. The output of the nth weak classifier to the input x is $y_n = h_n(x)$. If every y_n is weighted with an α_n which is created during the training process, the strong classifier is given by:

$$H(x) = sign\left(\sum_{n=1}^{N} \alpha_n h_n(x)\right) \quad (1)$$

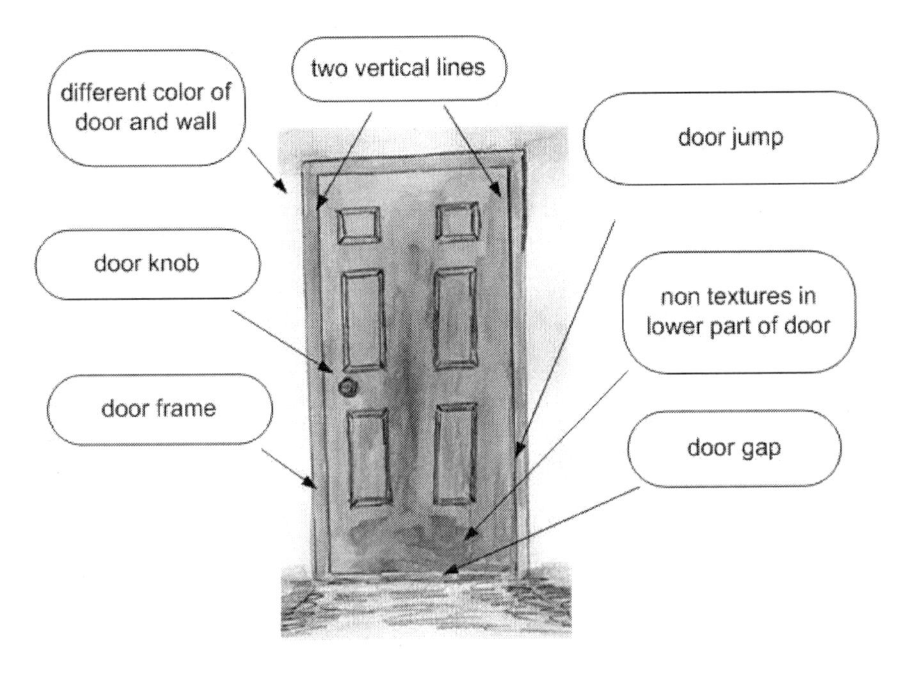

Fig. 2. Characterizing features of doors

3 Detection of Door Features

Our approach is illustrated in figure 3. As mentioned before, we use the robot's camera image and the laser-based distance signal for the door detection. Out of the camera image we extract vertical lines to find door candidates in the images. For this preselection we assume that each door has a vertical line on the right and left side. As a consequence, a door is not detected, if the door posts are not visible. In the next step we check each candidate for seven door features which represent the weak classifiers: a door can have a certain width *WidthClassifier*, the color of the door can be different from the color of the wall *ColorWallClassifier*, a door can have a texture at the bottom or not *TextureBottomClassifier*, a door may have a door frame *FrameClassifier* or a door knob *KnobClassifier*, also a door gap *GapClassifier* is possible. Finally, the door can stick out of the wall *JumpClassifier*. The buildup of the weak classifiers is described in the sections below. Each classifier resolves a binary decision. The best threshold for each classifier is measured with ROC curves by varying the threshold until the best one is found. The classifiers *GapClassifier*, *ColorWallClassifier* and *TextureBottomClassifier* are similarly implemented as in [6] and not further mentioned here.

3.1 Preselection

During the preselection vertical line pairs generated by the door frame represent door candidates for the AdaBoost algorithm. To receive vertical lines we apply the Contour Following Algorithm [11]. Compared to other transformations, this method has the advantage that we obtain the starting and end points of these lines.

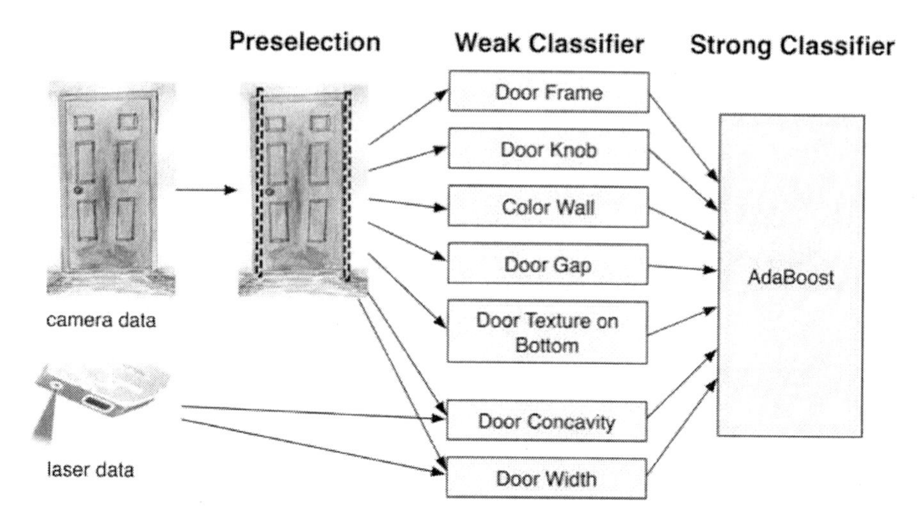

Fig. 3. This Image illustrates our approach to find doors. Several weak classifiers are combined using the AdaBoost algorithm in the training process. The door width and door concavity classifiers are using the laser distance data from the robot in addition to the camera image.

Not every combination of vertical line pairs in an image correspond to door candidates. The number of candidates can be drastically reduced by the following rules:

- The vertical lines need to have a minimal length.
- The horizontal distance between the two vertical lines has to be between a maximal and minimal value.
- The end points of the vertical lines have a minimal vertical shift.
- If there are lines close together, which all may represent a door candidate according to the earlier rules, only the inner lines are used. The outer lines are indicators for a door frame.

3.2 Weak Classifiers

To improve our AdaBoost algorithm in comparison to [6] we use four additional weak classifiers. At first the door knob classifier will be explained. For this classifier we used again the line image calculated during the preselection of the door candidates. However, for this classifier not the vertical lines are important, but the almost horizontal lines which result from the door knob. We use a height from about 0.9m based on the bottom end of the vertical lines to find the door knob areas. In these two areas (left and right side of a door) the classifier returns 'true' if at least two almost horizontal lines are found.

The second classifier is a door frame classifier. A door frame is required to install a door inside a wall. The frame can also be calculated during the determination of the vertical line pairs. A door frame in an image is characterized by duplicated vertical line pairs. If there is a close additional vertical line on both sides of the door, the door frame classifier is positive.

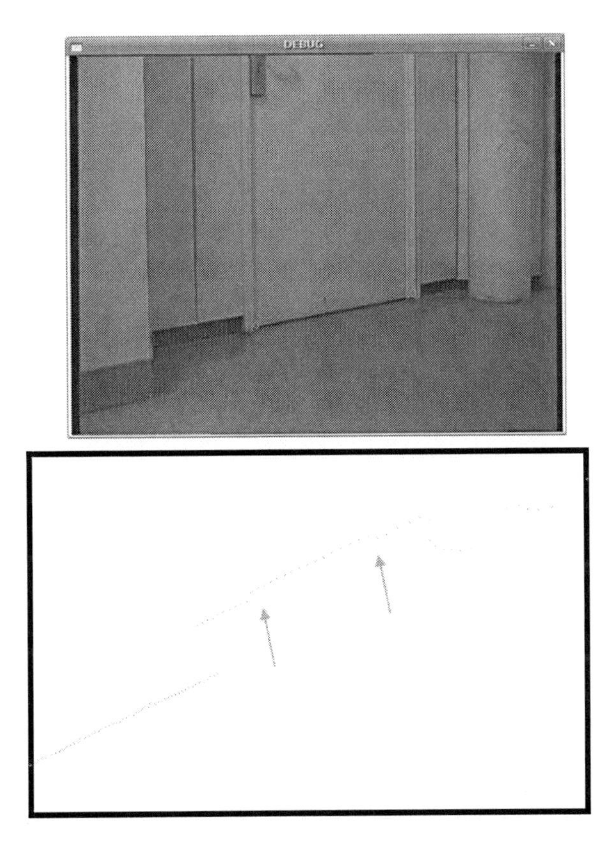

Fig. 4. The red arrows in the laser profile point to a door. The images show that the door is not running with the wall. It is receding or sticking out from the door.

Further, we use the door width to get one more weak classifier. There is a DIN standard (DIN18101) for door width. Unfortunately the DIN width values vary strongly. Even here it is not easy to find a strong classifier for a door. For a weak classifier we bordered the width in values between 0.735m and 1.110m. To calculate the distance between the two vertical lines we use the laser distance data provided from the laser range finder.

At last we consider that in many environments doors are receded into the wall, creating a concave shape for the doorway (see figure 4). This shape can be obtained using the robot laser distance data. For this the slope between each measured laser distance in the door candidate area is calculated. There exists a maximum and minimum slope value at the position of the door candidate (see figure 5). The *JumpClassifier* can be described by the following rules:

- If we calculate the slope between each measured laser distance point, without considering the door candidate, the standard deviation is almost zero.
- If we look at the slope at the door frame area we will find values which strongly vary from the calculated mean value.

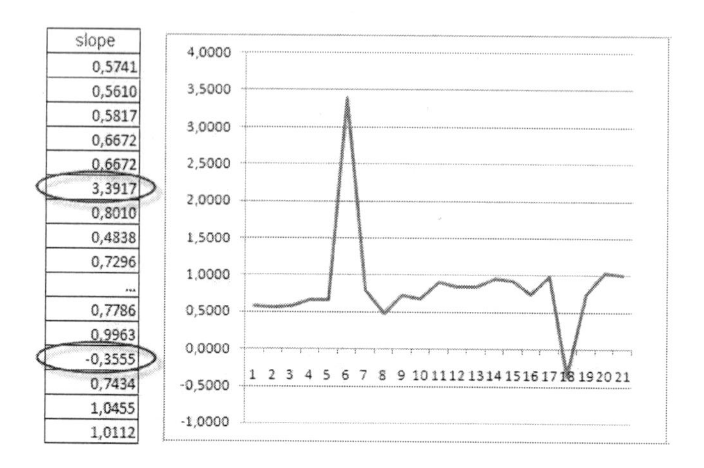

Fig. 5. Slopes between the measuring points from figure 4. We found turning points in the area of the door frame. It's used for the *JumpClassifier*.

4 Experimental Results

To test the performance of the system, a database of 210 test sets were taken with the Pioneer2DX robot. One test set consists of one camera image and one laser profile at a certain time. We considered pictures of doors and non-doors. From the 210 test sets we took 140 for the training process of the AdaBoost. The residual sets were taken to test our system. In these 70 test sets our preselection algorithm found overall 550 door candidates, of which 61 candidates correspond to real doors. The result of some examples of weak classifiers and the strong AdaBoost classifier is shown in a ROC curve diagram (see figure 6). As we can see in the ROC curves, the AdaBoost classifier reached the best detection rate. In our test the true-positive rate for the AdaBoost classifier reaches a value of 82% and a false-positive rate of 3%. We receive the same result if we take a look at the RPC methods (table 1). The best value of the F-score (combination from precision and recall) is obtained by the AdaBoost classifier.

Table 1. Results of the RPC methods. The F-score can be interpreted as a weighted average of precision and recall, where an F-score reaches its best value at 1 and worst score at 0.

	Recall	Fallout	Precision	F-score
WidthClassifier	0,73	0,13	0,45	0,56
JumpClassifier	0,64	0,19	0,35	0,45
TextureClassifier	0,56	0,06	0,56	0,56
ColorWallClassifier	0,07	0,05	0,18	0,10
GapClassifier	0,61	0,35	0,24	0,34
KnobClassifier	0,80	0,19	0,39	0,53
FrameClassifier	0,54	0,29	0,25	0,34
AdaBoostWithoutLaser	0,61	0,07	0,56	0,58
AdaBoost	0,82	0,03	0,79	0,81

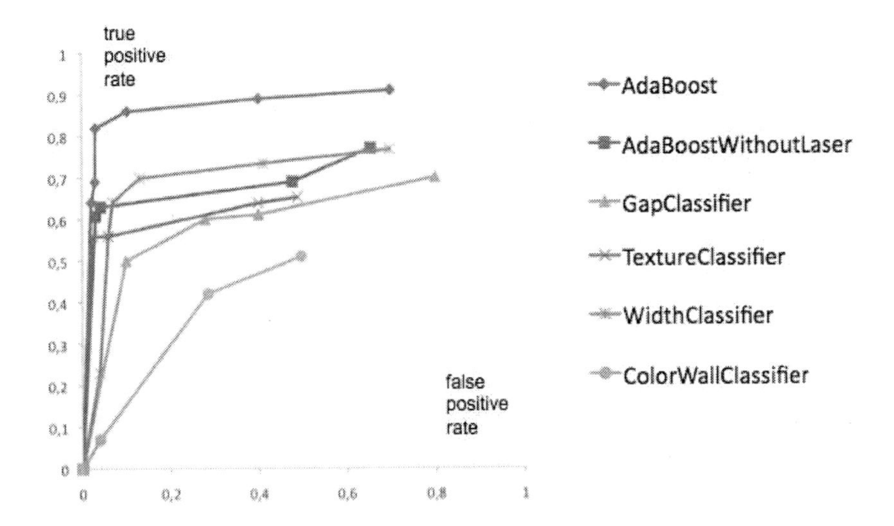

Fig. 6. ROC curves are showing the performance of our algorithm compared with the performance of some example single weak classifier

Typically detected doors are illustrated in figure 7. As can be seen, the algorithm is capable of detecting doors under different lighting situations and different viewpoints of the robot. It should be noted, that the absence of one or more door features does not cause a non-detection of the door. Figure 8 shows a false positive detection. They happen through walls or other objects which look very similar to doors.

As a next step, we looked at the result without the laser range finder (similar to [6], see table 1 and figure 6 *AdaBoostWithoutLaser*). This classifier combination (*TextureBottomClassifier*, *ColorWallClassifier*, *GapClassifier*, *KnobClassifier* and *FrameClassifier*) did not reach the same high result (detection rate 60% and false positive rate 7%). With this result we claim that in a strongly varying indoor environment with different kinds of doors a camera-based door detection is not strong enough to build up a powerful AdaBoost classifier. Further classifiers like the *JumpClassifier* and *WidthClassifier* can improve the result essentially.

Another advantage of the laser range finder is that the position of detected doors can be measured exactly. In combination with the robot position the doors can be marked in an existing map. The result is a map with doors as additional landmarks for improved robot localization and navigation.

We tested the system as a Player [12] driver on our Pioneer2DX robot. We used two different environments. In the first environment (basement of the university) all doors were detected (see figure 9). In the second environment (office environment) each door, except glass doors, was detected (see figure 10). The problem here is, that we received wrong laser distances, because the laser is going through the glass.

Fig. 7. These are typical examples of AdaBoost classifier detected doors. The pictures demonstrate that our approach is robust against different robot positions, reflection situation as well as different door features.

Fig. 8. This picture illustrates a sample false-positive error of the AdaBoost classifier. In the sample a wall, which looks similar to a door, is detected as door.

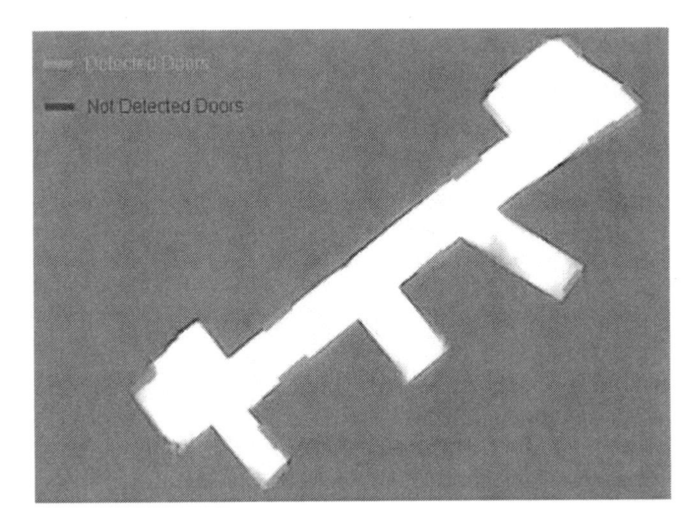

Fig. 9. The Image depicts the result of the robot first test run in the basement environment. Detected doors are marked green in the map. Each door was found.

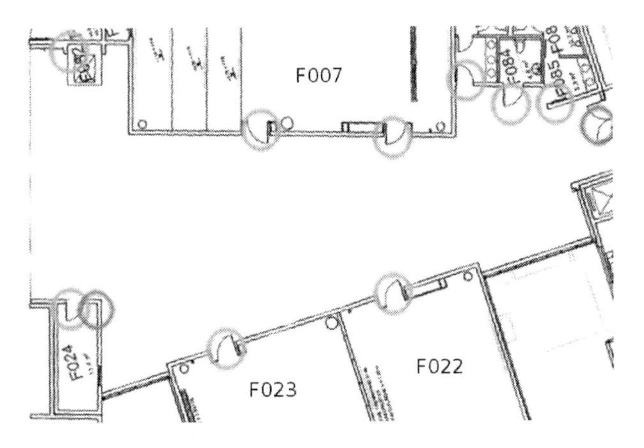

Fig. 10. Results of the second robot test run in the office environment. All detections are marked with green circles in the map. The non-detection (glass door) and false detection (see figure 8) are marked with red circles.

5 Conclusions and Future Work

In this paper we presented an approach for a laser- and camera-based door detection system. By using the AdaBoost algorithm we built a system with a detection rate of more than 82% and a very low error rate of 3%. It is a combination of several weak classifiers, e.g the color of the wall, door knob or door gap. We used the ROC and RPC methods to demonstrate that none of the other weak classifiers can replace the strong classifier created by the AdaBoost algorithm. Furthermore it was shown, that without the laser range finder, we could not reach the same high detection rate. The system has the ability to find doors in realtime. With an Intel Core Duo 2.4GHz processor we reached a performance of 12fps.

There are several possibilities to improve the system. Firstly, the training set can be enlarged. More test data would improve the alpha values for each weak classifier. When using the system in a new environment, it will provide a better result, if we add test data relating to this environment. Secondly, the weak classifiers can be modified and new weak classifiers can be added. E.g the *ColorWallClassifier* can be improved if the system automatically learns the wall color of the environment. New classifiers could use the door hinges or the light switch on the door side.

Currently, we are integrating this system with an autonomous map building system. That means, that the robot has the ability to create a map of an unknown environment and mark doors in it. In addition, we should look for new classifiers, which would allow it to detect open doors. Finally, the integration of a TOF-Camera could be a very promising approach for enhancing the door detection system.

References

1. Dedeoglu, G., Sukhatme, G.: Landmark-based matching algorithm for cooperative mapping by autonomous robots. In: Proc. 5th International Symp. Distributed Autonomous Robotic Systems, Citeseer, pp. 251–260 (2000)
2. Mozos, O., Rottmann, A., Triebel, R., Jensfelt, P., Burgard, W.: Semantic labeling of places using information extracted from laser and vision sensor data. In: IEEE/RSJ IROS Workshop: From Sensors to Human Spatial Concepts, Beijing, China, Citeseer (2006)
3. Stachniss, C., Martinez-Mozos, O., Rottmann, A., Burgard, W.: Semantic labeling of places. In: Proceedings of the International Symposium on Robotics Research (2005)
4. Vasudevan, S., Gchter, S., Nguyen, V., Siegwart, R.: Cognitive maps for mobile robots an object based approach. Robotics and Autonomous Systems 55, 359–371 (2007)
5. Murillo, A.C., Košecká, J., Guerrero, J.J., Sagüés, C.: Visual door detection integrating appearance and shape cues. Robot. Auton. Syst. 5, 512–521 (2008)
6. Chen, Z., Birchfield, S.T.: Visual detection of lintel-occluded doors from a single image. In: IEEE Computer Society Workshop on Visual Localization for Mobile Platforms, vol. 1, pp. 1–8 (2008)
7. Shi, W., Samarabandu, J.: Investigating the Performance of Corridor and Door Detection Algorithms in Different Environments. In: ICIA 2006, pp. 206–211 (2006)
8. Munoz-Salinas, R., Aguirre, E., Garcia-Silvente, M.: Detection of doors using a genetic visual fuzzy system for mobile robots. Autonomous Robots 21, 123–141 (2006)
9. Anguelov, D., Koller, D., Parker, E., Thrun, S.: Detecting and modeling doors with mobile robots. In: Proceedings of the IEEE International Conference on Robotics and Automation, ICRA (2004)
10. Freund, Y., Schapire, R.E.: A short introduction to boosting. J. Japan. Soc. for Artif. Intel. 14, 771–780 (1999)
11. Neira, J., Tardos, J.D.: Computer vision. Universidad de Zaragoza, Spain (2008)
12. Collett, T.H.J., MacDonald, B.A., Gerkey, B.: Player 2.0: Toward a practical robot programming framework. In: Australasian Conference on Robotics and Automation, Sydney (2005)

Combining Color and Spatial Color Distribution Information in a Fuzzy Rule Based Compact Composite Descriptor

Savvas A. Chatzichristofis[1], Yiannis S. Boutalis[1,2], and Mathias Lux[3]

[1]Department of Electrical and Computer Engineering,
Democritus University of Thrace, Xanthi 67100, Greece
[2]Department of Electrical, Electronic and Communication Engineering
Friedrich-Alexander University of Erlangen-Nuremberg, 91058 Erlangen, Germany
[3]Institute of Information Technology, Klagenfurt University, Austria
{schatzic,ybout}@ee.duth.gr
mlux@itec.uni-klu.ac.at

Abstract. In this paper, a novel low level feature for content based image retrieval is presented. The proposed feature structure combines color and spatial color distribution information. The combination of these two features in one vector classifies the proposed descriptor to the family of Composite Descriptors. In order to extract the color information, a fuzzy system is being used, which is mapping the number of colors that are included in the image into a custom palette of 8 colors. The way by which the vector of the proposed descriptor is being formed, describes the color spatial information contained in images. To be applicable in the design of large image databases, the proposed descriptor is compact, requiring only 48 bytes per image. Experiments presented in this paper demonstrate the effectiveness of the proposed technique especially for Hand-Drawn Sketches.

Keywords: Image retrieval, Compact composite descriptors, Spatial distribution, Hand-drawn sketches.

1 Introduction

As content based image retrieval (CBIR) is defined any technology, that in principle helps to organize digital image archives by their visual content. By this definition, anything ranging from an image similarity function to a robust image annotation engine falls under the purview of CBIR [1].

In CBIR systems, the visual content of the images is mapped into a new space named the feature space. The features, typically represented as vectors in this space, have to be discriminative and sufficient for retrieval purposes like e.g. the description of the objects, finding of duplicates or retrieval of visually similar scenes. The key to attaining a successful retrieval system is to choose the right features that represent the images as "strong" as possible [2]. A feature is a set of characteristics of the image, such as color, texture, and shape. In addition, a feature can further be enriched with information about the spatial distribution of the characteristic, that it describes.

J. Filipe, A. Fred, and B. Sharp (Eds.): ICAART 2010, CCIS 129, pp. 49–60, 2011.

Regarding CBIR schemes which rely on single features like color and/or color spatial information several schemes have been proposed. The algorithm proposed in [3] makes use of multiresolution wavelet decompositions of the images. In [4], each pixel as coherent or noncoherent based on whether the pixel and its neighbors have similar color. In [5][6][7] are presented the Spatial Color Histograms in which, in addition to the statistics in the dimensions of a color space, the distribution state of each single color in the spatial dimension is also taken into account. In [8] a color distribution entropy (CDE) method is proposed, which takes account of the correlation of the color spatial distribution in an image. In [9] a color correlograms method is proposed, which collects statistics of the co-occurrence of two colors. A simplification of this feature is the autocorrelogram, which only captures the spatial correlation between identical colors. The MPEG-7 standard [10] includes the Color Layout Descriptor [11], which represents the spatial distribution of color of visual signals in a very compact form.

The schemes which include more than one features in a compact vector can be regarded that they belong to the family of Compact Composite Descriptors (CCD). In [12] and [13] 2 descriptors are presented, that contain color and texture information at the same time in a very compact representation. In [14] a descriptor is proposed, that includes brightness and texture information in a vector with size of 48 bytes.

In this paper a new CCD is proposed, which combines color and spatial color distribution information. The descriptors of this type can be used for image retrieval by using hand-drawn sketch queries, since this descriptor captures the layout information of color feature. In addition, the descriptors of this structure are considered to be suitable for colored graphics, since such images contain relatively small number of color and less texture regions than the natural color images.

The rest of the paper is organized as follows: Section 2 describes how to extract the color information, which is embedded in the proposed descriptor, while Section 3 describes in details the descriptor's formation. Section 4 contains the experimental results of an image retrieval system that uses either the proposed desciptor or the MPEG-7 CLD descriptor on two benchmarking databases. Finally, the conclusions are given in Section 5.

2 Color Information

An easy way to extract color features from an image is by linking the color space channels. Linking is defined as the combination of more than one histogram to a single one. An example of color linking methods is the Scalable Color Descriptor [10], which is included in the MPEG-7 standard.

In the literature several methods are mentioned, that perform the linking process by using Fuzzy systems. In [15] the extraction of a fuzzy-linking histogram is presented based on the color space *CIE-L*a*b**. Their 3-input fuzzy system uses the L^*, a^* and b^* values from each pixel in an image to classify that pixel into one of 10 preset colors, transforming the image into a palette of the 10 preset colors. In this method, the defuzzyfication algorithm classifies the input pixel into one and only one output bin (color) of the system (crisp classification). Additionally, the required conversion of an image from the *RGB* color space to *CIEXYZ* and finally to *CIE-L*a*b** color space makes the method noticeably time-consuming. In [2], the technique is improved by replacing the color space and the defuzzyfication algorithm. In [12], a second fuzzy system is added, in order to replace the 10 colors palette with a new 24 color palette.

In this paper a new fuzzy linking system is proposed, that maps the colors of the image in a custom 8 colors palette. The system uses the three channels of HSV as inputs, and forms an 8-bin histogram as output. Each bin represents a present color as follows: (0) Black, (1) Red, (2) Orange/Yellow, (3) Green, (4) Cyan, (5) Blue, (6) Magenta and (7) White.

The system's operating principle is described as follows: Each pixel of the image is transformed to the HSV color space. The H, S and V values interacts with the fuzzy system. Depending on the activation value of the membership functions of the 3 system inputs, the pixel is classified by a participation value in one or more of the preset colors, that the system uses.

For the defuzzification process, a set of 28 TSK-like [16] rules is used, with fuzzy antecedents and crisp consequents.

The Membership Value Limits and the Fuzzy Rules are given in the Appendix.

3 Descriptor Formation

In order to incorporate the spatial distribution information of the color to the proposed descriptor, from each image, 8 Tiles with size of 4x4 pixels are created. Each Tile corresponds to one of the 8 preset colors described in Section 2. The Tiles are described as $T(c)$,$c \in [0,7]$, while the Tiles' pixel values are described as $T(c)_{M,N}$, $M,N \in [0,3]$. The Tiles' creation process is described as follows: A given image J is first sub-divided into 16 sub-images according to figure 2(a). Each sub image is denoted as $\acute{J}_{M,N}$. Each pixel that belongs to $\acute{J}_{M,N}$ is denoted as $P(i)_{\acute{J}_{M,N}}$, $i \in [0,I]$, I Number of pixels in $\acute{J}_{M,N}$. Each $\acute{J}_{M,N}$ contributes to the formation of the $T(c)_{M,N}$ for each c. Each $P(i)_{\acute{J}_{M,N}}$ is transformed to the HSV color space and the values of H, S and V constitute inputs to the Fuzzy-Linking system, which gives a participation value in space $[0,1]$ for each one of the 8 preset colors. The participation value of each color is defined as $MF(c)$, $c \in [0,7]$.

$$\sum_{c=0}^{7} MF(c) = 1 \qquad (1)$$

Each $T(c)_{M,N}$ increases by $MF(c)$. The process is repeated for all $P(i)_{\acute{J}_{M,N}}$ until $i = I$. Subsequently, the value of each $T(c)_{M,N}$ for each c is replaced according to the following formula:

$$T(c)_{M,N} = \frac{T(c)_{M,N}}{I} \qquad (2)$$

In this way, a normalization on the number I of the pixels, that participate in any $\acute{J}_{M,N}$ is achieved. On the completion of the process, the value of each $T(c)_{M,N}$ is quantized using a linear quantization table to an integer value within the interval $[0,7]$.

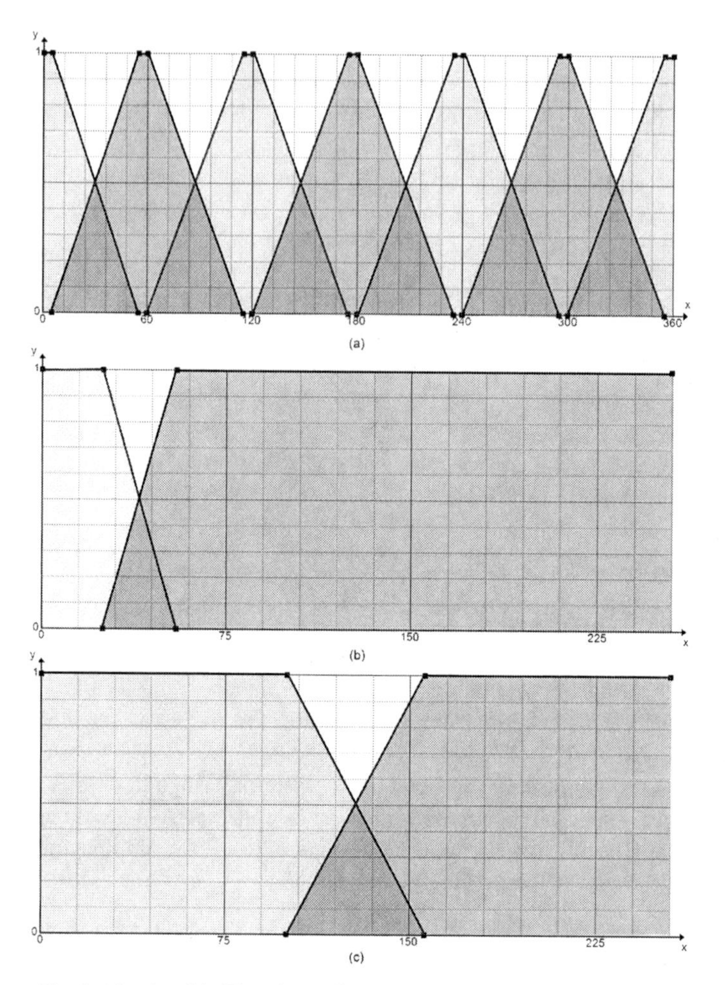

Fig. 1. Membership Functions of (a) Hue, (b) Saturation and (c) Value

The described process is repeated for each (M, N). On completion of the process 8 images of 3 bits are produced, where each one of them describes the distribution (quantitative and spatial) of each color. Figure 2(b) shows the resulting Tiles of Figure 2(a).

Scanning row-by-row each tile $T(c)$, a 16-dimensional vector can be formed, with each vector element requiring 3-bits for its representation. In this vector the element's index determines position (M, N) in the tile $T(c)_{M,N}$. The element's value determines the color quantity at the position (M, N) quantized in space $[0, 7]$. By repeating the process for all the Tiles, 8 such vectors can be produced, which if combined, by placing them successively, a 48 bytes descriptor is formed.

$$16(Elements) \times 8(Tiles) \times \frac{3}{8}(bytes) = 48bytes$$

This size compared with the size of other compact composite descriptors is considered satisfactory. The problem occurs in the length of the final vector which is set in 128

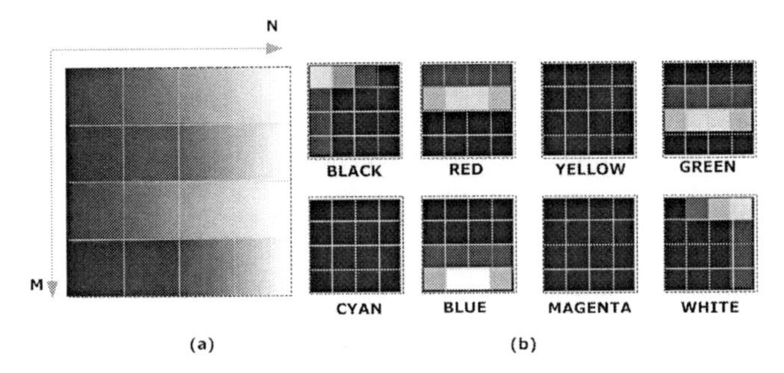

Fig. 2. (a) The image is divided into 16 sub images, (b) Result of the 8 Tiles production from the image (a)

elements. During data retrieval from databases, the length of the retrieved information is of great significance. In order to compress the length of the proposed descriptor the following information lossless procedure is applied. The 8 Tiles of 3 bits are combined in order to form a new 24-bit Tile ("color" Tile) of the same size. This Tile is defined as ϕ, while the values of the pixels are described as $\phi_{M,N}$, $M,N \in [0,3]$. In order to define each $\phi_{M,N}$, 3 values are needed. $R(\phi_{M,N})$ describes the amount of red, $G(\phi_{M,N})$ describes the amount of green and $B(\phi_{M,N})$ describes the amount of blue.

The value of each $T(c)_{M,N}$ is expressed in binary form. Given that the value of each $T(c)_{M,N}$ is described by 3 bits, 3 binary digits are needed for its description.

The binary form of $T(c)_{M,N}$ value is defined as a 3 places matrix, $T(c)_{M,N}[B]$, $B \in [0,2]$.

$\phi_{M,N}$ pixel is shaped using the following rules:

$$R(\phi_{M,N}) = \sum_{c=0}^{7} 2^c \times T(c)_{M,N}[0] \tag{3}$$

$$G(\phi_{M,N}) = \sum_{c=0}^{7} 2^c \times T(c)_{M,N}[1] \tag{4}$$

$$B(\phi_{M,N}) = \sum_{c=0}^{7} 2^c \times T(c)_{M,N}[2] \tag{5}$$

The process is repeated for each (M,N).

In order to make clear the production process of ϕ, the following example is given. The $T(c)_{M,N}$ values of the 8 Tiles for a given image J are presented in Table 1. Next to the 3-bit representation of each value, the binary form of the value appears. The first bit represents $T(c)_{M,N}[0]$, the second one represents $T(c)_{M,N}[1]$ and the third bit represents $T(c)_{M,N}[2]$. The value of each $\phi_{M,N}$ channel depends on the combination, using Eq. 3, Eq. 4 or Eq. 5, of all the $T(c)_{M,N}[B]$, $c \in [0,7]$. For example, The Blue channel of $\phi_{M,N}$ depends on the combination of all the $T(c)_{M,N}[2]$ (last column of the table). At the given example, the $B(\phi_{M,N})$ value is equal to $0 \times 2^1 + 1 \times 2^1 + 0 \times 2^2 + 1 \times 2^3 + 1 \times 2^4 + 0 \times 2^5 + 0 \times 2^6 + 1 \times 2^7 = 153$.

Table 1. Combining the 8 $T(c)_{M,N}$ values to create the $\phi_{M,N}$ value

c	3-Bit Value	$T(c)_{M,N}[0]$	$T(c)_{M,N}[1]$	$T(c)_{M,N}[2]$
0	1	0	0	1
1	0	0	0	0
2	0	0	0	0
3	3	0	1	1
4	1	0	0	1
5	0	0	0	0
6	0	0	0	0
7	1	0	0	1
		$R(\phi_{M,N})$	$G(\phi_{M,N})$	$B(\phi_{M,N})$
		0	8	153

After applying the procedure to each (M, N) the ϕ combined "color" tile is formed. Scanning row-by-row the ϕ, a 48-dimensional vector is formed.

$$16(Elements) \times 3(Channels) = 48Elements$$

where the values of each element are quantized in the interval $[0, 255]$ requiring 1 byte each. Therefore, the storage requirements of the descriptor remained the same (48 bytes), but its length in terms of number of elements was **losslessly** compressed by 62.5%. This obviously reduces the management overhead for implementation and allows for faster retrieval from the databases, in which it is stored.

4 Experimental Results

The extraction process of the proposed descriptor is applied to all images of the database and an XML file with the entries is created. Given that, the proposed descriptor is an MPEG-7 like descriptor, the schema of the SpCD as an MPEG-7 extension is described as follows.

```
<?xml version ="1.0" encoding="UTF-8"?>
<schema xmlns="http://www.w3.org/2001/XMLSchema"
        xmlns:mpeg7="urn:mpeg:mpeg7:schema :2004"
        xmlns:SpCDNS="SpCDNS" targetNamespace="SpCDNS">
  <import namespace="urn:mpeg:mpeg7:schema:2004"
                              schemaLocation="Mpeg7-2004.xsd"/>
  <complexType name="SpCDType" final="#all">
    <complexContent>
      <extension base="mpeg7:VisualDType">
       <sequence>
         <element name="value">
          <simpleType>
           <restriction>
            <simpleType>
              <listitemType="mpeg7:unsigned3"/>
```

```
          </simpleType>
           <length value="48"/>
         </restriction>
        </simpleType>
       </element>
     </sequence>
    </extension>
  </complexContent>
 </complexType>
</schema>
```

During the retrieval process, the user enters a query image in the form of Hand Drawn Color Sketch. From this image the 8 Tiles are exported and the 128-dimensional vector is formed. Next, each descriptor from the XML file is transformed from the 48-dimensional vector to the corresponding 128-dimensional vector, following the exact opposite procedure to that described in Section 3.

For the similarity matching we propose, in accordance to the other Compact Composite Descriptors, the Tanimoto coefficient. The distance T of the vectors X_i and X_j is defined as $T_{ij} \in [0,1]$ and is calculated as follows:

$$T_{ij} = t(X_i, X_j) = \frac{X_i^T \times X_j}{X_i^T \times X_i + X_j^T \times X_j - X_i^T \times X_j} \tag{6}$$

Where X^T is the transpose vector of X. In the absolute congruence of the vectors the Tanimoto coefficient takes the value 1, while in the maximum deviation the coefficient tends to zero.

After iteratively applying the process to all image descriptors contained in the XML file, the result list of best matching images is presented ranked by distance to the query image. The process is described in Figure 3.

In order to evaluate the performance of the proposed descriptor, experiments were carried out on 2 benchmarking image databases. The first database contains 1030 images, which they depict flags from around the world. The database consists of state flags, flags of unrecognized states, flags of formerly independent states, municipality flags, dependent territory flags and flags of first-level country subdivisions. All the images are taken from Wikipedia. We used 20 Hand Drawn Color Sketches as query images and as ground truth (list of similar images of each query) we considered only the flag that was attempted to be drawn with the Hand Drawn Color Sketch.

The "Paintings" database was also used. This database is incorporated in [17] and includes 1771 images of paintings by internationally well known artists. Also in this case, as query images were used 20 Hand Drawn Color Sketches and as ground truth was considered only the painting, which was attempted to be drawn with the Hand Drawn Color Sketch. Figure 4 shows a query sample from each database as well as the retrieval results.

The objective Averaged Normalized Modified Retrieval Rank (ANMRR)[10] is employed to evaluate the performance of the proposed descriptor.

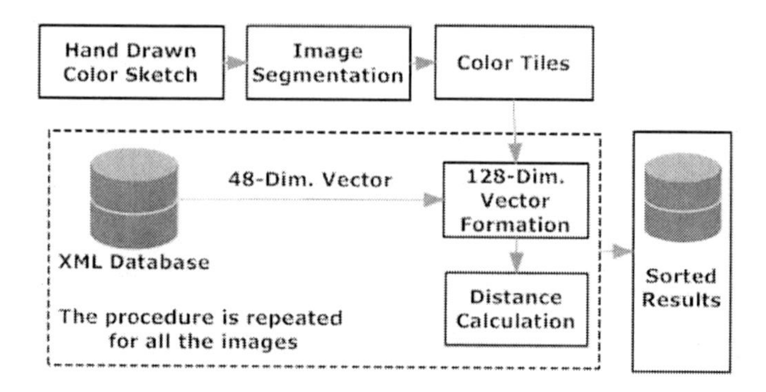

Fig. 3. Image Retrieval Procedure

The average rank AVR(q) for a given query q is:

$$AVR(q) = \sum_{k=1)}^{NG(q} \frac{Rank(k)}{NG(q)} \tag{7}$$

Where

- $NG(q)$ is the number of ground truth images for the query q. In our case $NG(q) = 1$
- K is the top ranked retrievals examined where:
  - $K = min(X \times NG(q), 2GMT)$, $GMT = max(NG(q))$, in our case $K = 2$
  - $NG(q) > 50$ then $X = 2$ else $X = 4$
- $Rank(k)$ is the retrieval rank of the ground truth image. Considering a query assume that as a result of the retrieval, the kth ground truth image for this query q is found at a position R. If this image is in the first K retrievals then $Rank(k) = R$ else $Rank(k) = K + 1$

The modified retrieval rank is:

$$MRR(q) = AVR(q) - 0.5 - 0.5 \times N(q) \tag{8}$$

The normalized modified retrieval rank is computed as follows:

$$NMRR(q) = \frac{MRR(q)}{K + 0.5 - 0.5 \times N(q)} \tag{9}$$

The average of NMRR over all queries defined as:

$$ANMRR(q) = \frac{1}{Q} \sum_{q=1}^{Q} NMRR(q) \tag{10}$$

Table 2 illustrates the ANMRR results in both databases. The results are compared with the corresponding results of the Color Layout Descriptor (CLD). The implementation

Table 2. ANMRR results

Database	SpCD ANMRR	CLD ANMRR
Flags	0.225	0.425
Paintings	0.275	0.45

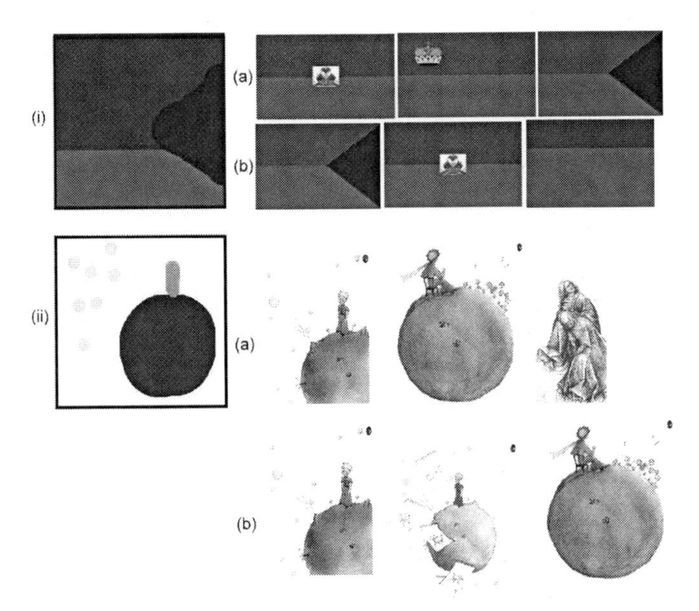

Fig. 4. Hand Drawn Color Sketch queries (i) for the Flags database and (ii) for the Paintings database. In each one of the queries (a) shows the first 3 results of the CLD, while (b) shows the first 3 results of the proposed descriptor.

of the CLD matches the implementation in [18]. The reason for this comparison is that only the CLD, from the descriptors that were referred in the introduction, is compact. Considering the results, it is evident that the proposed descriptor is able to achieve satisfactory retrieval results by using Hand Drawn Color Sketches as queries.

As mentioned in the introduction, the SpCD belongs to the family of the Compact Composite Descriptors, given that it combines more than one features in a compact vector (color and spatial color distribution information). The previous experiment demonstrates the effectiveness of the proposed descriptor on databases with artificially generated images. Similar, the quality of the other CCDs has so far been evaluated through retrieval from homogeneous benchmarking databases, containing images of only the type that each CCD is intended for. For example, Joint Composite Descriptor (JCD)[19] is tested on Wang[20], UCID[21] and NISTER[22] databases which contain natural color images while BTDH[14] on the IRMA[1] database consisting of grayscale medical radiology images. The JCD is the result of early fusion of the CEDD[12] and FCTH[13] which, as shown in [19] presents better retrieval results from the 2 descriptors which combined to result.

[1] IRMA is courtesy of TM Deserno, Dept.of Medical Informatics, RWTH Aachen, Germany.

Table 3. Experimental Results on Heterogeneous Database

Descriptor	40 Queries ANMRR
JCD	0.3554
BTDH	0.4015
SpCD	**0.3081**

In real-world image retrieval systems, databases have a much larger scale and may be heterogeneous. In this section, we evaluate the retrieval effectiveness of the CCDs on a heterogeneous database.

We created a heterogeneous database with 20230 images by joining the following: 9000 grayscale images from the IRMA 2005 database; 10200 natural color images from the NISTER database and the1030 artificially generated images from the Flags database as used in the previous experiment. We used 40 queries: The first 20 natural color image queries from the NISTER database and the first 20 grayscale queries of the IRMA 2005 database. The results are illustrated in Table 3. As one can see, the proposed descriptor performs better than the other CCDs.

The proposed descriptor is implemented in the image retrieval system img(Rummager) [23] and is available online[2] along with the image databases and the queries.

5 Conclusions

In this paper, a new compact composite descriptor has been presented. It combines color and spatial color distribution information. Characteristics of the proposed descriptor are its small storage requirements (48 bytes/ per image) and its small length (48 elements). Experiments were performed at 2 databases: one with artificial images, another one with natural and artificial images mixed in one data set. The proposed descriptor achieved very good results and outperformed the MPEG-7 CLD in both experiments.

References

1. Datta, R., Joshi, D., Li, J., Wang, J.: Image retrieval: Ideas, influences, and trends of the new age. ACM Computing Surveys 40(2), 1–60 (2008)
2. Chatzichristofis, S., Boutalis, Y.: A hybrid scheme for fast and accurate image retrieval based on color descriptors. In: IASTED International Conference on Artificial Intelligence and Soft Computing (2007)
3. Jacobs, C., Finkelstein, A., Salesin, D.: Fast multiresolution image querying. In: Proceedings of the 22nd Annual Conference on Computer Graphics and Interactive Techniques, pp. 277–286. ACM, New York (1995)
4. Pass, G., Zabih, R., Miller, J.: Comparing images using color coherence vectors. In: Proceedings of the Fourth ACM International Conference on Multimedia, pp. 65–73. ACM, New York (1997)

[2] `http://www.img-rummager.com`

5. Rao, A., Srihari, R., Zhang, Z.: Spatial color histograms for content-based image retrieval. In: 11th IEEE International Conference on Tools with Artificial Intelligence, Proceedings, pp. 183–186 (1999)
6. Cinque, L., Ciocca, G., Levialdi, S., Pellicano, A., Schettini, R.: Color-based image retrieval using spatial-chromatic histograms. Image and Vision Computing 19(13), 979–986 (2001)
7. Lim, S., Lu, G.: Spatial statistics for content based image retrieval. In: International Conference on Information Technology: Coding and Computing [Computers and Communications], Proceedings, ITCC 2003, pp. 155–159 (2003)
8. Sun, J., Zhang, X., Cui, J., Zhou, L.: Image retrieval based on color distribution entropy. Pattern Recognition Letters 27(10), 1122–1126 (2006)
9. Huang, J.: Image indexing using color correlograms. In: IEEE Computer Society Conference on Computer Vision and Pattern Recognition, pp. 762–768 (1997)
10. Manjunath, B., Ohm, J., Vasudevan, V., Yamada, A.: Color and texture descriptors. IEEE Transactions on Circuits and Systems for Video Technology 11(6), 703–715 (2001)
11. Kasutani, E., Yamada, A.: The mpeg-7 color layout descriptor: a compact image feature-description for high-speed image/video segment retrieval. In: Proceedings of 2001 International Conference on Image Processing, vol. 1 (2001)
12. Chatzichristofis, S., Boutalis, Y.: Cedd: Color and edge directivity descriptor: A compact descriptor for image indexing and retrieval. In: Gasteratos, A., Vincze, M., Tsotsos, J.K. (eds.) ICVS 2008. LNCS, vol. 5008, p. 312. Springer, Heidelberg (2008)
13. Chatzichristofis, S., Boutalis, Y.: Fcth: Fuzzy color and texture histogram-a low level feature for accurate image retrieval. In: Ninth International Workshop on Image Analysis for Multimedia Interactive Services, WIAMIS 2008, pp. 191–196 (2008)
14. Chatzichristofis, S.A., Boutalis, Y.S.: Content based radiology image retrieval using a fuzzy rule based scalable composite descriptor. Multimedia Tools and Applications 46, 493–519 (2010)
15. Konstantinidis, K., Gasteratos, A., Andreadis, I.: Image retrieval based on fuzzy color histogram processing. Optics Communications 248(4-6), 375–386 (2005)
16. Zimmermann, H.: Fuzzy sets, decision making, and expert systems. Kluwer Academic Pub., Dordrecht (1987)
17. Zagoris, K., Chatzichristofis, S., Papamrkos, N., Boutalis, Y.: Img(anaktisi): A web content based image retrieval system. In: 2nd International Workshop on Similarity Search and Applications (SISAP), pp. 154–155 (2009)
18. Lux, M., Chatzichristofis, S.: Lire: lucene image retrieval: an extensible java cbir library. ACM, New York (2008)
19. Chatzichristofis, S.A., Boutalis, Y.S., Lux, M.: Selection of the proper compact composite descriptor for improving content based image retrieval. In: SPPRA, pp. 134–140. ACTA Press (2009)
20. Wang, J., Li, J., Wiederhold, G.: Simplicity: Semantics-sensitive integrated matching for picture libraries. IEEE Transactions on Pattern Analysis and Machine Intelligence, 947–963 (2001)
21. Schaefer, G., Stich, M.: Ucid-an uncompressed colour image database. Storage and Retrieval Methods and Applications for Multimedia 5307, 472–480 (2004)
22. Nister, D., Stewenius, H.: Scalable recognition with a vocabulary tree. In: Proc. CVPR, Citeseer, vol. 5, pp. 2161–2168 (2006)
23. Chatzichristofis, S., Boutalis, Y., Lux, M.: Img(rummager): An interactive content based image retrieval system. In: 2nd International Workshop on Similarity Search and Applications (SISAP), pp. 151–153 (2009)

Appendix

Table 4. Fuzzy Interface Rules

If HUE is	And SATURATION is	And VALUE is	Then OUT is
RED 1	GRAY	WHITE	WHITE
RED 1	GRAY	BLACK	BLACK
RED 1	COLOR	WHITE	RED
RED 1	COLOR	BLACK	RED
YELLOW	GRAY	WHITE	WHITE
YELLOW	GRAY	BLACK	BLACK
YELLOW	COLOR	WHITE	YELLOW
YELLOW	COLOR	BLACK	YELLOW
GREEN	GRAY	WHITE	WHITE
GREEN	GRAY	BLACK	BLACK
GREEN	COLOR	WHITE	GREEN
GREEN	COLOR	BLACK	GREEN
CYAN	GRAY	WHITE	WHITE
CYAN	GRAY	BLACK	BLACK
CYAN	COLOR	WHITE	CYAN
CYAN	COLOR	BLACK	CYAN
BLUE	GRAY	WHITE	WHITE
BLUE	GRAY	BLACK	BLACK
BLUE	COLOR	WHITE	BLUE
BLUE	COLOR	BLACK	BLUE
MAGENTA	GRAY	WHITE	WHITE
MAGENTA	GRAY	BLACK	BLACK
MAGENTA	COLOR	WHITE	MAGENTA
MAGENTA	COLOR	BLACK	MAGENTA
RED 2	GRAY	WHITE	WHITE
RED 2	GRAY	BLACK	BLACK
RED 2	COLOR	WHITE	RED
RED 2	COLOR	BLACK	RED

Table 5. Fuzzy 8-bin Color System

Membership Function	Activation Value			
	0	1	1	0
HUE				
RED 1	0	0	5	55
YELLOW	5	55	60	115
GREEN	60	115	120	175
CYAN	120	175	180	235
BLUE	180	235	240	295
MAGENTA	240	295	300	355
RED 2	300	355	360	360
SATURATION				
GRAY	0	0	25	55
COLOR	25	55	255	255
VALUE				
WHITE	0	0	100	156
BLACK	100	156	255	255

Towards Min Max Generalization in Reinforcement Learning

Raphael Fonteneau[1], Susan A. Murphy[2], Louis Wehenkel[1], and Damien Ernst[1]

[1] University of Liège, Liège, Belgium
[2] University of Michigan, Ann Arbor, U.S.A.
{raphael.fonteneau,l.wehenkel,dernst}@ulg.ac.be
samurphy@umich.edu

Abstract. In this paper, we introduce a min max approach for addressing the generalization problem in Reinforcement Learning. The min max approach works by determining a sequence of actions that maximizes the worst return that could possibly be obtained considering any dynamics and reward function compatible with the sample of trajectories and some prior knowledge on the environment. We consider the particular case of deterministic Lipschitz continuous environments over continuous state spaces, finite action spaces, and a finite optimization horizon. We discuss the non-triviality of computing an exact solution of the min max problem even after reformulating it so as to avoid search in function spaces. For addressing this problem, we propose to replace, inside this min max problem, the search for the worst environment given a sequence of actions by an expression that lower bounds the worst return that can be obtained for a given sequence of actions. This lower bound has a tightness that depends on the sample sparsity. From there, we propose an algorithm of polynomial complexity that returns a sequence of actions leading to the maximization of this lower bound. We give a condition on the sample sparsity ensuring that, for a given initial state, the proposed algorithm produces an optimal sequence of actions in open-loop. Our experiments show that this algorithm can lead to more cautious policies than algorithms combining dynamic programming with function approximators.

1 Introduction

Since the late sixties, the field of Reinforcement Learning (RL) [27] has studied the problem of inferring from the sole knowledge of observed system trajectories, near-optimal solutions to optimal control problems. The original motivation was to design computational agents able to learn by themselves how to interact in a rational way with their environment. The techniques developed in this field have appealed researchers trying to solve sequential decision making problems in many fields such as Finance [15], Medicine [19,20] or Engineering [23].

RL algorithms are challenged when dealing with large or continuous state spaces. Indeed, in such cases they have to generalize the information contained in a generally sparse sample of trajectories. The dominating approach for generalizing this information is to combine RL algorithms with function approximators [2,16,9]. Usually, these approximators generalize the information contained in the sample to areas poorly covered by the sample by implicitly assuming that the properties of the system in those

J. Filipe, A. Fred, and B. Sharp (Eds.): ICAART 2010, CCIS 129, pp. 61–77, 2011.

areas are similar to the properties of the system in the nearby areas well covered by the sample. This in turn often leads to low performance guarantees on the inferred policy when large state space areas are poorly covered by the sample. This can be explained by the fact that when computing the performance guarantees of these policies, one needs to take into account that they may actually drive the system into the poorly visited areas to which the generalization strategy associates a favorable environment behavior, while the environment may actually be particularly adversarial in those areas. This is corroborated by theoretical results which show that the performance guarantees of the policies inferred by these algorithms degrade with the sample sparsity where, loosely speaking, the sparsity can be seen as the radius of the largest non-visited state space area.[1]

As in our previous work [12] from which this paper is an extended version, we assume a deterministic Lipschitz continuous environment over continuous state spaces, finite action spaces, and a finite time-horizon. In this context, we introduce a $\min\max$ approach to address the generalization problem. The $\min\max$ approach works by determining a sequence of actions that maximizes the worst return that could possibly be obtained considering any dynamics and reward functions compatible with the sample of trajectories, and a weak prior knowledge given in the form of upper bounds on the Lipschitz constants of the environment. However, we show that finding an exact solution of the $\min\max$ problem is far from trivial, even after reformulating the problem so as to avoid the search in the space of all compatible functions. To circumvent these difficulties, we propose to replace, inside this $\min\max$ problem, the search for the worst environment given a sequence of actions by an expression that lower bounds the worst return that can be obtained for a given sequence of actions. This lower bound is derived from [11] and has a tightness that depends on the sample sparsity. From there, we propose a Viterbi–like algorithm [28] for computing an open-loop sequence of actions to be used from a given initial state to maximize that lower bound. This algorithm is of polynomial computational complexity in the size of the dataset and the optimization horizon. It is named CGRL for Cautious Generalization (oriented) Reinforcement Learning since it essentially shows a cautious behaviour in the sense that it computes decisions that avoid driving the system into areas of the state space that are not well enough covered by the available dataset, according to the prior information about the dynamics and reward function. Besides, the CGRL algorithm does not rely on function approximators and it computes, as a byproduct, a lower bound on the return of its open-loop sequence of decisions. We also provide a condition on the sample sparsity ensuring that, for a given initial state, the CGRL algorithm produces an optimal sequence of actions in open-loop, and we suggest directions for leveraging our approach to a larger class of problems in RL.

The rest of the paper is organized as follows. Section 2 briefly discusses related work. In Section 3, we formalize the $\min\max$ approach to generalization, and we discuss its non trivial nature in Section 4. In Section 5, we exploit the results of [11] for lower

[1] Usually, these theoretical results do not give lower bounds per se but a distance between the actual return of the inferred policy and the optimal return. However, by adapting in a straightforward way the proofs behind these results, it is often possible to get a bound on the distance between the estimate of the return of the inferred policy computed by the RL algorithm and its actual return and, from there, a lower bound on the return of the inferred policy.

bounding the worst return that can be obtained for a given sequence of actions. Section 6 proposes a polynomial algorithm for inferring a sequence of actions maximizing this lower bound and states a condition on the sample sparsity for its optimality. Section 7 illustrates the features of the proposed algorithm and Section 8 discusses its interest, while Section 9 concludes.

2 Related Work

The $\min\max$ approach to generalization followed by the CGRL algorithm results in the output of policies that are likely to drive the agent only towards areas well enough covered by the sample. Heuristic strategies have already been proposed in the RL literature to infer policies that exhibit such a conservative behavior. As a way of example, some of these strategies associate high negative rewards to trajectories falling outside of the well covered areas. Other works in RL have already developped $\min\max$ strategies when the environment behavior is partially unknown [17,4,24]. However, these strategies usually consider problems with finite state spaces where the uncertainities come from the lack of knowledge of the transition probabilities [7,5]. In model predictive control (MPC) where the environment is supposed to be fully known [10], $\min\max$ approaches have been used to determine the optimal sequence of actions with respect to the "worst case" disturbance sequence occuring [1]. The CGRL algorithm relies on a methodology for computing a lower bound on the worst possible return (considering any compatible environment) in a deterministic setting with a mostly unknown actual environment. In this, it is related to works in the field of RL which try to get from a sample of trajectories lower bounds on the returns of inferred policies [18,22].

3 Problem Statement

We consider a discrete-time system whose dynamics over T stages is described by a time-invariant equation

$$x_{t+1} = f(x_t, u_t) \quad t = 0, 1, \ldots, T-1,$$

where for all t, the state x_t is an element of the compact state space $\mathcal{X} \subset \mathbb{R}^{d_{\mathcal{X}}}$ where $\mathbb{R}^{d_{\mathcal{X}}}$ denotes the $d_{\mathcal{X}}-$dimensional Euclidian space and u_t is an element of the finite (discrete) action space $\mathcal{U}$. $T \in \mathbb{N}_0$ is referred to as the optimization horizon. An instantaneous reward $r_t = \rho(x_t, u_t) \in \mathbb{R}$ is associated with the action u_t taken while being in state x_t. For every initial state $x \in \mathcal{X}$ and for every sequence of actions $(u_0, \ldots, u_{T-1}) \in \mathcal{U}^T$, the cumulated reward over T stages (also named return over T stages) is defined as

$$J^{u_0,\ldots,u_{T-1}}(x) = \sum_{t=0}^{T-1} \rho(x_t, u_t) \; ,$$

where $x_{t+1} = f(x_t, u_t)\;, \forall t \in \{0, \ldots, T-1\}$ and $x_0 = x$. We assume that the system dynamics f and the reward function ρ are Lipschitz continuous, i.e. that there exist finite constants $L_f, L_\rho \in \mathbb{R}$ such that: $\forall x', x'' \in \mathcal{X}, \forall u \in \mathcal{U}$,

$$\|f(x', u) - f(x'', u)\|_{\mathcal{X}} \leq L_f \|x' - x''\|_{\mathcal{X}} ,$$
$$|\rho(x', u) - \rho(x'', u)| \leq L_\rho \|x' - x''\|_{\mathcal{X}} ,$$

where $\|.\|_{\mathcal{X}}$ denotes the Euclidian norm over the space $\mathcal{X}$. We further suppose that: (i) the system dynamics f and the reward function ρ are unknown, (ii) a set of one-step transitions $\mathcal{F}_n = \{(x^l, u^l, r^l, y^l)\}_{l=1}^n$ is known where each one-step transition is such that $y^l = f(x^l, u^l)$ and $r^l = \rho(x^l, u^l)$, (iii) $\forall a \in \mathcal{U}, \ \exists (x, u, r, y) \in \mathcal{F}_n \ : u = a$ (each action $a \in \mathcal{U}$ appears at least once in $\mathcal{F}_n$) and (iv) two constants L_f and L_ρ satisfying the above-written inequalities are known.[2] We define the set of functions $\mathcal{L}^f_{\mathcal{F}_n}$ (resp. $\mathcal{L}^\rho_{\mathcal{F}_n}$) from $\mathcal{X} \times \mathcal{U}$ into $\mathcal{X}$ (resp. into $\mathbb{R}$) as follows :

$$\mathcal{L}^f_{\mathcal{F}_n} = \left\{ f' : \mathcal{X} \times \mathcal{U} \to \mathcal{X} \left| \begin{cases} \forall x', x'' \in \mathcal{X}, \forall u \in \mathcal{U} , \\ \|f'(x', u) - f'(x'', u)\|_{\mathcal{X}} \leq L_f \|x' - x''\|_{\mathcal{X}} , \\ \forall l \in \{1, \ldots, n\}, f'(x^l, u^l) = f(x^l, u^l) = y^l \end{cases} \right. \right\} ,$$

$$\mathcal{L}^\rho_{\mathcal{F}_n} = \left\{ \rho' : \mathcal{X} \times \mathcal{U} \to \mathbb{R} \left| \begin{cases} \forall x', x'' \in \mathcal{X}, \forall u \in \mathcal{U} , \\ |\rho'(x', u) - \rho'(x'', u)| \leq L_\rho \|x' - x''\|_{\mathcal{X}} , \\ \forall l \in \{1, \ldots, n\}, \rho'(x^l, u^l) = \rho(x^l, u^l) = r^l \end{cases} \right. \right\} .$$

In the following, we call a "compatible environment" any pair $(f', \rho') \in \mathcal{L}^f_{\mathcal{F}_n} \times \mathcal{L}^\rho_{\mathcal{F}_n}$. Given a compatible environment (f', ρ'), a sequence of actions $(u_0, \ldots, u_{T-1}) \in \mathcal{U}^T$ and an initial state $x \in \mathcal{X}$, we introduce the $(f', \rho')-$return over T stages when starting from $x \in \mathcal{X}$:

$$J^{u_0, \ldots, u_{T-1}}_{(f', \rho')}(x) = \sum_{t=0}^{T-1} \rho'(x'_t, u_t) ,$$

where $x'_0 = x$ and $x'_{t+1} = f'(x'_t, u_t)$, $\forall t \in \{0, \ldots, T-1\}$. We introduce $L^{u_0, \ldots, u_{T-1}}_{\mathcal{F}_n}(x)$ such that

$$L^{u_0, \ldots, u_{T-1}}_{\mathcal{F}_n}(x) = \min_{(f', \rho') \in \mathcal{L}^f_{\mathcal{F}_n} \times \mathcal{L}^\rho_{\mathcal{F}_n}} \left\{ J^{u_0, \ldots, u_{T-1}}_{(f', \rho')}(x) \right\} .$$

The existence of $L^{u_0, \ldots, u_{T-1}}_{\mathcal{F}_n}(x)$ is ensured by the following arguments: (i) the space $\mathcal{X}$ is compact, (ii) the set $\mathcal{L}^f_{\mathcal{F}_n} \times \mathcal{L}^\rho_{\mathcal{F}_n}$ is closed and bounded considering the $\|.\|_\infty$ norm ($\|(f', \rho')\|_\infty = \sup_{(x,u) \in \mathcal{X} \times \mathcal{U}} \|(f'(x, u), \rho'(x, u))\|_{\mathbb{R}^{d_\mathcal{X}+1}}$ where $\|.\|_{\mathbb{R}^{d_\mathcal{X}+1}}$ is the Euclidian norm over $\mathbb{R}^{d_\mathcal{X}+1}$) and (iii) one can show that the mapping $\mathcal{M}^{u_0, \ldots, u_{T-1}}_{\mathcal{F}_n, x} : \mathcal{L}^f_{\mathcal{F}_n} \times \mathcal{L}^\rho_{\mathcal{F}_n} \to \mathbb{R}$ such that $\mathcal{M}^{u_0, \ldots, u_{T-1}}_{\mathcal{F}_n, x}(f', \rho') = J^{u_0, \ldots, u_{T-1}}_{(f', \rho')}(x)$ is a continuous mapping. Furthermore, this also proves that

$$\forall (u_0, \ldots, u_{T-1}) \in \mathcal{U}^T, \forall x \in \mathcal{X}, \exists (f^{u_0, \ldots, u_{T-1}}_{\mathcal{F}_n, x}, \rho^{u_0, \ldots, u_{T-1}}_{\mathcal{F}_n, x}) \in \mathcal{L}^f_{\mathcal{F}_n} \times \mathcal{L}^\rho_{\mathcal{F}_n} :$$
$$J^{u_0, \ldots, u_{T-1}}_{(f^{u_0, \ldots, u_{T-1}}_{\mathcal{F}_n, x}, \rho^{u_0, \ldots, u_{T-1}}_{\mathcal{F}_n, x})}(x) = L^{u_0, \ldots, u_{T-1}}_{\mathcal{F}_n}(x). \quad (1)$$

[2] These constants do not necessarily have to be the smallest ones satisfying these inequalities (i.e., the Lispchitz constants).

Our goal is to compute, given an initial state $x \in \mathcal{X}$, an open-loop sequence of actions $(\dot{u}_0(x), \ldots, \dot{u}_{T-1}(x)) \in \mathcal{U}^T$ that gives the highest return in the least favorable compatible environment. This problem can be formalized as the min max problem:

$$(\dot{u}_0(x), \ldots, \dot{u}_{T-1}(x)) \in \underset{(u_0,\ldots,u_{T-1}) \in \mathcal{U}^T}{\arg\max} \left\{ L_{\mathcal{F}_n}^{u_0,\ldots,u_{T-1}}(x) \right\} .$$

4 Reformulation of the min max Problem

Since $\mathcal{U}$ is finite, one could solve the min max problem by computing for each $(u_0, \ldots, u_{T-1}) \in \mathcal{U}^T$ the value of $L_{\mathcal{F}_n}^{u_0,\ldots,u_{T-1}}(x)$. As the latter computation is posed as an infinite-dimensional minimization problem over the function space $\mathcal{L}_{\mathcal{F}_n}^f \times \mathcal{L}_{\mathcal{F}_n}^\rho$, we first show that it can be reformulated as a finite-dimensional problem over $\mathcal{X}^{T-1} \times \mathbb{R}^T$. This is based on the observation that $L_{\mathcal{F}_n}^{u_0,\ldots,u_{T-1}}(x)$ is actually equal to the lowest sum of rewards that could be collected along a trajectory compatible with an environment from $\mathcal{L}_{\mathcal{F}_n}^f \times \mathcal{L}_{\mathcal{F}_n}^\rho$, and is precisely stated by the following Theorem (see Appendix A for the proof).

Theorem 1 (Equivalence). *Let* $(u_0, \ldots, u_{T-1}) \in \mathcal{U}^T$ *and* $x \in \mathcal{X}$. *Let* $K_{\mathcal{F}_n}^{u_0,\ldots,u_{T-1}}(x)$ *be the solution of the following optimization problem:*

$$K_{\mathcal{F}_n}^{u_0,\ldots,u_{T-1}}(x) = \min_{\substack{\hat{r}_0 \ldots \hat{r}_{T-1} \in \mathbb{R} \\ \hat{x}_0 \ldots \hat{x}_{T-1} \in \mathcal{X}}} \left\{ \sum_{t=0}^{T-1} \hat{r}_t \right\},$$

where the variables $\hat{x}_0, \ldots, \hat{x}_{T-1}$ *and* $\hat{r}_0, \ldots, \hat{r}_{T-1}$ *satisfy the constraints*

$$\left.\begin{array}{l} |\hat{r}_t - r^{l_t}| \leq L_\rho \|\hat{x}_t - x^{l_t}\|_{\mathcal{X}} , \\ \|\hat{x}_{t+1} - y^{l_t}\|_{\mathcal{X}} \leq L_f \|\hat{x}_t - x^{l_t}\|_{\mathcal{X}} \end{array}\right\} \forall l_t \in \{0, \ldots, T-1 | u^{l_t} = u_t\} ,$$

$$\left.\begin{array}{l} |\hat{r}_t - \hat{r}_{t'}| \leq L_\rho \|\hat{x}_t - \hat{x}_{t'}\|_{\mathcal{X}} ; \\ \|\hat{x}_{t+1} - \hat{x}_{t'+1}\|_{\mathcal{X}} \leq L_f \|\hat{x}_t - \hat{x}_{t'}\|_{\mathcal{X}} \end{array}\right\} \forall t, t' \in \{0, \ldots, T-1 | u_t = u_{t'}\} ,$$

$$\hat{x}_0 = x .$$

Then,

$$K_{\mathcal{F}_n}^{u_0,\ldots,u_{T-1}}(x) = L_{\mathcal{F}_n}^{u_0,\ldots,u_{T-1}}(x) .$$

Unfortunately, this latter minimization problem turns out to be non-convex in its generic form and, hence "off the shelf" algorithms will only be able to provide upper bounds on its value. Furthermore, the overall complexity of an algorithm that would be based on the enumeration of $\mathcal{U}^T$, combined with a local optimizer for the inner loop, may be intractable as soon as the cardinality of the action space $\mathcal{U}$ and/or the optimization horizon T become large.

We leave the exploration of the above formulation for future research. Instead, in the following subsections, we use the results from [11] to define a maximal lower bound $B_{\mathcal{F}_n}^{u_0,\ldots,u_{T-1}}(x) \leq L_{\mathcal{F}_n}^{u_0,\ldots,u_{T-1}}(x)$ for a given initial state $x \in \mathcal{X}$ and a sequence

$(u_0, \ldots, u_{T-1}) \in \mathcal{U}^T$. Furthermore, we show that the maximization of this lower bound $B^{u_0,\ldots,u_{T-1}}_{\mathcal{F}_n}(x)$ with respect to the choice of a sequence of actions lends itself to a dynamic programming type of decomposition. In the end, this yields a polynomial algorithm for the computation of a sequence of actions $(\hat{u}^*_{\mathcal{F}_n,0}(x), \ldots, \hat{u}^*_{\mathcal{F}_n,T-1}(x))$ maximizing a lower bound of the original $\min\max$ problem, i.e. $(\hat{u}^*_{\mathcal{F}_n,0}(x), \ldots, \hat{u}^*_{\mathcal{F}_n,T-1}(x)) \in \underset{(u_0,\ldots,u_{T-1}) \in \mathcal{U}^T}{\arg\max} \left\{ B^{u_0,\ldots,u_{T-1}}_{\mathcal{F}_n}(x) \right\}$.

5 Lower Bound on the Return of a Given Sequence of Actions

In this section, we present a method for computing, from a given initial state $x \in \mathcal{X}$, a sequence of actions $(u_0, \ldots, u_{T-1}) \in \mathcal{U}^T$, a dataset of transitions, and weak prior knowledge about the environment, a lower bound on $L^{u_0,\ldots,u_{T-1}}_{\mathcal{F}_n}(x)$. The method is adapted from [11]. In the following, we denote by $\mathcal{F}^T_{n,(u_0,\ldots,u_{T-1})}$ the set of all sequences of one-step system transitions $[(x^{l_0}, u^{l_0}, r^{l_0}, y^{l_0}), \ldots, (x^{l_{T-1}}, u^{l_{T-1}}, r^{l_{T-1}}, y^{l_{T-1}})]$ that may be built from elements of $\mathcal{F}_n$ and that are compatible with $u_0, \ldots, u_{T-1}$, i.e. for which $u^{l_t} = u_t, \forall t \in \{0, \ldots, T-1\}$. First, we compute a lower bound on $L^{u_0,\ldots,u_{T-1}}_{\mathcal{F}_n}(x)$ from any given element τ from $\mathcal{F}^T_{n,(u_0,\ldots,u_{T-1})}$. This lower bound $B(\tau, x)$ is made of the sum of the T rewards corresponding to τ ($\sum_{t=0}^{T-1} r^{l_t}$) and T negative terms. Every negative term is associated with a one-step transition. More specifically, the negative term corresponding to the transition $(x^{l_t}, u^{l_t}, r^{l_t}, y^{l_t})$ of τ represents an upper bound on the variation of the cumulated rewards over the remaining time steps that can occur by simulating the system from a state x^{l_t} rather than $y^{l_{t-1}}$ (with $y^{l_{-1}} = x$) and considering any compatible environment (f', ρ') from $\mathcal{L}^f_{\mathcal{F}_n} \times \mathcal{L}^\rho_{\mathcal{F}_n}$. By maximizing $B(\tau, x)$ over $\mathcal{F}^T_{n,(u_0,\ldots,u_{T-1})}$, we obtain a maximal lower bound on $L^{u_0,\ldots,u_{T-1}}_{\mathcal{F}_n}(x)$. Furthermore, we prove that the distance from the maximal lower bound to the actual return $J^{u_0,\ldots,u_{T-1}}(x)$ can be characterized in terms of the sample sparsity.

5.1 Computing a Bound from a Given Sequence of One-Step Transitions

We have the following lemma.

Lemma 1. *Let* $(u_0, \ldots, u_{T-1}) \in \mathcal{U}^T$ *be a sequence of actions and* $x \in \mathcal{X}$ *an initial state. Let* $\tau = [(x^{l_t}, u^{l_t}, r^{l_t}, y^{l_t})]_{t=0}^{T-1} \in \mathcal{F}^T_{n,(u_0,\ldots,u_{T-1})}$. *Then,*

$$B(\tau, x) \leq L^{u_0,\ldots,u_{T-1}}_{\mathcal{F}_n}(x) \leq J^{u_0,\ldots,u_{T-1}}(x) ,$$

with

$$B(\tau, x) \doteq \sum_{t=0}^{T-1} \left[r^{l_t} - L_{Q_{T-t}} \| y^{l_{t-1}} - x^{l_t} \|_{\mathcal{X}} \right], y^{l_{-1}} = x , L_{Q_{T-t}} = L_\rho \sum_{i=0}^{T-t-1} (L_f)^i .$$

The proof is given in Appendix B. The lower bound on $L^{u_0,\ldots,u_{T-1}}_{\mathcal{F}_n}(x)$ derived in this lemma can be interpreted as follows. Given any compatible environment $(f', \rho') \in$

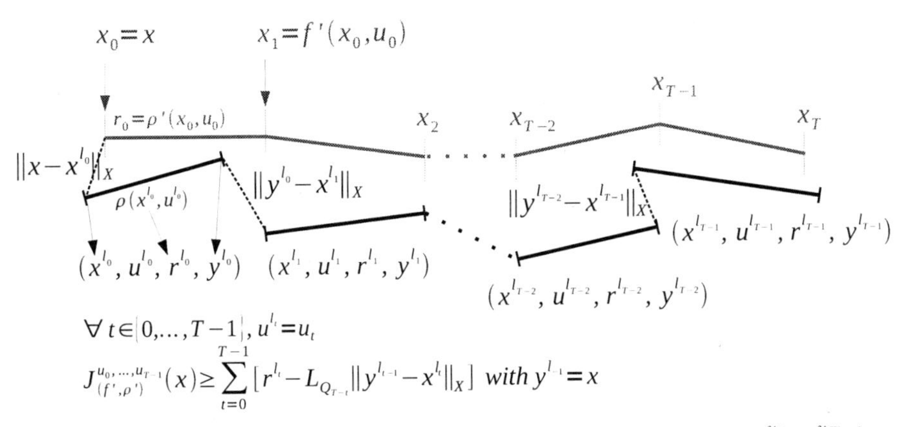

Fig. 1. A graphical interpretation of the different terms composing the bound on $J^{u_0,\ldots,u_{T-1}}_{(f',\rho')}(x)$ computed from a sequence of one-step transitions

$\mathcal{L}^f_{\mathcal{F}_n} \times \mathcal{L}^\rho_{\mathcal{F}_n}$, the sum of the rewards of the "broken" trajectory formed by the sequence of one-step system transitions τ can never be greater than $J^{u_0,\ldots,u_{T-1}}_{(f',\rho')}(x)$, provided that every reward r^{l_t} is penalized by a factor $L_{Q_{T-t}} \|y^{l_{t-1}} - x^{l_t}\|_{\mathcal{X}}$. This factor is in fact an upper bound on the variation of the $(T-t)$-state-action value function given any compatible environment (f', ρ') (see Appendix B) that can occur when "jumping" from $(y^{l_{t-1}}, u_t)$ to (x^{l_t}, u_t). An illustration of this is given in Figure 1.

5.2 Tightness of Highest Lower Bound over All Compatible Sequences of One-Step Transitions

We define

$$B^{u_0,\ldots,u_{T-1}}_{\mathcal{F}_n}(x) = \max_{\tau \in \mathcal{F}^T_{n,(u_0,\ldots,u_{T-1})}} B(\tau, x)$$

and we analyze in this subsection the distance from the lower bound $B^{u_0,\ldots,u_{T-1}}_{\mathcal{F}_n}(x)$ to the actual return $J^{u_0,\ldots,u_{T-1}}(x)$ as a function of the sample sparsity. The sample sparsity is defined as follows: let $\mathcal{F}_{n,a} = \{(x^l, u^l, r^l, y^l) \in \mathcal{F}_n | u^l = a\}$ ($\forall a,\ \mathcal{F}_{n,a} \neq \emptyset$ according to assumption (iii) given in Section 3). Since $\mathcal{X}$ is a compact subset of $\mathbb{R}^{d_\mathcal{X}}$, it is bounded and there exists $\alpha \in \mathbb{R}^+$:

$$\forall a \in \mathcal{U}\ ,\ \sup_{x' \in X} \left\{ \min_{(x^l,u^l,r^l,y^l) \in \mathcal{F}_{n,a}} \{\|x^l - x'\|_{\mathcal{X}}\} \right\} \leq \alpha\ . \tag{2}$$

The smallest α which satisfies equation (2) is named the sample sparsity and is denoted by $\alpha^*_{\mathcal{F}_n}$. We have the following theorem.

Theorem 2 (Tightness of Highest Lower Bound)

$$\exists\, C > 0 :\ \forall (u_0, \ldots, u_{T-1}) \in \mathcal{U}^T, J^{u_0,\ldots,u_{T-1}}(x) - B^{u_0,\ldots,u_{T-1}}_{\mathcal{F}_n}(x) \leq C\alpha^*_{\mathcal{F}_n}.$$

The proof of Theorem 2 is given in the Appendix of [12]. The lower bound $B^{u_0,\ldots,u_{T-1}}_{\mathcal{F}_n}(x)$ thus converges to the $T-$stage return of the sequence of actions $(u_0, \ldots, u_{T-1}) \in \mathcal{U}^T$ when the sample sparsity $\alpha^*_{\mathcal{F}_n}$ decreases to zero.

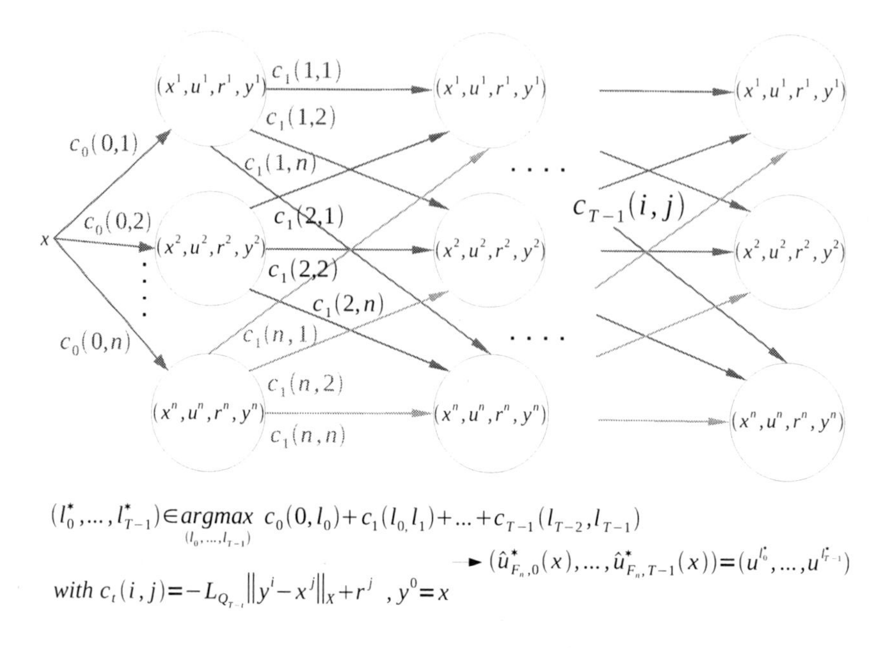

Fig. 2. A graphical interpretation of the CGRL algorithm

6 Computing a Sequence of Actions Maximizing the Highest Lower Bound

Let $\mathfrak{B}^*_{\mathcal{F}_n}(x) = \underset{(u_0,\dots,u_{T-1})\in\mathcal{U}^T}{\arg\max} \left\{B^{u_0,\dots,u_{T-1}}_{\mathcal{F}_n}(x)\right\}$. The CGRL algorithm computes for each initial state $x \in \mathcal{X}$ a sequence of actions $(\hat{u}^*_{\mathcal{F}_n,0}(x),\dots,\hat{u}^*_{\mathcal{F}_n,T-1}(x))$ that belongs to $\mathfrak{B}^*_{\mathcal{F}_n}(x)$. From what precedes, it follows that the actual return $J^{\hat{u}^*_{\mathcal{F}_n,0}(x),\dots,\hat{u}^*_{\mathcal{F}_n,T-1}(x)}(x)$ of this sequence is lower-bounded by the quantity $\underset{(u_0,\dots,u_{T-1})\in\mathcal{U}^T}{\max} B^{u_0,\dots,u_{T-1}}_{\mathcal{F}_n}(x)$. Due to the tightness of the lower bound $B^{u_0,\dots,u_{T-1}}_{\mathcal{F}_n}(x)$, the value of the return which is guaranteed will converge to the true return of the sequence of actions when $\alpha^*_{\mathcal{F}_n}$ decreases to zero. Additionaly, we prove in Section 6.1 that when the sample sparsity $\alpha^*_{\mathcal{F}_n}$ decreases below a particular threshold, the sequence $(\hat{u}^*_{\mathcal{F}_n,0}(x),\dots,\hat{u}^*_{\mathcal{F}_n,T-1}(x))$ is optimal. To identify a sequence of actions that belongs to $\mathfrak{B}^*_{\mathcal{F}_n}(x)$ without computing for all sequences $(u_0,\dots,u_{T-1})\in\mathcal{U}^T$ the value $B^{u_0,\dots,u_{T-1}}_{\mathcal{F}_n}(x)$, the CGRL algorithm exploits the fact that the problem of finding an element of $\mathfrak{B}^*_{\mathcal{F}_n}(x)$ can be reformulated as a shortest path problem.

6.1 Convergence of $(\hat{u}^*_{\mathcal{F}_n,0}(x),\dots,\hat{u}^*_{\mathcal{F}_n,T-1}(x))$ towards an Optimal Sequence of Actions

We prove hereafter that when $\alpha^*_{\mathcal{F}_n}$ gets lower than a particular threshold, the CGRL algorithm can only output optimal policies.

Theorem 3 (Convergence of the CGRL Algorithm)
Let

$$\mathfrak{J}^*(x) = \left\{(u_0, \ldots, u_{T-1}) \in \mathcal{U}^T | J^{u_0,\ldots,u_{T-1}}(x) = J^*(x)\right\} ,$$

and let us suppose that $\mathfrak{J}^*(x) \neq \mathcal{U}^T$ *(if* $\mathfrak{J}^*(x) = \mathcal{U}^T$*, the search for an optimal sequence of actions is indeed trivial). We define*

$$\epsilon(x) = \min_{(u_0,\ldots,u_{T-1}) \in \mathcal{U}^T \setminus \mathfrak{J}^*(x)} \{J^*(x) - J^{u_0,\ldots,u_{T-1}}(x)\} .$$

Then

$$C\alpha^*_{\mathcal{F}_n} < \epsilon(x) \implies (\hat{u}^*_{\mathcal{F}_n,0}(x), \ldots, \hat{u}^*_{\mathcal{F}_n,T-1}(x)) \in \mathfrak{J}^*(x) .$$

The proof of Theorem 3 is given in the Appendix of [12].

6.2 Cautious Generalization Reinforcement Learning Algorithm

The CGRL algorithm computes an element of the set $\mathfrak{B}^*_{\mathcal{F}_n}(x)$ defined previously. Let $\mathcal{D} : \mathcal{F}_n^T \to \mathcal{U}^T$ be the operator that maps a sequence of one-step system transitions $\tau = [(x^{l_t}, u^{l_t}, r^{l_t}, y^{l_t})]_{t=0}^{T-1}$ into the sequence of actions $(u^{l_0}, \ldots, u^{l_{T-1}})$. Using this operator, we can write

$$\mathfrak{B}^*_{\mathcal{F}_n}(x) = \left\{ (u_0, \ldots, u_{T-1}) \in \mathcal{U}^T \left| \begin{array}{l} \exists \tau \in \arg\max_{\tau \in \mathcal{F}_n^T} \{B(\tau, x)\} , \\ \mathcal{D}(\tau) = (u_0, \ldots, u_{T-1}) \end{array} \right. \right\} .$$

Or, equivalently

$$\mathfrak{B}^*_{\mathcal{F}_n}(x) = \left\{ (u_0, \ldots, u_{T-1}) \in \mathcal{U}^T \left| \begin{array}{l} \exists \tau \in \arg\max_{\tau \in \mathcal{F}_n^T} \sum_{t=0}^{T-1} \left[r^{l_t} - L_{Q_{T-t}} \|y^{l_{t-1}} - x^{l_t}\|_{\mathcal{X}}\right] , \\ \mathcal{D}(\tau) = (u_0, \ldots, u_{T-1}) \end{array} \right. \right\} .$$

From this expression, we can notice that a sequence of one-step transitions τ such that $\mathcal{D}(\tau)$ belongs to $\mathfrak{B}^*_{\mathcal{F}_n}(x)$ can be obtained by solving a shortest path problem on the graph given in Figure 2. The CGRL algorithm works by solving this problem using the Viterbi algorithm and by applying the operator $\mathcal{D}$ to the sequence of one-step transitions τ corresponding to its solution. Its complexity is quadratic with respect to the cardinality n of the input sample $\mathcal{F}_n$ and linear with respect to the optimization horizon T.

7 Illustration

In this section, we illustrate the CGRL algorithm on a variant of the puddle world benchmark introduced in [26]. In this benchmark, a robot whose goal is to collect high cumulated rewards navigates on a plane. A puddle stands in between the initial position of the robot and the high reward area. If the robot is in the puddle, it gets highly negative rewards. An optimal navigation strategy drives the robot around the puddle to reach the high reward area. Two datasets of one-step transitions have been used in our example. The first set $\mathcal{F}$ contains elements that uniformly cover the area of the state space that can be reached within T steps. The set $\mathcal{F}'$ has been obtained by removing from $\mathcal{F}$ the

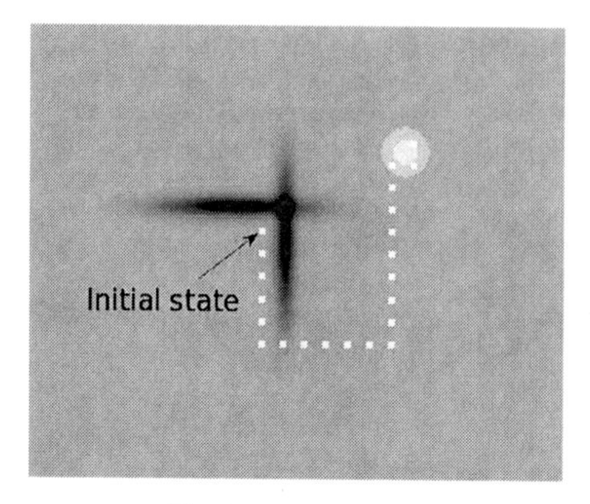

Fig. 3. CGRL with $\mathcal{F}$

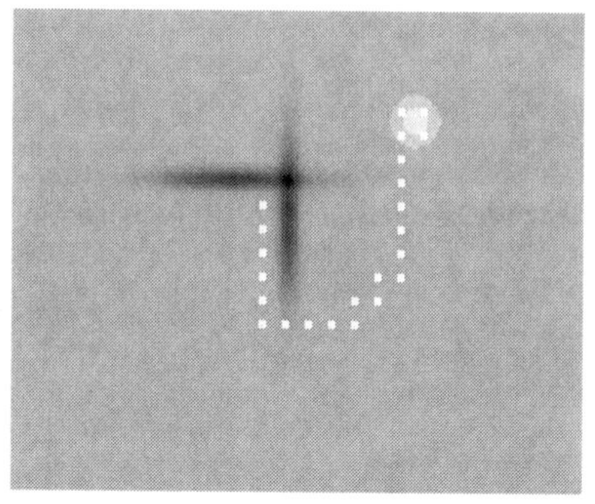

Fig. 4. FQI with $\mathcal{F}$

elements corresponding to the highly negative rewards.[3] The full specification of the benchmark and the exact procedure for generating $\mathcal{F}$ and $\mathcal{F}'$ are given in [12].

On Figure 3, we have drawn the trajectory of the robot when following the sequence of actions computed by the CGRL algorithm. Every state encountered is represented by a white square. The plane upon which the robot navigates has been colored such that the darker the area, the smaller the corresponding rewards are. In particular, the puddle area is colored in dark grey/black. We see that the CGRL policy drives the robot around the puddle to reach the high-reward area – which is represented by the light-grey circles. The CGRL algorithm also computes a lower bound on the cumulated rewards obtained by this action sequence. Here, we found out that this lower bound was rather conservative.

Figure 4 represents the policy inferred from $\mathcal{F}$ by using the (finite-time version of the) Fitted Q Iteration algorithm (FQI) combined with extremely randomized trees as function approximators [9]. The trajectories computed by the CGRL and FQI algorithms are very similar and so are the sums of rewards obtained by following these two trajectories. However, by using $\mathcal{F}'$ rather that $\mathcal{F}$, the CGRL and FQI algorithms do not lead to similar trajectories, as it is shown on Figures 5 and 6. Indeed, while the CGRL policy still drives the robot around the puddle to reach the high reward area, the FQI policy makes the robot cross the puddle. In terms of optimality, this latter navigation strategy is much worse. The difference between both navigation strategies can be explained as follows. The FQI algorithm behaves as if it were associating to areas of the state space that are not covered by the input sample, the properties of the elements of this sample that are located in the neighborhood of these areas. This in turn explains why it computes a policy that makes the robot cross the puddle. The same behavior could probably be observed by using other algorithms that combine dynamic programming strategies with kernel-based approximators or averagers [3,14,21]. The CGRL algorithm generalizes the information contained in the dataset, by assuming, given the intial state, the most adverse behavior for the environment according to its weak prior

[3] Although this problem might be treated by on-line learning methods, in some settings - for whatever reason - on-line learning may be impractical and all one will have is a batch of trajectories.

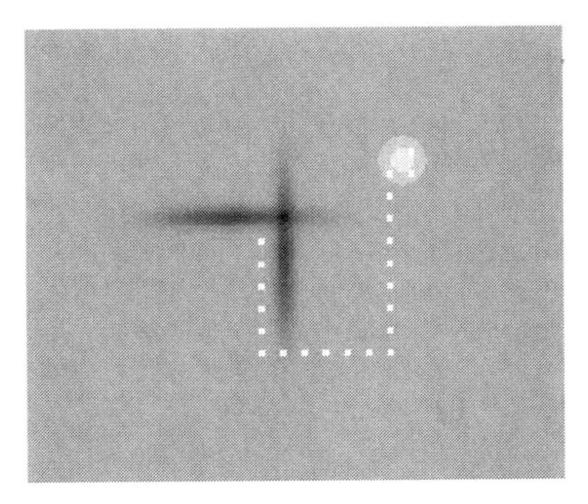

Fig. 5. CGRL with $\mathcal{F}'$

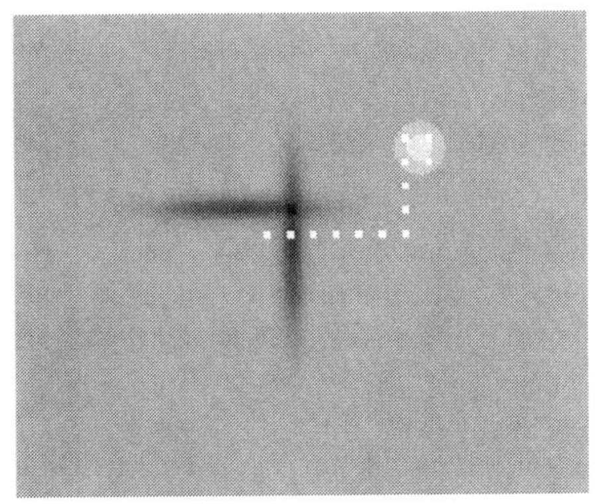

Fig. 6. FQI with $\mathcal{F}'$

knowledge about the environment. This results in the fact that the CGRL algorithm penalizes sequences of decisions that could drive the robot in areas not well covered by the sample, and this explains why the CGRL algorithm drives the robot around the puddle when run with $\mathcal{F}'$.

8 Discussion

The CGRL algorithm outputs a sequence of actions as well as a lower bound on its return. When $L_f > 1$ (e.g. when the system is unstable), this lower bound will decrease exponentially with T. This may lead to very low performance guarantees when the optimization horizon T is large. However, one can also observe that the terms $L_{Q_{T-t}}$ – which are responsible for the exponential decrease of the lower bound with the optimization horizon – are multiplied by the distance between the end state of a one-step transition and the beginning state of the next one-step transition of the sequence τ ($\|y^{l^*_{t-1}} - x^{l^*_t}\|_{\mathcal{X}}$) solution of the shortest path problem of Figure 2. Therefore, if these states $y^{l^*_{t-1}}$ and $x^{l^*_t}$ are close to each other, the CGRL algorithm can lead to good performance guarantees even for large values of T. It is also important to notice that this lower bound does not depend explicitly on the sample sparsity $\alpha^*_{\mathcal{F}_n}$, but depends rather on the initial state for which the sequence of actions is computed. Therefore, this may lead to cases where the CGRL algorithm provides good performance guarantees for some specific initial states, even if the sample does not cover every area of the state space well enough.

Other RL algorithms working in a similar setting as the CGRL algorithm, while not exploiting the weak prior knowledge about the environment, do not output a lower bound on the return of the policy h they infer from the sample of trajectories $\mathcal{F}_n$. However, some lower bounds on the return of h can still be computed. For instance, this can be done by exploiting the results of [11] upon which the CGRL algorithm is based. However, one can show that following the strategy described in [11] would necessarily lead to a bound lower than the lower bound associated to the sequence of actions computed by the CGRL algorithm. Another strategy would be to design global lower bounds on their policy by adapting proofs used to establish the consistency of these algorithms.

As a way of example, by proceeding like this, we can design a lower bound on the return of the policy given by the FQI algorithm when combined with some specific approximators which have, among others, Lipschitz continuity properties. These algorithms compute a sequence of state-action value functions $\hat{Q}_1, \hat{Q}_2, \ldots, \hat{Q}_T$ and compute the policy $h : \{0, 1, \ldots, T-1\} \times X$ defined as follows : $h(t, x_t) \in \arg\max_{u \in \mathcal{U}} \hat{Q}_{T-t}(x_t, u)$. For instance when using kernel-based approximators [21], we have as result that the return of h when starting from a state x is larger than $\hat{Q}_T(x, h(0, x)) - (C_1 T + C_2 T^2) \cdot \alpha^*_{\mathcal{F}_n}$ where C_1 and C_2 depends on L_f, L_ρ, the Lipschtiz constants of the class of approximation and an upper bound on ρ (the proof of this result can be found in [13]). The explicit dependence of this lower bound on $\alpha^*_{\mathcal{F}_n}$ as well as the large values of C_1 and C_2 tend to lead to a very conservative lower bound, especially when $\mathcal{F}_n$ is sparse.

9 Conclusions

In this paper, we have considered min max based approaches for addressing the generalization problem in RL. In particular, we have proposed and studied an algorithm that outputs a policy that maximizes a lower bound on the worst return that may be obtained with an environment compatible with some observed system transitions. The proposed algorithm is of polynomial complexity and avoids regions of the state space where the sample density is too low according to the prior information. A simple example has illustrated that this strategy can lead to cautious policies where other batch-mode RL algorithms fail because they unsafely generalize the information contained in the dataset.

From the results given in [11], it is also possible to derive in a similar way tight upper bounds on the return of a policy. In this respect, it would also be possible to adopt a "max max" generalization strategy by inferring policies that maximize these tight upper bounds. We believe that exploiting together the policy based on a min max generalization strategy and the one based on a max max generalization strategy could offer interesting possibilities for addressing the exploitation-exploration tradeoff faced when designing intelligent agents. For example, if the policies coincide, it could be an indication that further exploration is not needed.

When using batch mode reinforcement learning algorithms to design autonomous intelligent agents, a problem arises. After a long enough time of interaction with their environment, the sample the agents collect may become so large that batch mode RL-techniques may become computationally impractical, even with small degree polynomial algorithms. As suggested by [8], a solution for addressing this problem would be to retain only the most "informative samples". In the context of the proposed algorithm, the complexity for computing the optimal sequence of decisions is quadratic in the size of the dataset. We believe that it would be interesting to design lower complexity algorithms based on subsampling the dataset based on the initial state information.

The work reported in this paper has been carried out in the particular context of deterministic Lipschtiz continuous environments. We believe that extending this work to environments which satisfy other types of properties (for instance, Hölder continuity assumptions or properties that are not related with continuity) or which are possibly also stochastic is a natural direction for further research.

Acknowledgements. Raphael Fonteneau acknowledges the financial support of the FRIA (Fund for Research in Industry and Agriculture). Damien Ernst is a research associate of the FRS-FNRS. This paper presents research results of the Belgian Network BIOMAGNET (Bioinformatics and Modeling: from Genomes to Networks), funded by the Interuniversity Attraction Poles Programme, initiated by the Belgian State, Science Policy Office. We also acknowledge financial support from NIH grants P50 DA10075 and R01 MH080015. We also thank Bertrand Cornélusse and Quentin Louveaux for their insightful comments on the min max formulation. The scientific responsibility rests with its authors.

References

1. Bemporad, A., Morari, M.: Robust model predictive control: A survey. Robustness in Identification and Control 245, 207–226 (1999)
2. Bertsekas, D., Tsitsiklis, J.: Neuro-Dynamic Programming. Athena Scientific, Belmont (1996)
3. Boyan, J., Moore, A.: Generalization in reinforcement learning: Safely approximating the value function. In: Advances in Neural Information Processing Systems (NIPS 1995), vol. 7, pp. 369–376. MIT Press, Denver (1995)
4. Chakratovorty, S., Hyland, D.: Minimax reinforcement learning. In: Proceedings of AIAA Guidance, Navigation, and Control Conference and Exhibit, San Francisco, CA, USA (2003)
5. Csáji, B.C., Monostori, L.: Value function based reinforcement learning in changing Markovian environments. Journal of Machine Learning Research 9, 1679–1709 (2008)
6. De Berg, M., Cheong, O., Van Kreveld, M., Overmars, M.: Computational Geometry: Algorithms and Applications. Springer, Heidelberg (2008)
7. Delage, E., Mannor, S.: Percentile optimization for Markov decision processes with parameter uncertainty. Operations Research (2006)
8. Ernst, D.: Selecting concise sets of samples for a reinforcement learning agent. In: Proceedings of the Third International Conference on Computational Intelligence, Robotics and Autonomous Systems (CIRAS 2005), Singapore (2005)
9. Ernst, D., Geurts, P., Wehenkel, L.: Tree-based batch mode reinforcement learning. Journal of Machine Learning Research 6, 503–556 (2005)
10. Ernst, D., Glavic, M., Capitanescu, F., Wehenkel, L.: Reinforcement learning versus model predictive control: a comparison on a power system problem. IEEE Transactions on Systems, Man, and Cybernetics - Part B: Cybernetics 39, 517–529 (2009)
11. Fonteneau, R., Murphy, S., Wehenkel, L., Ernst, D.: Inferring bounds on the performance of a control policy from a sample of trajectories. In: Proceedings of the 2009 IEEE Symposium on Adaptive Dynamic Programming and Reinforcement Learning (IEEE ADPRL 2009), Nashville, TN, USA (2009)
12. Fonteneau, R., Murphy, S., Wehenkel, L., Ernst, D.: A cautious approach to generalization in reinforcement learning. In: Proceedings of the Second International Conference on Agents and Artificial Intelligence (ICAART 2010), Valencia, Spain (2010)
13. Fonteneau, R., Murphy, S.A., Wehenkel, L., Ernst, D.: Computing bounds for kernel-based policy evaluation in reinforcement learning. Tech. rep., Arxiv (2010)
14. Gordon, G.: Approximate Solutions to Markov Decision Processes. Ph.D. thesis, Carnegie Mellon University (1999)
15. Ingersoll, J.: Theory of Financial Decision Making. Rowman and Littlefield Publishers, Inc. (1987)

16. Lagoudakis, M., Parr, R.: Least-squares policy iteration. Jounal of Machine Learning Research 4, 1107–1149 (2003)
17. Littman, M.L.: Markov games as a framework for multi-agent reinforcement learning. In: Proceedings of the Eleventh International Conference on Machine Learning (ICML 1994), New Brunswick, NJ, USA (1994)
18. Mannor, S., Simester, D., Sun, P., Tsitsiklis, J.: Bias and variance in value function estimation. In: Proceedings of the Twenty-first International Conference on Machine Learning (ICML 2004), Banff, Alberta, Canada (2004)
19. Murphy, S.: Optimal dynamic treatment regimes. Journal of the Royal Statistical Society, Series B 65(2), 331–366 (2003)
20. Murphy, S.: An experimental design for the development of adaptive treatment strategies. Statistics in Medicine 24, 1455–1481 (2005)
21. Ormoneit, D., Sen, S.: Kernel-based reinforcement learning. Machine Learning 49(2-3), 161–178 (2002)
22. Qian, M., Murphy, S.: Performance guarantees for individualized treatment rules. Tech. Rep. 498, Department of Statistics, University of Michigan (2009)
23. Riedmiller, M.: Neural fitted Q iteration - first experiences with a data efficient neural reinforcement learning method. In: Gama, J., Camacho, R., Brazdil, P.B., Jorge, A.M., Torgo, L. (eds.) ECML 2005. LNCS (LNAI), vol. 3720, pp. 317–328. Springer, Heidelberg (2005)
24. Rovatous, M., Lagoudakis, M.: Minimax search and reinforcement learning for adversarial tetris. In: Konstantopoulos, S., Perantonis, S., Karkaletsis, V., Spyropoulos, C.D., Vouros, G. (eds.) SETN 2010. LNCS, vol. 6040, pp. 417–422. Springer, Heidelberg (2010)
25. Rudin, W.: Real and Complex Analysis. McGraw-Hill, New York (1987)
26. Sutton, R.: Generalization in reinforcement learning: Successful examples using sparse coding. In: Advances in Neural Information Processing Systems (NIPS 1996), vol. 8, pp. 1038–1044. MIT Press, Denver (1996)
27. Sutton, R., Barto, A.: Reinforcement Learning. MIT Press, Cambridge (1998)
28. Viterbi, A.: Error bounds for convolutional codes and an asymptotically optimum decoding algorithm. IEEE Transactions on Information Theory 13(2), 260–269 (1967)

A Proof of Theorem 1

- Let us first prove that $L_{\mathcal{F}_n}^{u_0,\ldots,u_{T-1}}(x) \leq K_{\mathcal{F}_n}^{u_0,\ldots,u_{T-1}}(x)$. Let us assume that we know a set of variables $\hat{x}_0,\ldots,\hat{x}_{T-1}$ and $\hat{r}_0,\ldots,\hat{r}_{T-1}$ that are solution of the optimization problem. To each action $u \in \mathcal{U}$, we associate the sets $A_u = \{x^l \in \{x^1,\ldots,x^n\} | u^l = u\}$ and $B_u = \{\hat{x}_t \in \{\hat{x}_0,\ldots,\hat{x}_{T-1}\} | u_t = u\}$. Let $S_u = A_u \cup B_u$. For simplicity in the proof, we assume that the points of S_u are in *general position*, i.e., no $(d_{\mathcal{X}}+1)$ points from S_u lie in a $(d_{\mathcal{X}}-1)-$dimensional plane (the points are affinely independent). This allows to compute a $d_{\mathcal{X}}-$dimensional triangulation $\{\Delta^1,\ldots,\Delta^p\}$ of the convex hull $H(S_u)$ defined by the set of points S_u [6]. We introduce for every value of $u \in \mathcal{U}$ two Lipschitz continuous functions $\tilde{f}_u : \mathcal{X} \to \mathcal{X}$ and $\tilde{\rho}_u : \mathcal{X} \to \mathbb{R}$ defined as follows:
  - *Inside the convex hull* $H(S_u)$
    Let $g_u^f : S_u \to \mathcal{X}$ and $g_u^\rho : S_u \to \mathbb{R}$ be such that:

$$\forall x^l \in A_u\,, \begin{cases} g_u^f(x^l) = f(x^l,u) \\ g_u^\rho(x^l) = \rho(x^l,u) \end{cases} \text{and } \forall \hat{x}_t \in B_u \backslash A_u\,, \begin{cases} g_u^f(\hat{x}_t) = \hat{x}_{t+1} \\ g_u^\rho(\hat{x}_t) = \hat{r}_t \end{cases} .$$

Then, we define the functions $\tilde{f}_u$ and $\tilde{\rho}_u$ inside $H(S_u)$ as follows:

$$\forall k \in \{1, \ldots, p\}, \forall x' \in \Delta^k \ , \tilde{f}_u(x') = \sum_{i=1}^{d_{\mathcal{X}}+1} \lambda_i^k(x') g_u^f(s_i^k) \ ,$$
$$\tilde{\rho}_u(x') = \sum_{i=1}^{d_{\mathcal{X}}+1} \lambda_i^k(x') g_u^\rho(s_i^k) \ ,$$

where $s_i^k \quad i = 1 \ldots (d_{\mathcal{X}} + 1)$ are the vertices of Δ^k and $\lambda_i^k(x)$ are such that $x' = \sum_{i=1}^{d_{\mathcal{X}}+1} \lambda_i^k(x') s_i^k$ with $\sum_{i=1}^{d_{\mathcal{X}}+1} \lambda_i^k(x') = 1$ and $\lambda_i^k(x') \geq 0, \ \forall i$.

- *Outside the convex hull* $H(S_u)$
  According the Hilbert Projection Theorem [25], for every point $x'' \in \mathcal{X}$, there exists a unique point $y'' \in H(S_u)$ such that $\|x'' - y''\|_{\mathcal{X}}$ is minimized over $H(S_u)$. This defines a mapping $t_u : \mathcal{X} \to H(S_u)$ which is $1-$Lipschitzian. Using the mapping t_u, we define the functions $\tilde{f}_u$ and $\tilde{\rho}_u$ outside $H(S_u)$ as follows:

$$\forall x'' \in \mathcal{X} \backslash H(S_u), \tilde{f}_u(x'') = \tilde{f}_u(t_u(x'')) \text{ and } \tilde{\rho}_u(x'') = \tilde{\rho}_u(t_u(x'')) \ .$$

We finally introduce the functions $\tilde{f}$ and $\tilde{\rho}$ over the space $\mathcal{X} \times \mathcal{U}$ as follows:

$$\forall (x', u) \in \mathcal{X} \times \mathcal{U}, \tilde{f}(x', u) = \tilde{f}_u(x') \text{ and } \tilde{\rho}(x', u) = \tilde{\rho}_u(x') \ .$$

One can easily show that the pair $(\tilde{f}, \tilde{\rho})$ belongs to $\mathcal{L}^f_{\mathcal{F}_n} \times \mathcal{L}^\rho_{\mathcal{F}_n}$ and satisfies

$$J^{u_0, \ldots, u_{T-1}}_{(\tilde{f}, \tilde{\rho})}(x) = \sum_{t=0}^{T-1} \tilde{\rho}(\hat{x}_t, u_t) = \sum_{t=0}^{T-1} \hat{r}_t$$

with $\hat{x}_{t+1} = \tilde{f}(\hat{x}_t, u_t)$ and $\hat{x}_0 = x$. This proves that

$$L^{u_0, \ldots, u_{T-1}}_{\mathcal{F}_n}(x) \leq K^{u_0, \ldots, u_{T-1}}_{\mathcal{F}_n}(x) \ .$$

(Note that one could still build two functions $(\tilde{f}, \tilde{\rho}) \in \mathcal{L}^f_{\mathcal{F}_n} \times \mathcal{L}^\rho_{\mathcal{F}_n}$ even if the sets of points $(S_u)_{u \in \mathcal{U}}$ are not in general position)

- Then, let us prove that $K^{u_0, \ldots, u_{T-1}}_{\mathcal{F}_n}(x) \leq L^{u_0, \ldots, u_{T-1}}_{\mathcal{F}_n}(x)$. We consider the environment $(f^{u_0, \ldots, u_{T-1}}_{\mathcal{F}_n, x}, \rho^{u_0, \ldots, u_{T-1}}_{\mathcal{F}_n, x})$ introduced in Equation (1) at the end of Section 3. One has

$$L^{u_0, \ldots, u_{T-1}}_{\mathcal{F}_n}(x) = J^{u_0, \ldots, u_{T-1}}_{(f^{u_0, \ldots, u_{T-1}}_{\mathcal{F}_n, x}, \rho^{u_0, \ldots, u_{T-1}}_{\mathcal{F}_n, x})}(x) = \sum_{t=0}^{T-1} \tilde{r}_t \ ,$$

with $\forall t \in \{0, \ldots, T-1\}$,

$$\tilde{r}_t = \rho^{u_0, \ldots, u_{T-1}}_{\mathcal{F}_n, x}(\tilde{x}_t, u_t) \ , \tilde{x}_{t+1} = f^{u_0, \ldots, u_{T-1}}_{\mathcal{F}_n, x}(\tilde{x}_t, u_t) \ , \tilde{x}_0 = x \ .$$

The variables $\tilde{x}_0, \ldots, \tilde{x}_{T-1}$ and $\tilde{r}_0, \ldots, \tilde{r}_{T-1}$ satisfy the constraints introduced in Theorem (1). This proves that

$$K^{u_0, \ldots, u_{T-1}}_{\mathcal{F}_n}(x) \leq L^{u_0, \ldots, u_{T-1}}_{\mathcal{F}_n}(x)$$

and completes the proof.

B Proof of Lemma 1

Before proving Lemma 1, we prove a preliminary result related to the Lipschitz continuity of state-action value functions. Let $(f', \rho') \in \mathcal{L}^f_{\mathcal{F}_n} \times \mathcal{L}^\rho_{\mathcal{F}_n}$ be a compatible environment. For $N = 1, \ldots, T$, let us define the family of (f', ρ')–state-action value functions $Q^{u_0,\ldots,u_{T-1}}_{N,(f',\rho')} : \mathcal{X}' \times \mathcal{U} \to \mathbb{R}$ as follows:

$$Q^{u_0,\ldots,u_{T-1}}_{N,(f'\rho')}(x', u) = \rho'(x', u) + \sum_{t=T-N+1}^{T-1} \rho'(x'_t, u_t),$$

where $x'_{T-N+1} = f'(x', u)$ and $x'_{t+1} = f'(x'_t, u_t)$, $\forall t \in \{T - N + 1, \ldots, T - 1\}$. $Q^{u_0,\ldots,u_{T-1}}_{N,(f',\rho')}(x', u)$ gives the sum of rewards from instant $t = T - N$ to instant $T - 1$ given the compatible environment (f', ρ') when (i) the system is in state $x' \in \mathcal{X}$ at instant $T - N$, (ii) the action chosen at instant $T - N$ is u and (iii) the actions chosen at instants $t > T - N$ are u_t. The value $J^{u_0,\ldots,u_{T-1}}_{(f',\rho')}(x)$ can be deduced from the value of $Q^{u_0,\ldots,u_{T-1}}_{T,(f',\rho')}(x, u_0)$ as follows:

$$\forall x \in \mathcal{X},\ J^{u_0,\ldots,u_{T-1}}_{(f',\rho')}(x) = Q^{u_0,\ldots,u_{T-1}}_{T,(f',\rho')}(x, u_0). \tag{3}$$

We also have $\forall x' \in \mathcal{X}, \forall u \in \mathcal{U}, \forall N \in \{1, \ldots, T - 1\}$

$$Q^{u_0,\ldots,u_{T-1}}_{N+1,(f',\rho')}(x', u) = \rho'(x', u) + Q^{u_0,\ldots,u_{T-1}}_{N,(f',\rho')}(f'(x', u), u_{T-N}) \tag{4}$$

Lemma 2 (Lipschitz continuity of $Q^{u_0,\ldots,u_{T-1}}_{N,(f',\rho')}$)
$\forall N \in \{1, \ldots, T\}, \forall (x', x'') \in \mathcal{X}^2, \forall u \in \mathcal{U}$,

$$|Q^{u_0,\ldots,u_{T-1}}_{N,(f',\rho')}(x', u) - Q^{u_0,\ldots,u_{T-1}}_{N,(f',\rho')}(x'', u)| \le L_{Q_N} \|x' - x''\|_{\mathcal{X}},$$

with $L_{Q_N} = L_\rho \sum_{i=0}^{N-1} (L_f)^i$.

Proof. We consider the statement $\mathcal{H}(N)$: $\forall (x', x'') \in \mathcal{X}^2, \forall u \in \mathcal{U}$,

$$|Q^{u_0,\ldots,u_{T-1}}_{N,(f',\rho')}(x', u) - Q^{u_0,\ldots,u_{T-1}}_{N,(f',\rho')}(x'', u)| \le L_{Q_N} \|x' - x''\|_{\mathcal{X}}.$$

We prove by induction that $\mathcal{H}(N)$ is true $\forall N \in \{1, \ldots, T\}$. For the sake of clarity, we denote $|Q^{u_0,\ldots,u_{T-1}}_{N,(f',\rho')}(x', u) - Q^{u_0,\ldots,u_{T-1}}_{N,(f',\rho')}(x'', u)|$ by Δ_N.

- *Basis* $(N = 1)$ *:* We have $\Delta_N = |\rho'(x', u) - \rho'(x'', u)|$, and since $\rho' \in \mathcal{L}^\rho_{\mathcal{F}_n}$, we can write $\Delta_N \le L_\rho \|x' - x''\|_{\mathcal{X}}$. This proves $\mathcal{H}(1)$.
- *Induction step:* We suppose that $\mathcal{H}(N)$ is true, $1 \le N \le T - 1$. Using equation (4), we can write $\Delta_{N+1} = \left|Q^{u_0,\ldots,u_{T-1}}_{N+1,(f',\rho')}(x', u) - Q^{u_0,\ldots,u_{T-1}}_{N+1,(f',\rho')}(x'', u)\right|$ $= \left|\rho'(x', u) - \rho'(x'', u) + Q^{u_0,\ldots,u_{T-1}}_{N,(f',\rho')}(f'(x', u), u_{T-N}) - Q^{u_0,\ldots,u_{T-1}}_{N,(f',\rho')}(f'(x'', u), u_{T-N})\right|$ and, from there, $\Delta_{N+1} \le \left|\rho'(x', u) - \rho'(x'', u)\right| + \left|Q^{u_0,\ldots,u_{T-1}}_{N,(f',\rho')}(f'(x', u), u_{T-N}) - Q^{u_0,\ldots,u_{T-1}}_{N,(f',\rho')}(f'(x'', u), u_{T-N})\right|$. $\mathcal{H}(N)$ and the Lipschitz continuity of ρ' give

$$\Delta_{N+1} \le L_\rho \|x' - x''\|_{\mathcal{X}} + L_{Q_N} \|f'(x', u) - f'(x'', u)\|_{\mathcal{X}}.$$

Since $f' \in \mathcal{L}^f_{\mathcal{F}_n}$, the Lipschitz continuity of f' gives $\Delta_{N+1} \le L_\rho \|x' - x''\|_{\mathcal{X}} + L_{Q_N} L_f \|x' - x''\|_{\mathcal{X}}$, then $\Delta_{N+1} \le L_{Q_{N+1}} \|x' - x''\|_{\mathcal{X}}$ since $L_{Q_{N+1}} = L_\rho + L_{Q_N} L_f$. This proves $\mathcal{H}(N + 1)$ and ends the proof.

Proof of Lemma 1

- The inequality $L_{\mathcal{F}_n}^{u_0,\ldots,u_{T-1}}(x) \leq J^{u_0,\ldots,u_{T-1}}(x)$ is trivial since (f,ρ) belongs to $\mathcal{L}_{\mathcal{F}_n}^f \times \mathcal{L}_{\mathcal{F}_n}^\rho$.
- Let $(f',\rho') \in \mathcal{L}_{\mathcal{F}_n}^f \times \mathcal{L}_{\mathcal{F}_n}^\rho$ be a compatible environment. By assumption we have $u^{l_0} = u_0$, then we use equation (3) and the Lipschitz continuity of $Q_{T,(f',\rho')}^{u_0,\ldots,u_{T-1}}$ to write

$$|J_{(f',\rho')}^{u_0,\ldots,u_{T-1}}(x) - Q_{T,(f',\rho')}^{u_0,\ldots,u_{T-1}}(x^{l_0},u_0)| \leq L_{Q_T}\|x - x^{l_0}\|_{\mathcal{X}}.$$

It follows that $Q_{T,(f',\rho')}^{u_0,\ldots,u_{T-1}}(x^{l_0},u_0) - L_{Q_T}\|x - x^{l_0}\|_{\mathcal{X}} \leq J_{(f',\rho')}^{u_0,\ldots,u_{T-1}}(x)$.
According to equation (4), we have
$Q_{T,(f',\rho')}^{u_0,\ldots,u_{T-1}}(x^{l_0},u_0) = \rho'(x^{l_0},u_0) + Q_{T-1,(f'\rho')}^{u_0,\ldots,u_{T-1}}(f'(x^{l_0},u_0),u_1)$ and from there

$$Q_{T,(f',\rho')}^{u_0,\ldots,u_{T-1}}(x^{l_0},u_0) = r^{l_0} + Q_{T-1,(f',\rho')}^h(y^{l_0},u_1).$$

Thus, $Q_{T-1,(f',\rho')}^{u_0,\ldots,u_{T-1}}(y^{l_0},u_1) + r^{l_0} - L_{Q_T}\|x - x^{l_0}\|_{\mathcal{X}} \leq J_{(f',\rho')}^{u_0,\ldots,u_{T-1}}(x)$.
The Lipschitz continuity of $Q_{T-1,(f',\rho')}^{u_0,\ldots,u_{T-1}}$ with $u_1 = u^{l_1}$ gives

$$|Q_{T-1,(f',\rho')}^{u_0,\ldots,u_{T-1}}(y^{l_0},u_1) - Q_{T-1,(f',\rho')}^{u_0,\ldots,u_{T-1}}(x^{l_1},u^{l_1})| \leq L_{Q_{T-1}}\|y^{l_0} - x^{l_1}\|_{\mathcal{X}}.$$

This implies that

$$Q_{T-1,(f',\rho')}^{u_0,\ldots,u_{T-1}}(x^{l_1},u_1) - L_{Q_{T-1}}\|y^{l_0} - x^{l_1}\|_{\mathcal{X}} \leq Q_{T-1,(f',\rho')}^{u_0,\ldots,u_{T-1}}(y^{l_0},u_1).$$

We have therefore

$$Q_{T-1,(f',\rho')}^{u_0,\ldots,u_{T-1}}(x^{l_1},u_1) + r^{l_0} - L_{Q_T}\|x - x^{l_0}\|_{\mathcal{X}} - L_{Q_{T-1}}\|y^{l_0} - x^{l_1}\|_{\mathcal{X}} \leq J_{(f',\rho')}^{u_0,\ldots,u_{T-1}}(x).$$

By developing this iteration, we obtain

$$J_{(f',\rho')}^{u_0,\ldots,u_{T-1}}(x) \geq \sum_{t=0}^{T-1}\left[r^{l_t} - L_{Q_{T-t}}\|y^{l_{t-1}} - x^{l_t}\|_{\mathcal{X}}\right]. \tag{5}$$

The right side of Equation (5) does not depend on the choice of $(f',\rho') \in \mathcal{L}_{\mathcal{F}_n}^f \times \mathcal{L}_{\mathcal{F}_n}^\rho$; Equation (5) is thus true for $(f',\rho') = (f_{\mathcal{F}_n,x}^{u_0,\ldots,u_{T-1}}, \rho_{\mathcal{F}_n,x}^{u_0,\ldots,u_{T-1}})$ (cf. Equation (1) in Section 3). This finally gives

$$L_{\mathcal{F}_n}^{u_0,\ldots,u_{T-1}}(x) \geq \sum_{t=0}^{T-1}\left[r^{l_t} - L_{Q_{T-t}}\|y^{l_{t-1}} - x^{l_t}\|_{\mathcal{X}}\right]$$

since $L_{\mathcal{F}_n}^{u_0,\ldots,u_{T-1}}(x) = J_{(f_{\mathcal{F}_n,x}^{u_0,\ldots,u_{T-1}},\rho_{\mathcal{F}_n,x}^{u_0,\ldots,u_{T-1}})}^{u_0,\ldots,u_{T-1}}(x)$.

Designing an Epigenetic Approach in Artificial Life: The EpiAL Model

Jorge A.B. Sousa and Ernesto Costa

Evolutionary and Complex Systems Group
Center of Informatics and Systems of the University of Coimbra, Coimbra, Portugal
pelaio@student.dei.uc.pt, ernesto@dei.uc.pt

Abstract. Neo-Darwinist concepts of evolution are being questioned by new approaches, and one of these sources of debate is the epigenetic theory. Epigenetics focus the relation between phenotypes and their environment, studying how this relation contributes for the regulation of the genetic expression, while producing inheritable traits. In this work, an approach for designing epigenetic concepts, including regulation and inheritance, in an Artificial Life model are presented. The model makes use of a dynamic environment, which influences these epigenetic actions. It is possible to perceive evolutionary differences regarding the epigenetic phenomena, both at individual and population levels, as the epigenetic agents achieve a regulatory state and adapt to dynamic environments. The persistence of acquired traits is also observable in the experiments, despite the absence of the signal that induces those same traits.

Keywords: Artificial life, Epigenetics, Regulation, Dynamic environments, Environmental influence.

1 Introduction

150 years after the publication of *"On the Origin of Species"* [10], Darwin's theory of evolution by means of natural selection is still a source of inspiration, but also of debate. Neo-Darwinism [12], the idea that evolution is gene centric and that the organisms' structures are untouchable by the environment, is due for a revision, as defended by some authors [17]. Different theories claim that the modern synthesis provides for an incomplete view of evolution, regarding the separation between organism and nature [11]. Genetic structures are far more complex than previously thought, due to the non linear nature of the mapping between genotype and phenotype. Epigenetics posit the existence of environmentally based regulatory operations, and the possibility for inheritance of structural marks [17]. Several approaches in Artificial Life (ALife) attempt to model biological phenomena, mostly focusing on the Neo-Darwinist point of view in evolution. Most models perceive the environment solely as a factor of selection, discarding interrelations between agents and their world, such as the influence of the environment in the developmental processes [18], which could provide for an alternative evolutionary approach [17]. Here, we present an approach for an ALife model that considers epigenetic concepts, focusing on the regulation of organisms and the possible inheritance of epigenetic acquired marks. By including regulatory elements in the

J. Filipe, A. Fred, and B. Sharp (Eds.): ICAART 2010, CCIS 129, pp. 78–90, 2011.

agents' structures, our model enables the study of the evolutionary significance of epigenetic concepts, while contributing to an enriched ALife approach.

2 Epigenetics

Epigenetics is conceived as a set of genetic mechanisms and operations involved in the regulation of gene activity, allowing the creation of phenotypic variation without a modification in the genes' nucleotides. Some of these variations are also possibly inheritable between generations of individuals [1]. The focus is on the major importance of development in organisms, but also on the relationship between the mechanisms of development, the genetic system and, ultimately, with evolution itself [17]. In terms of phylogeny, epigenetics refers to the traits that are, or can be, inherited by means other than DNA nucleotides. Regarding ontogeny, it refers to the influence, through epigenetic effects, of structural genetic parts of an individual during its lifetime [13]. Although epigenetic marks can be inheritable, their frequency of inheritance is lower than nucleotides [9]. This is due to the fact that epigenetic signals are much easier to alter through environmental disturbances and, therefore, it could result in a high and undesirable variability [13]. There are different types of (cellular) epigenetic inheritance systems (EIS) [17], and our model makes use of the chromatin marking system, in which methylation is a possible epigenetic mark [17]. Methylation of DNA refers to the addition of a methyl group to the base sequence, that although does not change the coding properties of the base, can influence its gene expression [8,7]. Stress responses (tolerance, resistance, avoidance or escape) from the organisms, induced from environmental conditions, can lead to responses of methylation or demethylation of a binding site of chromosomes [13]. The effects of these responses can both be heritable and remain present during more than one generation [13]. Most individuals, during development, possess mechanisms that erase the methylation marks from the parents, sometimes resetting the genome to the original state [6]. Nevertheless, there are cases in which these erasure operations are not fully accomplished, with epigenetic variation persisting through meiosis, and being retained by the offspring [17].

3 State of the Art

ALife is a scientific area whose main goal is the study of life and life-like phenomena by means of computational models, aiming at a better understanding of those phenomena. As a side effect, ALife has also produced nature-inspired solutions to different engineering problems. At the core of ALife activity, we find the development of models that can be simulated with computers. Although some works use epigenetic theory (or related concepts) for problem solving techniques, to the best of our knowledge, none tackles the question of experimenting with epigenetic ideas regarding the evolutionary questions posed by the concept itself. This is a flaw that is pointed out by other authors as well [18]. There are some approaches for problems solving model that take inspiration in epigenetic, or epigenetic related ideas. In [16], the authors model an agent structure that is able to edit the genotype, allowing the same genotype to produce different phenotypic expressions. This edition, however, is not influenced by environmental conditions. A dynamic approach for the environment is presented in [5], where an Avida [4] based model promotes a time based symmetric environment in order to induce the agents to produce phenotypic plasticity [3]. In [15], epigenetic theory is used in order to model a different sort of genetic programming, with the modelling of different life phases (development, adult life) being used to adapt the agents to the environment. During the simulations, the agents adapt to the environment using epigenetic based processes, that are separated between somatic and germ line structures. Finally, in [14], epigenetic concepts are used for the formulation of an Epigenetic Evolutionary Algorithm (EGA). The algorithm is used in order to attempt an optimization for the internal structures of organizations, with a focus on the autopoietic behavior of the systems.

4 EpiAL Model

EpiAL aims at studying the plausibility for the existence of epigenetic phenomena and its relevance to an evolutionary system, from an ALife point of view. In this section, we first describe the conceptual design and notions used in *EpiAL*, focusing in the agent, the regulatory mechanisms and the environment. Then we present the dynamics of the system, explaining how to evaluate the agents and the mechanisms of inheritance of *EpiAL*.

4.1 Conceptual Design

In our model, epigenetics is considered as the ability for an agent to modify its phenotypic expression due to environmental conditions. This means that an agent has regulatory structures that, given an input from the environment, can act upon the genotype, regulating its expression. We also consider the possibility for the epigenetic marks to be inherited between generations, through the transmission of partial or full epigenetic marks (methylation values), allowing the existence of acquired traits (methyl marks) to be transmitted through generations of agents. The proposed model is based on two fundamental entities, the **agent(s)** and the **environment**. Their relation and constitution are conceptually depicted on figure 1. The environment provides inputs that are perceived by the sensors of the agent (S). These sensors are connected to the elements that compose the epigenotype (EG), i.e., the epigenetic code of the agent. The epigenotype is composed of structures that can act over the agents' genotype (G), performing regulatory functions. This regulation occurs in methylation sites that are assigned to each of the genes, controlling their expression. The methylation value for a gene is a real value comprehended between 0 (not methylated at all) and 1 (fully methylated). It is this value that determines, stochastically, the type of expression for each of the genes. Finally, genes are expressed, originating the phenotype of the agents, which is composed of a set of traits (T). This mapping from expressed genotype to phenotype is performed according to a function (**f**) that relates sets of genes with the traits. In figure 2, we show an example where each trait is dependent of 3 genes.

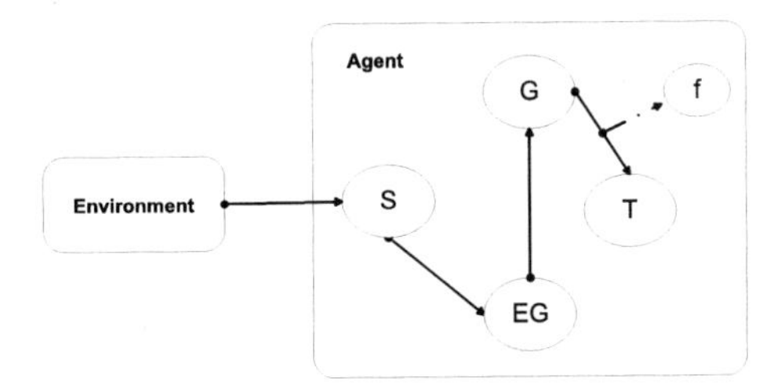

Fig. 1. EpiAL conceptual model

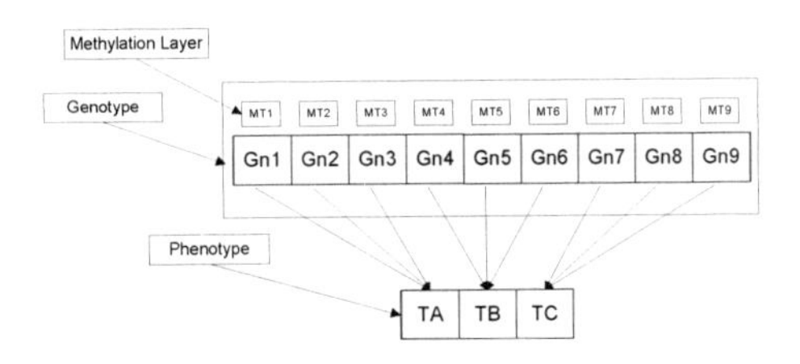

Fig. 2. Genotype to phenotype mapping

4.2 Regulation

The regulatory actions are taken under control of the epigenes. An agent can contain one or several epigenes, that have the structure shown in figure 3. Epigenes are composed of two main parts, sensory and regulatory (figure 3). The sensory section is a tuple that states (i) the sensor to which the epigene is connected and (ii) the reference value that is going to be used to compare with an environmental effect. The possible reference values (for instance, the mean values for each range of environmental effects) are previously setup by the experimenter. The other constituent of the epigenes, the regulatory section, contains a list of tuples, with each tuple representing a gene and a regulatory operation. During a regulatory phase, the reference value encoded in the epigene is compared against the environmental value perceived by the sensor corresponding to the epigene. If the epigene detects an activation, it acts on the genotype by firing the rules that it encodes. Formulas 1 and 2 are used to define the update of the methylation sites, with $m(g)$ representing the methylation value of gene g, $BaseMethyl$ a methylation constant and $AcValue$ the activation value perceived from the sensory operation, proportional to the level of activation perceived. The rules can be to *methylate* (formula 1) or *demethylate* (formula 2) a certain gene, which means that a methylation value is either increased or decreased for that gene.

$$m(g)_{new} = max(1, m(g)_{old} + (BaseMethyl * AcValue)) \tag{1}$$

$$m(g)_{new} = min(0, m(g)_{old} - (BaseMethyl * AcValue)) \tag{2}$$

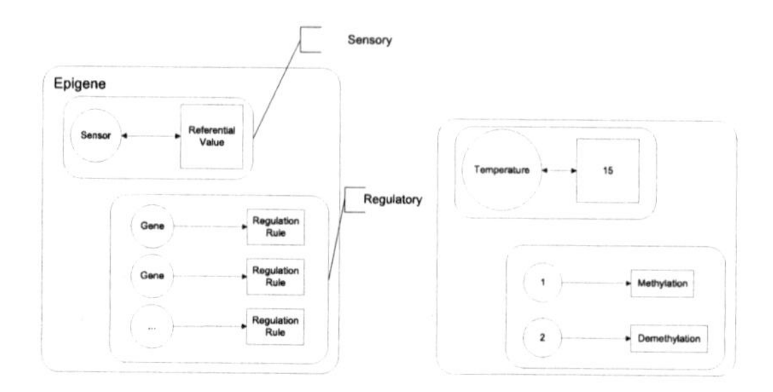

Fig. 3. Agent epigene structure. In the left, the general structure. In the right, a concrete example of an epigene.

4.3 Environment

The environment is modelled as a 2D grid with each location being transposable or not (a wall). Each of the locations also has different attributes - temperature, light and food - that can vary along time. These attributes are used to define the favoured traits for the locations. A certain location in the world favours agents which have the traits more adequate for the current environmental conditions. Therefore, a change in an environmental condition might also imply a change in the trait that is favoured. This is shown in figure 4, where a modification in the temperature state also implies a modification in the set of favoured traits. However, environmental dynamics are also used to promote the regulatory expression of agents. This allows for an influence of the environment over the agents not only by performing selection, but also by inducing possible structural changes (either adaptive or disruptive) within the agent. Because the agents sense environmental conditions, if there is a change in the environment then it is possible that regulatory actions are undertaken in the agents structures.

4.4 System Dynamics

The simulation of the agents in the environment is performed in evolutionary steps, as shown in figure 5. Agents are born and, either during their development phase or their adult life, are subject to regulatory phases. Here, by sensing environmental conditions, they regulate their expression. At the core of the evolutionary system is a Darwinian evolution and, as such, agents are subject to selection, with the best ones being chosen for reproduction and new agents being subject to mutation. The agents live for a number of generations, originate offspring periodically and die of old age, with the possibility for different generations to live at the same time in the environment. The time step of the agents is independent from the one in the environment. An agent can perform several internal cycles of regulation or food digestion during only one environmental cycle.

4.5 Evaluation

As agents do possibly have different performances during their lifetime, due to their different regulation status, it is relevant to possess means for this evaluation to take into

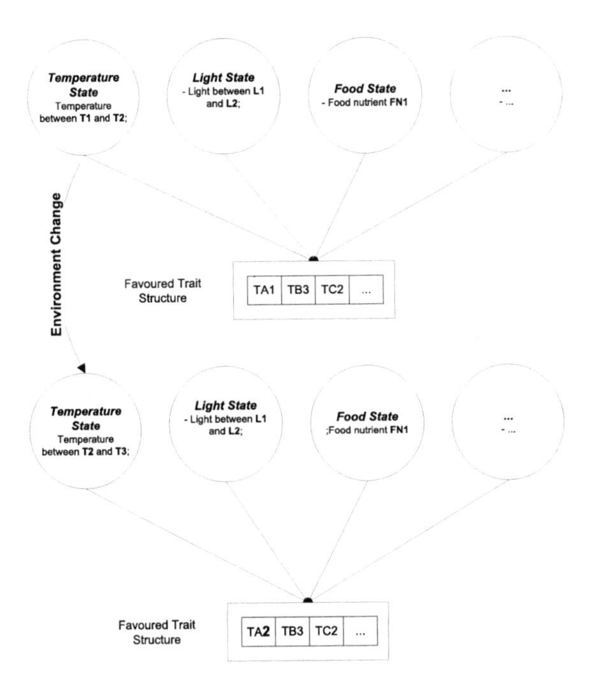

Fig. 4. Environmental traits setup of favoured traits

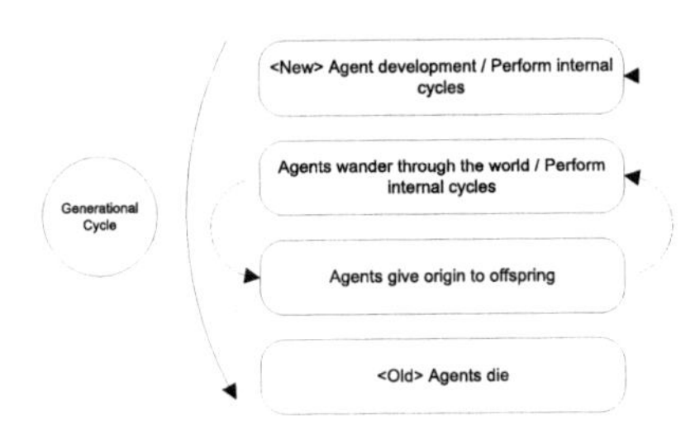

Fig. 5. EpiAL system cycles

account not only the immediate performance, but also an account of the past behaviour of the agent. Moreover, recent values are more important to evaluate an agent. In order to evaluate the agents in *EpiAL*, two measures are used: *direct* fitness evaluation, i.e., the immediate fitness of the agents. This value is obtained by comparing the agents' trait values with the favoured ones for the location (cell) in which the agent currently stands (formula 3). From the formula, it is clear that the lower the sum of the module value of the differences, the better. Therefore, agents attempt to minimize this direct fitness. During the life of the agents, these evaluations are stored. When one desires to evaluate the global performance of an agent, the *weighted* evaluation is performed. This calculation is based on the exponential moving average method, proposed in [2]. In simple terms, this method assigns more weight to recent evaluations. The weighted sum calculated in formula 4 is used to obtain the weighted fitness, as depicted in formula 5

($f(t)$ is the direct fitness evaluation for time t). This allows to perceive if the current regulatory state of the agent is beneficial or detrimental. At the same time, allows agents that are born later to be evaluated without prejudice from the older agents.

$$Fitness = \sum_{i=1}^{3} |Phenotype_i - CellFavouredTrait_i| \tag{3}$$

$$wSum(t) = w(1)f(1) + w(2)f(2) + \ldots + w(t)f(t) \tag{4}$$

$$WeightedFit = \begin{cases} \frac{1}{wSum} & \text{if } wSum > 0.2 \\ 5 & \text{if } wSum <= 0.2 \end{cases} \tag{5}$$

4.6 Inheritance

During reproduction, crossover operations are used, so the genetic material of the parents is combined to form the offspring genetic material. The genotype is immutable during an agent's life, but there is, however, the possibility for the methylation marks to be inherited by the offspring. In *EpiAL*, there are three different mechanisms of epigenetic methylation mark transmission: (i) faithful transmission, in which the methylation marks are transmitted entirely to the offspring; (ii) complete erasure, where all the marks are deleted from parent to offspring, being reset in the new organism; or (iii) partial and stochastic erasure, in which some of the marks may be partially erased. This partial erasure is either performed uniformly over the whole genotype or independently for each gene, according to formulas 6 ($DeltaMax$ being the maximum erasure value) and 7. Partial inheritance of methyl marks is exemplified in figure 6. Here, methylation values of genes 1, 4 and 9, for the first offspring, and genes 3, 7 and 9, for the second offspring, are slightly decreased, compared to the original parents' marks. They have suffered a partial, stochastic erasure.

$$\Delta Erase = Rand(0, DeltaMax) \tag{6}$$

$$\forall g\colon methyl(g)_{new} = methyl(g)_{current} - \Delta Erase \tag{7}$$

5 Experimental Results

Several types of experiments were performed with the *EpiAL* model, in order to study the mechanisms that influence the evolution of the agents. Agents were subject to environments where the three conditions encoded in the world (temperature, light and food) were modified, either periodically or non periodically. Populations without epigenetic mechanisms are compared against epigenetic ones. The epigenetic mechanisms - encoded in the agents' epigenotype - remain static and are not subject to evolution. In table 1 are shown the main parameters used or tested, while table 2 presents the different major mechanisms that were also experimented with in our simulations. These values were applied in several different simulations in which the environmental setup would vary from modifying one or more conditions, periodically or non periodically, with

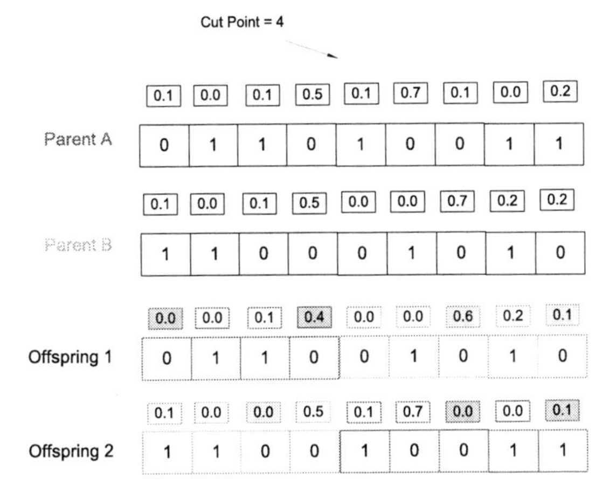

Fig. 6. Partial transfer of methylation marks

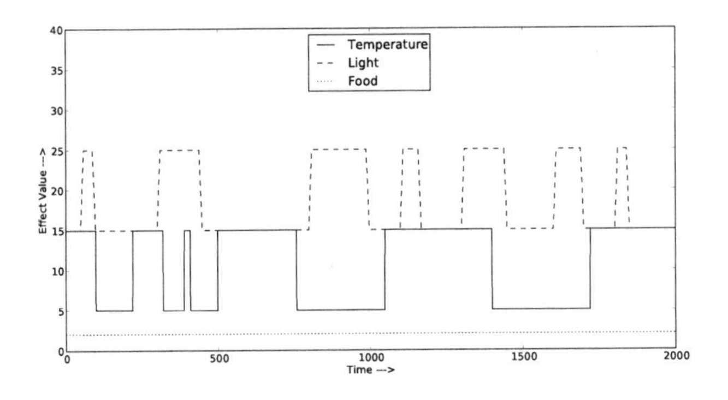

Fig. 7. Dynamic, non periodic temperature and light

higher or lower modification values. As available space does not permit the exhaustive presentation of all these results, we present two types of experiments that demonstrate the typical behavior of the *EpiAL* platform.

Figure 7 shows a dynamic environment where the temperature and light are modified, non periodically, along time.

The results (average fitness values for sets of 30 runs) for this environment are shown in figure 8. Non epigenetic populations, using tournament selection with a mutation rate of 0.01, perform poorly in this sort of environment, whereas the epigenetic populations, subject to the same evolutionary operations of selection, crossover and mutation, are able to endure such an environment. Epigenetic populations adapt to the dynamic environment by adapting also to the epigenetic structures they possess. Agents which evolve to take advantage of the epigenetic layer are preferentially selected to reproduce. This is performed indirectly, by selecting the agents with the best weighted fitness - i.e., during the agents' lifetime, the ones that performed better. As such, along time, the populations increase their adaptation to the dynamic environment. One can observe

Table 1. Parameters for the simulations

Parameter	Value
Agent Step	1
Fitness Smooth Factor	0.5
Aging Step	1
Aging Value	0.5/1/1.5
Dying Age Base	100
World Step	1
Mating Step	10
Initial Agents	15
Offsprings (per mating)	6
Agents Initial Age	[-10;-5;-1]
Epigenetic Sensing Factor	[3;4]
Methyl Base Value	[0.1;0.3;0.5;0.7]
Genetic Mutation Rate	[0.01; 0.1]
Losers' Hype (Tournament)	[0.05; 0.15]

Table 2. Methods for the simulations

Action	Method
Agent Aging	Fitness Related
Agent Death	Stochastic
Selection	Tournament (2 Pair)
Reproduction	Sexual (Crossover)
Genetic Crossover	Trait Based (Gaps of 3)
Partial Methyl Erasure	Overall

different evolutionary behaviors, regarding the different epigenetic inheritance mechanisms. In the case where the methylation marks are not inherited, a 'spiked' behavior originates due to the fact that the new agents are born from parents that are regulated with methyl marks for different environments, with no change on the genotype. The epigenetic marks are not transmitted and, as such, the agents are born with the genotype for warmer/darker environments - the ones actually coded in the genotype -, but are not regulated to colder/brighter ones. The case where there is no methyl mark inheritance is the worst of the epigenetic variants, while the cases where the inheritance occurs in full terms, or only partially, are very similar in terms of performance.

The apparently significant difference in the evolutionary behavior of the agents regarding the possession or absence of epigenetic mechanisms poses some interesting questions that can be tackled with the *EpiAL* model. For instance, we experimented with a basic movement behavior: some of the agents would be able to move into different locations, while others would be fixed into the place where they are born. The environment changes not as a whole, but only in some specific locations (the environmental modification in these areas occurs as depicted in figure 9). As such, some of the agents can move out of unfavorable locations, while others have to endure the conditions from their birth location. The results for this simulation is presented in figure 10.

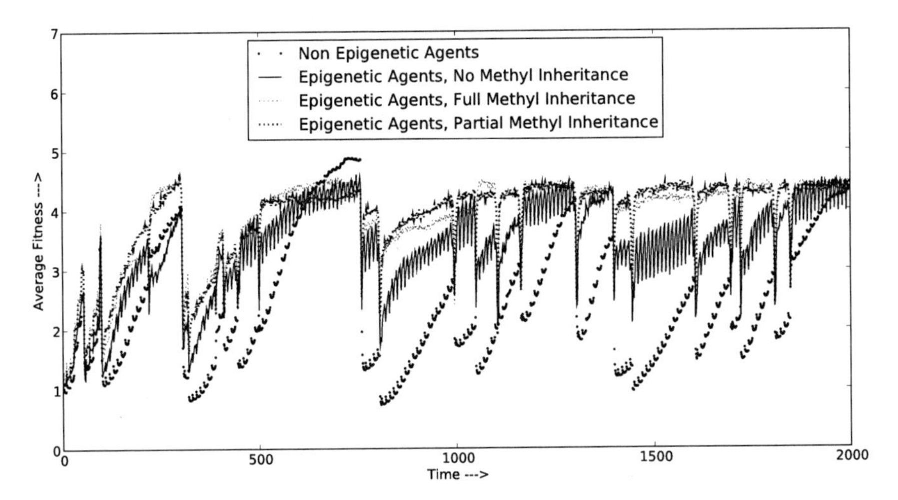

Fig. 8. Average fitness results for different populations, regarding the environment shown in 7

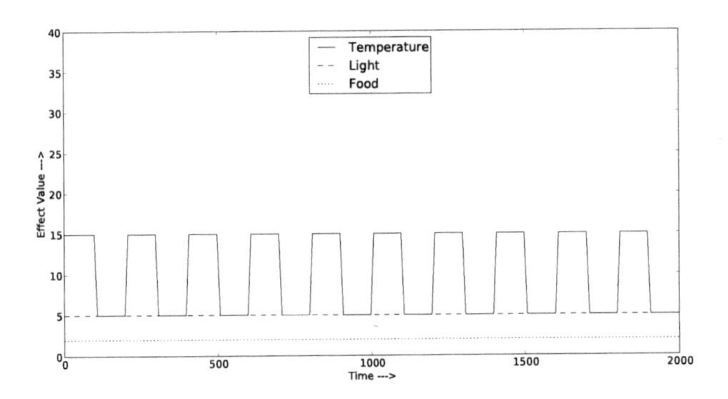

Fig. 9. Dynamic, periodic temperature

Results indicate that the impact of either possessing epigenetic mechanisms or not, regarding populations that can move is not nearly as significant as for populations that cannot move. The moving populations have a fairly similar performance whether they are epigenetic or not. In the case of the grounded populations, as they cannot move from the modified areas, they are subject to harsher environments, unless they regulate their genotypes to the new conditions. This is in agreement with the epigenetic theory, in which plants, organisms that, indeed, cannot move, possess biological mechanisms that are much more plastic than the ones in animals. Some of these mechanisms are considered epigenetic [7].

The epigenetic elements of the agents act as evolutionary and regulatory tools, but they can also be subject to mutational events. In order to experiment with these effects, populations were exposed to a mutation in their methylation sites, starting at iteration 1000, while evolving in the context of the environment shown in 9. Here, agents that possess a full methyl mark inheritance become increasingly worse with the mutational events. This is a sort of *accumulation* effect of mutational events, which is not perceived where there is an erasure, complete or partial, of the methyl marks. As the methylation marks, in living systems, are very easily manipulated by the environment, many errors also originate from this process [13,9]. In evolutionary terms, the dynamics observed in these experiments from the model could pose as a possible reason for the existence of an erasure mechanism at all.

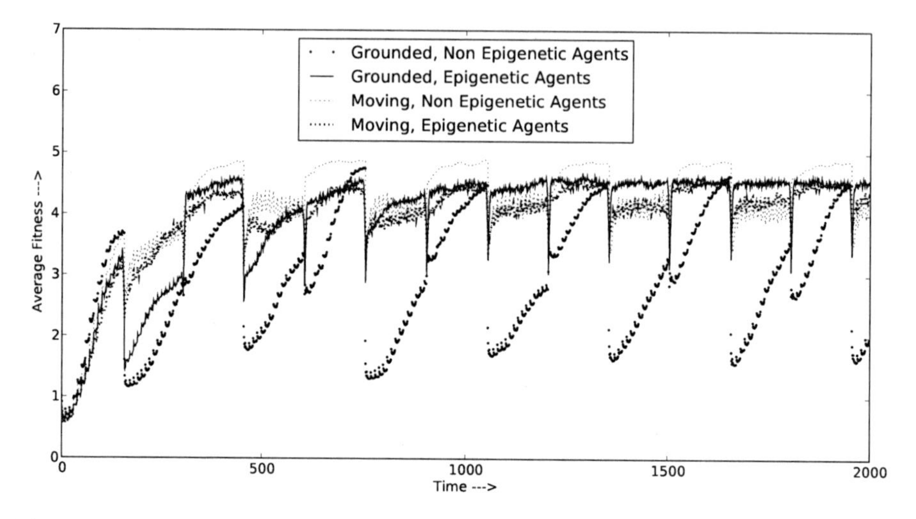

Fig. 10. Average fitness results for different populations, regarding an environment modified in some of the areas according to dynamics in figure 9

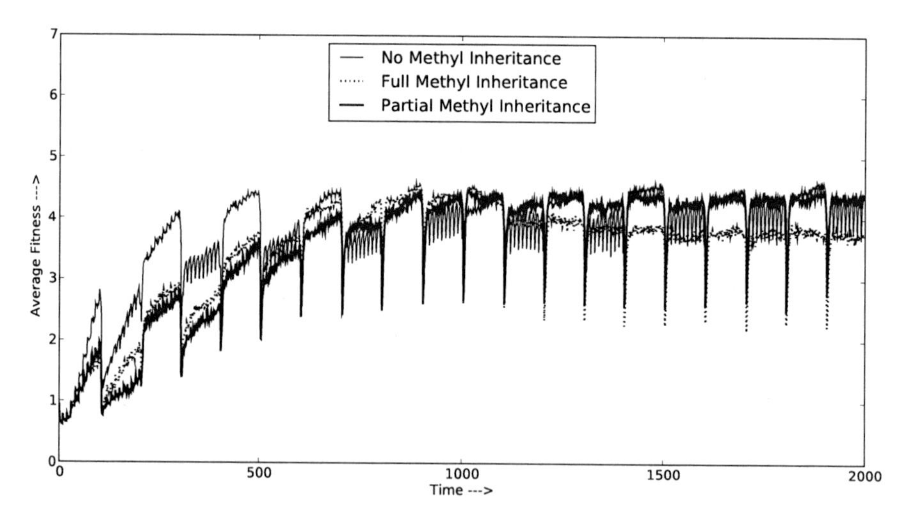

Fig. 11. Average fitness results for different populations, regarding a mutational epigenetic effect starting at iteration 1000

Regarding the results of the model in general, a legitimate question arises, related to whether there is a *memory role* at play, a way in which the agents remember the environmental modifications and inherit those memories. There are no hard coded mechanisms that the agents can take advantage of, in order to remember timed, cyclic conditions in the environment and transmit that information to their offspring. Nevertheless, it can be stated that the main role is played by both a *regulatory* and a *genetic memory* factor. Considering that the construction of the methylation marks are a mechanism of life-time adaptation, and if those marks are inherited, it is only fair to state that there is some 'knowledge' inherited by the offspring. However, that sort of 'memory' is derived from epigenetic regulatory mechanisms. The agents have no means to remember a specific state from the environment, unless the epigenetic components allow them to mark their genotypes with such knowledge. And, even then, such knowledge has to evolve (genetically, at least) in order to be of any use. Methylation, by itself, is worthless if the genetic elements are not evolved to cooperate with the epigenetic mechanisms. Despite some evidence regarding the partial inheritance of these acquired marks (as studied in [13], for instance), the *EpiAL* model enables the possibility to consider that this genetic memory does not operate. The results obtained from the experiments have shown that, although these results are usually worse than the simulations with methyl mark inheritance, there can be improved results from this solely regulatory behavior, compared with the absence of epigenetic mechanisms at all. As such, we consider that the key role of the epigenetic dimension modelled in *EpiAL* is mostly regulatory, with the genetic memory variants improving on the results obtained by regulatory mechanisms.

6 Conclusions and Future Work

EpiAL, the model hereby presented, is designed with a biological enhancement in ALife regarding epigenetic mechanisms. The dynamics of the environment are able to handle multiple conditions, in cyclic or non periodic modifications. Agents possess mechanisms to incorporate and process these environmental conditions, allowing an epigenetic influence in their coding structures. This enables the simulation with different mechanics regarding the epigenetic effects, ranging from the regulatory aspects to the inheritance methods. Using an abstraction for the phenomena of methylation in the genome, *EpiAL* is able to represent regulatory operations for adapting to dynamic environments and allows diverse inheritance patterns to be experimented with. The results show that there could be a significant difference regarding the agents' possession of epigenetic mechanisms. The non epigenetic populations find it hard to thrive in dynamic environments, while the epigenetic populations are able to regulate themselves to dynamic conditions. Although we ventured in experiment with movement and mutations, different sorts of experimentations can be performed with the actual implementation of the model. It would also be interesting to perceive if stronger nocive effects of the epigenetic mechanisms can be found. This work is but a first step, in which we attempted to show that there is a difference, in evolutionary terms, in considering agents with an epigenetic variant. By pursuing this work further, we can hope to achieve a better understanding of the field of epigenetics. The subject of epigenetics is young and still maturing, but, as a hot subject in biology, it provides an example of an excellent opportunity to capitalize over the knowledge achieved in the past years, regarding the modelling of biological phenomena. We intend to pursue the continuous development of the model, with a focus on two dimensions: the first is regarding the biological knowledge that we believe the *EpiAL* model can assist in studying and better understanding. The emergent fields of developmental biology (evo-devo) tackle these issues and have use for simulation tools that can assist in theoretical speculation. The other dimension is related to problem solving techniques. Problems with dynamic environments are tackled by several algorithmic approaches, and we believe that the conceptual basis of the *EpiAL* model has some evolutionary and adaptation mechanisms that could provide a different approach in tackling these problems. This twofold path is but another hint that ALife models can be used concurrently with the discoveries found on the field of epigenetics and developmental biology, providing an actual relation of biomutualism between the fields of computation and biology.

References

1. Achelis, S.: Technical Analysis from A to Z. McGraw-Hill, New York (2000)
2. Gilbert, S.F., Epel, D.: Ecological Developmental Biology. Sinauer Associates, Inc. (2009)
3. Pigliucci, M.: Phenotypic Plasticity: Beyond Nature and Nurture. The Johns Hopkins University Press, Baltimore (2001)
4. Ofria, C., Wilke, C.O.: Avida: a software platform for research in computational evolutionary biology. Artif. Life 10, 191–229 (2004)
5. Clune, J., Ofria, C., Pennock, R.T.: Investigating the Emergence of Phenotypic Plasticity in Evolving Digital Organisms. In: Almeida e Costa, F., Rocha, L.M., Costa, E., Harvey, I., Coutinho, A. (eds.) ECAL 2007. LNCS (LNAI), vol. 4648, pp. 74–83. Springer, Heidelberg (2007)
6. Youngson, N.A., Whitelaw, E.: Transgenerational Epigenetic Effects. Annual Review of Genomics and Human Genetics 9, 233–257 (2008)
7. Boyko, A., Kovalchuk, I.: Epigenetic Control of Plant Stress Response. Environmental and Molecular Mutagenesis 49, 61–72 (2008)
8. Bender, J.: DNA Methylation and Epigenetics. Annual Reviews Plant Biology 55, 41–68 (2004)
9. Holliday, R.: Epigenetics: an overview. Developmental Genetics 15, 453–457 (1994)
10. Darwin, C.R.: The Origin of Species by Means of Natural Selection. John Murray, London (1859)
11. Waddington, C.H.: The Epigenotype. Endeavour (1942)
12. Dawkins, R.: The Selfish Gene Selection. Oxford University Press, Oxford (1976)
13. Gorelick, R.: Evolutionary Epigenetic Theory. PhD Thesis @ Arizona State University (2004), `http://http-server.carleton.ca/~rgorelic/Publications/Root%20Gorelick's%20PhD%20Dissertation%202004.pdf` (accessed)
14. Periyasamy, S., Gray, A., Kille, P.: The Epigenetic Algorithm. In: IEEE World Congress on Computational Intelligence, pp. 3228–3236. IEEE Press, New York (2008)
15. Tanev, I., Yuta, K.: Epigenetic Programming: Genetic programming incorporating epigenetic learning through modification of histones. Information Sciences 178, 4469–4481 (2008)
16. Rocha, L.M., Kaur, J.: Genotype Editing and the Evolution of Regulation and Memory. In: Almeida e Costa, F., Rocha, L.M., Costa, E., Harvey, I., Coutinho, A. (eds.) ECAL 2007. LNCS (LNAI), vol. 4648, pp. 63–73. Springer, Heidelberg (2007)
17. Jablonka, E., Lamb, M.J.: Evolution in Four Dimensions: Genetic, Epigenetic, Behavioral, and Symbolic Variation in the History of Life. MIT Press, Cambridge (2005)
18. Rocha, L.M.: Reality is Stranger than Fiction - What can Artificial Life do about Advances in Biology? In: Talk @ the 9th European Conference on Artificial Life Thesis (2007), `http://informatics.indiana.edu/rocha/discuss_ecal07.html` (accessed)

A Spontaneous Topic Change of Dialogue for Conversational Agent Based on Human Cognition and Memory

Sungsoo Lim, Keunhyun Oh, and Sung-Bae Cho

Department of Computer Science, Yonsei University
262 Seongsanno, Seodaemun-gu, Seoul 120-749, Korea
{lss,ocworld}@sclab.yonsei.ac.kr, sbcho@cs.yonsei.ac.kr

Abstract. Mixed-initiative interaction (MII) plays an important role for the flexible dialogues in conversational agent. Since conventional research on MII process dialogues based on the predefined methodologies, they only provide simple and static dialogues rather than complicated and dynamic dialogues through context-aware themselves. In this paper, we proposed a spontaneous conversational agent that provides MII and can change topics of dialogue dynamically based on human cognitive architecture and memory structure. Based on the global workspace theory, one of the simple cognitive architecture models, the proposed agent is aware of the context of dialogue in conscious level and chooses the topic in unconscious level which is the most relevant to the current context as the next topic of dialogues. We represent the unconscious part of memory using semantic network which is a popular representation for storing knowledge in the field of cognitive science, and retrieve the semantic network according to the spreading activation theory which is proven to be efficient for inferring in semantic networks. It is verified that the proposed method spontaneously changes the topics of dialogues through some dialogue examples on the domain of schedule management.

Keywords: Mixed-initiative interaction, Global workspace theory, Semantic network, Spreading activation theory, Conversational agent.

1 Introduction

Conversational agent can be classified into user-initiative, system-initiative, and mixed-initiative agent with the subject who plays a leading role when solving problems. In the user-initiative conversational agent, the user takes a leading role when continuing the conversation, requesting necessary information and services to the agent with the web searching engine and question-answer systems. On the other hand, in the system-initiative conversational agent, the agent calls on users to provide with information by answering the predefined questions. Although the various conversational agents have been suggested with the user-initiative or system-initiative way, these techniques still have significant limitations for efficient problem solving.

J. Filipe, A. Fred, and B. Sharp (Eds.): ICAART 2010, CCIS 129, pp. 91–100, 2011.

In conversation, people often use ambiguous expressions and background knowledge or the context of dialogue is often presupposed so that missing or spurious sentences appear frequently [1, 2]. In order to resolve ambiguities and uncertainties, the mixed-initiative (MI) approach has been presented in the field of HCI [3]. Contrary to a tight interaction mechanism, the MI approach attempts to solve a problem by allowing the human and the system to collaborate in incremental stages [4].

Mixed-initiative conversational agent is defined as the process of the user and system; which both have the initiative; and solve problems effectively by continuing identification of each other's intention through mutual interaction when needed [5, 6]. Macintosh, Ellis, and Allen showed that mixed-initiative interaction can provide better satisfaction to the user, comparing to system-based interface and mixed-initiative interface through ATTAIN (Advanced Traffic and Travel Information System) [7].

The research to implement mixed-initiative interaction has been studied widely. Hong used hierarchical Bayesian network to embody mixed-initiative interaction [5], and Bohus and Rundnicky utilized dialogue stack [8]. However, these methods, which utilize predefined methodology depending on working memory, can process only static conversation and cannot change topics naturally.

In this paper, we focus on the question: how can the conversational agent naturally change the topics of dialogue? We assume that the changed topics in human-human dialogue are related with their own experiences and the semantics which are presented in the current dialogues.

We apply the global workspace theory (GWT) on the process of changing topics. GWT is a simple cognitive architecture that was developed to account qualitatively for a large set of matched pairs of conscious and unconscious processes. On the view of memory, we define the consciousness part as a working memory and the unconsciousness part as a long-term memory. By the broadcasting process of GWT, the most related unconscious process is called and becomes a conscious process which means one of the candidates of the next topics of dialogues. We model the unconsciousness part (or long-term memory) using semantic network and the broadcasting process using the spreading activation theory.

2 Related Works

2.1 Global Workspace Theory

Global workspace theory models the problem solving cognitive process of the human being. As Fig. 1 shows, there is independent and different knowledge in the unconscious area. This theory defines the independent knowledge as the processor. Simple work such as listening to the radio is possible in the unconscious but complex work is not possible only in the unconscious. Hence, he calls the necessary processors through the consciousness and solves the faced problem by combining processors [9].

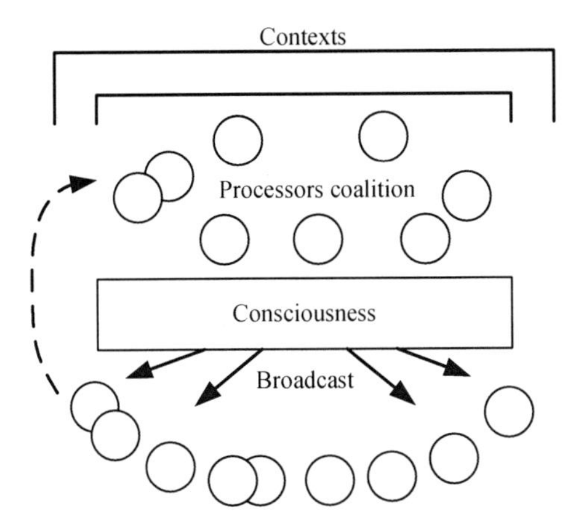

Fig. 1. Global workspace theory

An easy way to understand about GWT is in terms of a "theatre metaphor". In the "theatre of consciousness" a "spotlight of selective attention" shines a bright spot on stage. The bright spot reveals the contents of consciousness, actors moving in and out, making speeches or interacting with each other. In this paper, the bright spot could be interpreted as current topic of dialogue, the dark part on the stage as the candidate topics of next dialogue, and the outside of the stage represents the unconsciousness part (long-term memory).

2.2 Structure of Human Memory and Semantic Network

Fig. 2 shows the structure of human memory. Sensory memory receives the information or stimuli from the outside environment, and the working memory solves problems with the received information. The working memory cannot contain the received memory in the long-term since it store the information only when the sensory memory is in the cognition process of the present information. Hence, the working memory calls the necessary information from the long-term memory when the additional memory needed [10].

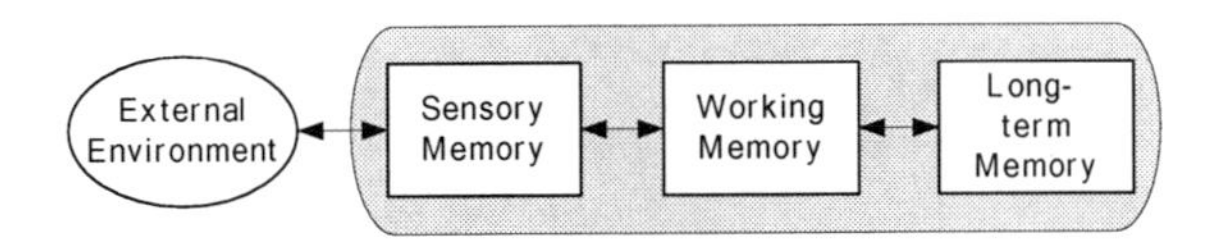

Fig. 2. Human memory structure

The long-term memory of human can be classified into non-declarative memory which cannot be described with a certain language and declarative memory which can be. In the conversation agent, we deal the declarative memory and construct the declarative memory in the long-term memory. The declarative memory is divided into semantic memory and episodic memory. Semantic memory is the independent data which contain only the relationship, and episodic memory is the part which stores data related with a certain event [11].

In this paper, for the domain conversation, we transfer the needed keywords and the past conversation record to semantic memory and episodic memory respectively, and express descriptive memory as semantic network. In the field of cognitive science, semantic network is a popular representation for storing knowledge in long-term memory. Semantic network is the directional graph which consists of nodes connected with edges. Each node represents the concept, and the edge represents the relationship between concepts the nodes mean. Semantic network is mainly used as a form for knowledge symbol, and it is simple, natural, clear, and significant [12]. Semantic network is utilized to measure the relationship between the created keywords during the present conversation and the past conversation, and generate system-initiative interaction when the past conversation with the significant level of relationship is discovered.

2.3 Spreading Activation Theory

Spreading activation serves as a fundamental retrieval mechanism across a wide variety of cognitive tasks [13]. It could be applied as a method for searching semantic networks. In the spreading activation theory, the activation value of each node spreads to its neighboring nodes. As the first step of the search process, a set of source nodes (e.g. concepts in a semantic network) is labeled with weights or activation values and then iteratively propagating or spreading that activation out to other nodes linked to the source nodes.

3 Changing Topics of Dialogue

In this paper, we propose a method for changing topic of dialogue in conversational agent. Fig. 3 represents the overview of proposed method adapting memory structure of human and global workspace theory.

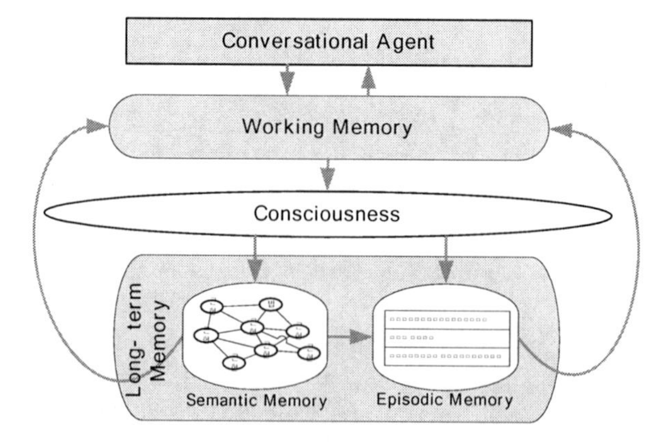

Fig. 3. Overview of proposed method

In the working memory, the information needed to process the present conversation and the candidates of next topics of dialogues are stored. Long-term memory is composed of semantic memory and episodic memory. Semantic memory expresses the relationship between important keywords in the conversation, and episodic memory stores the past conversation which is not completed. The proposed method is to represent such two types of memory into a semantic network.

Conversational agent processes the current dialogue using the information in working memory. When the current dialogue is ended, it selects some candidates of next topics in the long-term memory by broadcasting using spreading activation. If there are some topics which have more activation value than threshold, they are called and become candidates of next topics. Finally, the conversational agent selects the most proper topics from the candidate topics.

3.1 Conversational Agent

In this paper, we adjust and adopt the conversation agent using CAML (Conversational Agent Markup Language) to our experiment [14]. CAML was designed in order to reduce efforts on system construction when applying the conversation interface to a certain domain. It helps set up the conversation interface easily by building several necessary functions of domain services and designing the conversation scripts without modifying the source codes of conversational agent. The agent using CAML works as follows.

1) Analyze the user's answers and choose the scripts to handle them
2) Confirm the necessary factors to offer service which is provided by the chosen scripts. (If there is no information of the factors, system asks the user about them.)
3) Provide services

This agent use both stack and queue of the conversation topics to manage the stream of conversation, and provide different actions to the identical input by the faced situation. System initiative conversation with working memory is already constructed; we focus on the system-initiative conversation using long-term memory.

3.2 Semantic Networks and Spreading Activation

Fig. 4 shows the structure of semantic network and a spreading activation process of this research. The nodes in the network consist of semantic memories and the episodic memories, and the edges show the intensity of connections.

Formally, a semantic networks model can be defined as a structure $N = (C, R, \delta, W)$ consisting of

- A set of concepts $C = \{C_s, C_e\}$ which represents the semantics of keywords and episodic memory, respectively,
- A set of semantic relation $R = \{R_{s \to s}, R_{s \to e}\}$ which represents the semantic relations between keywords, and between keyword and episodic memory, respectively,

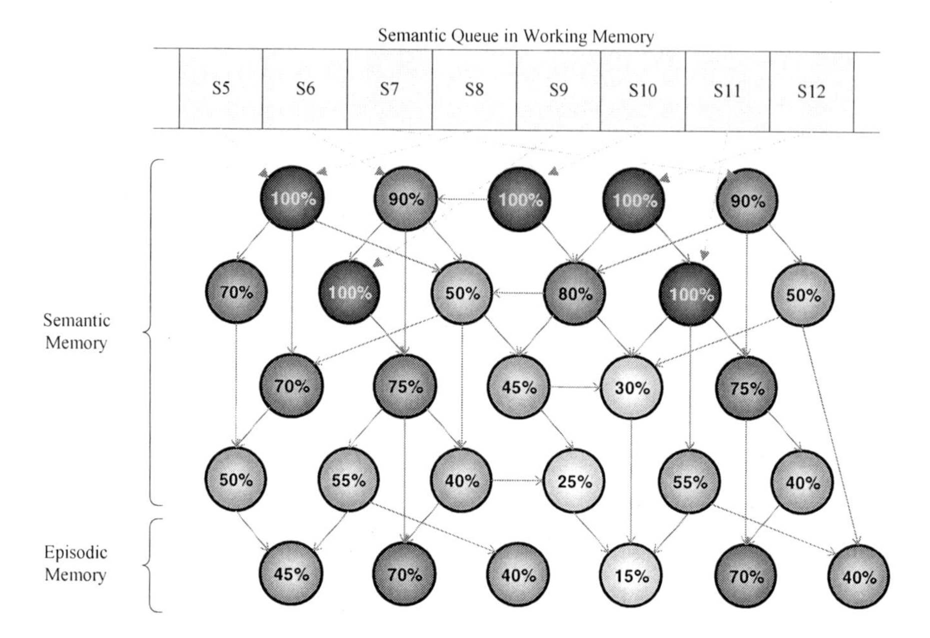

Fig. 4. Semantic networks and spreading activation process

- Function δ: C x $C \rightarrow R$, which associates a pair of concepts with a particular semantic relation, and
- A set of weight functions $W = \{W_c(c_x), W_r(c_x, c_y)\}$ which assign weights to concepts and relations, respectively.

The past conversation stored in episodic memory C_e goes up to working memory by broadcasting process, and we customize this process with applying spreading activation theory. Searching network is done by BFS (breath first search) algorithm using priority queue, and calculates the level of relationship between the corresponding node and the working memory when searching from the node to the next node. Finally, it calculates the level of relationship of episodic memory, the leaf node, and if it has relationship over a certain level, corresponding information goes up to working memory.

Fig. 5 shows the pseudo code of semantic network searching process. For the first stage of spreading activation, it gets the initial semantics from semantic queue in working memory which contains the latest keywords of dialogues. The activation values for the initial concepts are set by 1. Then the activation values are spread through the semantic relations. The spreading values are calculated as follows:

$$c_y.value = c_x.value * W_r(c_x, c_y) * W_c(c_y)$$

The weight function $W_c(c_x)$ returns 1 when $c_x \in C_s$, and returns the priority of episodic memory when $c_x \in C_e$.

```
procedure SpreadingActivation
  input:  Cs  // a set of semantics in working memory
  output: E // a set of episodic memory
begin
  Q.clear() // Q = priority queue
  for each cx in Cs
    Q.push(cx)
  end for
  while Q is not Empty
    cx = Q.pop()
    for each linked concept cy with cx
      cy.value = cx.value*Wr(cx,cy)*Wc(cy)
      if cy.value < threshold1
         then continue
      if isVisitedConcept(cy) is false
         then Q.push(cy)
      if cy is episodic memory
         then E.insert(cy)
    end for
  end while
end proc
```

Fig. 5. Pseudo code for Spreading Activation

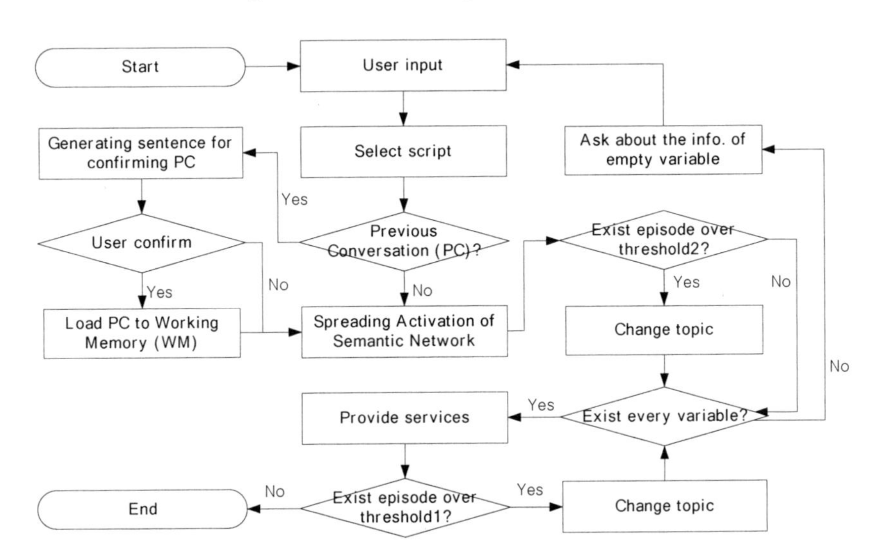

Fig. 6. Flow chart of the conversation process

Fig. 6 represents a flow chart of the conversation process in conversation agents. In this paper, we use the two levels of the threshold values. If the level of relationship is over the threshold value 1, the information has a significant link. If the level is over the threshold value 2, it means that it has both a significant link and urgency. If there is episodic memory over the threshold value 2, the agent stops the current conversation to process the past conversation. And if there is episodic memory with only threshold value 1, the agent waits for the current conversation finished and then it starts the next topic with the past conversation.

4 Dialogue Examples

We use the scheduling program domain to test the utility of the proposed method. Agent needs 4 factors: the type of schedule, subject, time and location. It leaves the conversation unhandled if the factors are unknown because of user's situation. Hence, semantic network for this scheduling conversation agent has four types of internal nodes in the network: the type of schedule, subject, time, and location. The leaf nodes have episodes with 4 values. In this paper, we show how the scheduling program works on 4 situations, when the episodic memory is constructed.

1) Register the schedule

U: I got a dinner appointment.
A: Who will you meet?
U: I'm going to meet Gina.
A: What time will you meet her?
U: At 7 P.M.
A: Where do you meet her?
U: We decided to meet at Yeouido.
A: Okay. The new dinner schedule had registered.

2) Register unprocessed schedule

U: I got a meeting appointment with Jaenam.
A: Is that a dinner appointment at Shinchon you told me on June 13.
U: Yes, that's right.
A: Did you arrange the time to meet?
U: Yes. I decided to meet him at 6 P.M.
A: Okay. The dinner schedule had registered.

3) Process the related conversation

U: I and Kangseok will go out to drink.
A: What time will you meet him?
U: Tomorrow. At 8 P.M.
A: Where do you meet him?
U: At Apgujeong.
A: Okay. The new schedule had registered.
Did you arrange the place of tomorrow's dinner with Gina?
U: Not yet. I'll tell you when I decide.
A: Okay.

4) Process the related conversation 2

U: I will study with Gina on this weekend.
A: Did you decide where you and Gina have dinner tonight?
U: Oh! Let it be Shinchon.
A: Okay. The location had registered. When do you study?
U: At 2 P.M.
A: Where do you study?
U: At Gongdeok.
A: Okay. The study schedule had registered.

The first scenario shows the general schedule registration process. The agent asks the necessary information and registers the corresponding schedule when all information is obtained. The second scenario shows the past schedule reprocessing. Whenever the schedule information gathered, the agent confirms that there is any correspondence, and continue the past conversation if there is the same data as shown in the scenario. In the third scenario, there exists related conversation which has activation value between threshold value 1 and threshold value 2, so that the agent continues the related conversation after the present conversation was finished. If the activation value of related conversation is bigger than threshold value 2, it stops the present conversation and processes the past conversation ahead because it means the related conversation has high priority.

5 Conclusions

Former conversation agent provides mixed-initiative conversation according to the predefined methodology and it only depends on the working memory so that only static conversation can be processed. Hence, in this research, we have studied to give an active function to conversational agent that can change the topic of dialogues. The proposed method models the declarative memory of long-term memory with the semantic network, and implements the broadcasting process in global workspace theory using spreading activation. By searching relevant episode memory with current dialogue, the conversational agent can change the topic of dialogue by itself.

As shown in the dialogue examples, the proposed method works according to the relationship between the present conversation and long-term memory so that the various mixed-initiative conversations can be occurred.

Hereafter, it is necessary to form the semantic network automatically by using the frequency of appeared keywords during the conversation and coherence of keywords. Also, the adaptation of memory reduction function is required to calculate the relationship smoothly.

Acknowledgements. This research was supported by the Conversing Research Center Program through the National Research Foundation of Korea(NRF) funded by the Ministry of Education, Science and Technology (2009-0093676).

References

1. Meng, H., et al.: The use of belief networks for mixed-initiative dialog modeling. IEEE Trans. Speech Audio Process. 11(6), 757–773 (2003)
2. Sammut, C.: Managing context in a conversational agent. Electron. Trans. Artif. Intell. 5, 189–202 (2001)
3. Allen, J.: Mixed initiative interaction. IEEE Intell. Syst. 15(4), 14–23 (1999)
4. Horvitz, E.: Uncertainty, action, and interaction: In pursuit of mixed initiative computing. IEEE Intell. Syst. 14(5), 17–20 (1999)
5. Hong, J.-H., et al.: Mixed-initiative human-robot interaction using hierarchical Bayesian networks. IEEE Trans. Systems, Man and Cybernetics 37(6), 1158–1164 (2007)

6. Tecuci, G., et al.: Seven aspects of mixed-initiative reasoning: An introduction to this special issue on mixed-initiative assistants. AI Magazine 28(2), 111–118 (2007)
7. Macintosh, A., et al.: Evaluation of a mixed-initiative dialogue multimodal interface. In: The 24th SGAI International Conf. on Innovative Techniques and Applications of Artificial Intelligence, pp. 265–278 (2005)
8. Bohus, D., Rudnicky, A.I.: The RavenClaw dialog management framework: Architecture and systems. Computer Speech & Language 23(3), 332–361 (2009)
9. Moura, I.: A model of agent consciousness and its implementation. Neurocomputing 69, 1984–1995 (2006)
10. Atkinson, R.C., Shiffrin, R.M.: Human memory: A proposed system and its control processes. In: The Psychology of Learning and Motivation, pp. 89–195. Academic Press, New York (1968)
11. Squire, L.R., Zola-Morgan, S.: The medial temporal lobe memory system. Science 253(5026), 1380–1386 (1991)
12. Sowa, J.F.: Semantic networks revised and extended for the second edition. In: Encyclopedia of Artificial Intelligence. John Wiley & Sons, Chichester (1992)
13. Anderson, J.R.: A spreading activation theory of memory. Journal of Verbal Learning and Verbal Behavior 22, 261–295 (1983)
14. Lim, S., Cho, S.-B.: CAML: Conversational agent markup language for application-independent dialogue system. In: 9th China-India-Japan-Korean Joint Workshop on Neurobiology and Neuroinformatics, pp. 70–71 (2007)

Clustering Data with Temporal Evolution: Application to Electrophysiological Signals

Liliana A.S. Medina and Ana L.N. Fred

Instituto de Telecomunicações, Instituto Superior Técnico
Lisbon, Portugal
{lmedina,afred}@lx.it.pt
http://www.it.pt

Abstract. Electrocardiography signals (ECGs) are typically analyzed for medical diagnosis of pathologies and are relatively unexplored as physiological behavioral manifestations. In this work we analyze these signals by employing unsupervised learning methods with the intent of assessing the existence of significant changes of their features related to stress occurring in the performance of a computer-based cognitive task. In the clustering context, this continuous change of the signal means that it is difficult to assign signal samples to clusters such that each cluster corresponds to a differentiated signal state.

We propose a methodology based on clustering algorithms, clustering ensemble methods and evolutionary computation for detection of patterns in data with continuous temporal evolution. The obtained results show the existence of differentiated states in the data sets that represent the ECG signals, thus confirming the adequacy and validity of the proposed methodology in the context of the exploration of these electrophysiological signals for emotional states detection.

Keywords: Unsupervised learning, Temporal data, Genetic algorithm, Electrocardiogram, ECG, Stress detection.

1 Introduction

Of the existing classification methods, unsupervised learning is especially appealing to organize data which has little or no labeling information associated to it [3]. A clustering algorithm organizes the patterns into k groups or clusters, based on the similarity or dissimilarity values between pairs of objects, such that objects in the same cluster are more similar than objects of different clusters [5][3]. The adopted similarity might be statistical or geometrical, such as a proximity measure based on a distance metric in the d-dimensional representation space of the d features that characterize the data [5]. The result will be a partition of the analyzed data set.

The work presented here is centered on the analysis of time series of electrophysiological signals, from an unsupervised learning perspective, to assess in particular the existence of differentiated emotional states. Given that typically the signal is characterized by a continuous temporal evolution, this means that the values of the features that represent it will also change gradually with time and that will possibly reflect transient emotional states present in the structure of the signal. In the clustering context, the fact

J. Filipe, A. Fred, and B. Sharp (Eds.): ICAART 2010, CCIS 129, pp. 101–115, 2011.

that such transient states occur, means that it is difficult to assign signal samples to clusters such that each cluster corresponds to a differentiated state.

This represents a challenge in the implementation of clustering methods to analyze time series, like the aforementioned electrophysiological signals, because the clusters are not well separated, which in turn introduces ambiguities in the observation of differentiated emotional states. In order to assess and evaluate the existence of these emotional states, we propose an analysis methodology based on a genetic algorithm combined with clustering techniques. The goal of this work methodology is to eliminate transient states in order to clarify the existence of well separated clusters, each corresponding to differentiated states present in the data time series. This methodology can be applied to electrophysiological signals acquired during the performance of cognitive tasks, such as electrocardiography signals (ECG) or electroencephalography signals (EEG). In this paper we specifically address the identification of stress from ECG signals.

2 Proposed Methodology

We propose a methodology for analysis of temporal data series, represented in Fig. 1. It is based on unsupervised learning techniques in order to unveil similarity relations between the temporal patterns that represent the data, and also to detect differentiated states in the temporal sequences that represent the data, by applying a genetic algorithm specifically conceived for this purpose.

After the acquisition and preprocessing of electrophysiological signals, these are represented by a set of j samples. Each sample corresponds to a given segment of the signal, therefore being associated to a time stamp, and is characterized by a d-dimensional feature vector, $f = [f_1 ... f_d]$.

The proposed methodology encompasses steps of learning similarities between temporal patterns, and the detection of states from these. These two main steps are described in the following subsections. The overall process consists of refining the state detection by means of a genetic algorithm that uses the output of these clustering and state detection procedures.

2.1 Learning Similarities with Evidence Accumulation

Different clustering algorithms lead in general to different clustering results. A recent approach in unsupervised learning consists of producing more robust clustering results by combining the results of different clusterings. Groups of partitions of a data set are called clustering ensembles and can be generated by choice of clustering algorithms or algorithmic parameters, as described in [2]. Evidence Accumulation (EAC) is a clustering ensemble method that deals with partitions with different number of clusters by employing a voting mechanism to combine the clustering results, leading to a new measure of similarity between patterns represented by a co-association matrix. The underlying assumption is that patterns belonging to a natural cluster are very likely to be assigned in the same cluster in different partitions. Taking the co-occurrences of pairs

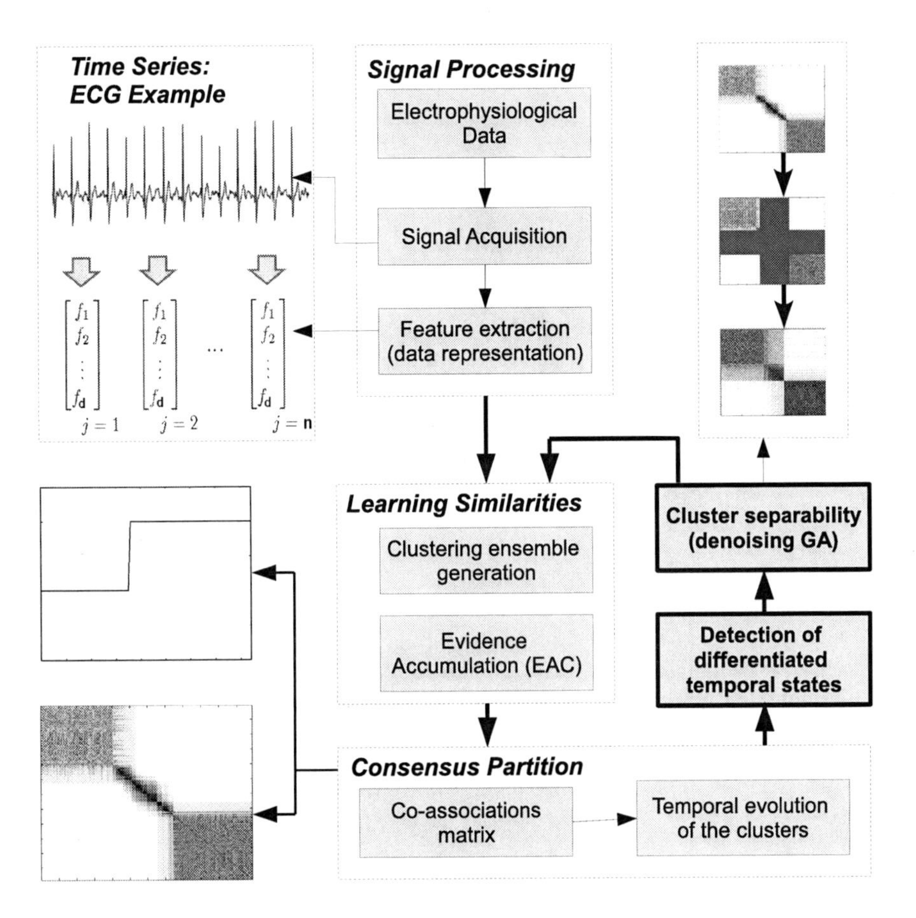

Fig. 1. Proposed work methodology for analysis of temporal series. The time series illustrated corresponds to ECG signals.

of patterns in the same cluster as votes for their association, the N data partitions of n patterns are mapped into a $n \times n$ co-association matrix:

$$C(i,j) = \frac{n_{ij}}{N} \tag{1}$$

where n_{ij} is the number of times the pattern pair (i, j) is assigned to the same cluster among the N partitions.

Graphically, the clusters can be visualized in the representation of the co-association matrix: if contiguous patterns belong to the same cluster, then quadrangular shapes will be present in this representation [8]. A co-association matrix is illustrated in Fig. 2(a). The chosen color scheme ranges from white to black (grayscale), corresponding to the gradient of similarity. Pure black corresponds to the highest similarity. Given that our major goal is to test that the temporal evolution of emotional states corresponds to a temporal evolution of the analyzed signal, the graphical representation of the co-association matrix is a powerful tool to assess the relationships of signal samples ordered by instant of occurrence.

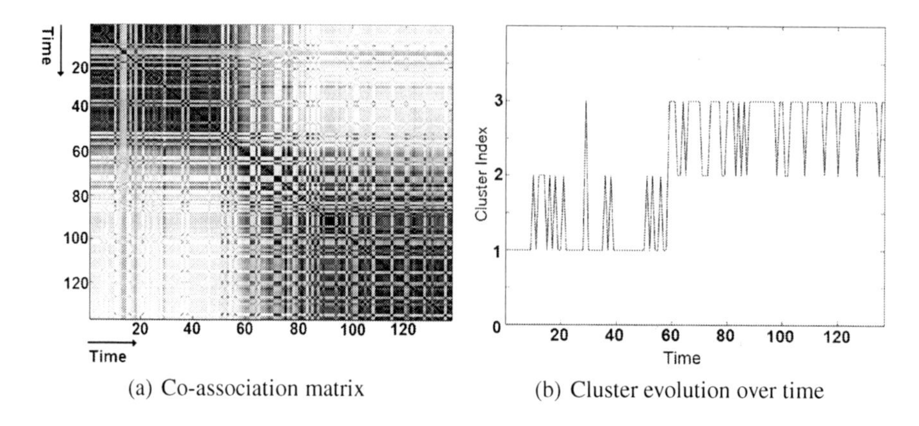

(a) Co-association matrix (b) Cluster evolution over time

Fig. 2. Graphical representation examples: co-association matrix and temporal evolution of the clusters extracted from it

A consensus partition can be extracted from the co-association matrix by applying an hierarchical clustering method [2][8]. Hierarchical algorithms are either divisive or agglomerative based on whether the partitioning process is top-down or bottom-up [4]: agglomerative methods initially treat each pattern as a single cluster and will agglomerate these clusters based on proximity values, represented by a proximity matrix, while divisive methods assume initially that the entire data set is a single cluster [4]. These processes of agglomerating or dividing clusters are represented graphically by a dendrogram. A partition with k clusters is obtained by cutting the dendrogram at the k-th level. On the other hand, the decision on the number of clusters might be based on specific criteria, such as the cluster lifetime criterion: the k cluster lifetime is defined as the range of threshold values on the dendrogram that lead to the identification of k clusters [2].

An example of an extracted partition is depicted in Fig. 2(b), where the relationship between the temporally sequenced samples (x-axis) and the cluster to which they are assigned (y-axis), is plotted. It is possible to observe that cluster transitions generally occur between adjacent clusters: cluster 1 evolves to cluster 2, cluster 2 evolves between clusters 2 and 3, etc [8]. This is a meaningful result for the testing of the hypothesis of temporal evolution of emotional states.

2.2 Detection of Temporal States and Cluster Separability

The detection of temporal states is performed by comparing and examining the temporal evolution of clusters of one or more partitions produced from the learned similarity matrix. The goal of this analysis is the assessment of underlying structures that might correspond to the temporal evolution of differentiated states. The proposed criteria for this assessment are illustrated in Fig. 3. Each criterion considers sample segments of the temporal evolution of the clusters. Differentiated states are detected if: **(1)** there are segments such that all the samples of each segment belong to a single cluster (Fig. 3(a)); **(2)** each segment is comprised of samples belonging to different clusters, such that each

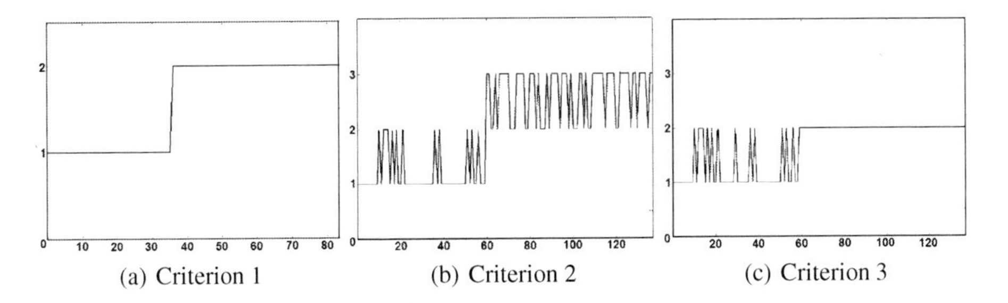

Fig. 3. Visualization of the proposed state detection criteria

Algorithm 1. State detection method

input : Cluster labels.
output: Consensus partition.

1. Split the n acquired samples into w windows $(L_1, L_2, ..., L_w)$ each with n samples.
2. Create the cluster indicator matrix M_b with w rows and i columns. $M_b(s,i) = 1$ if cluster label i is present in the L_s window. Repeat this procedure to all windows.
3. Identify meta-clusters by clustering over M_b with EAC, by examining the graphical representation of the state evolution of the consensus partition.

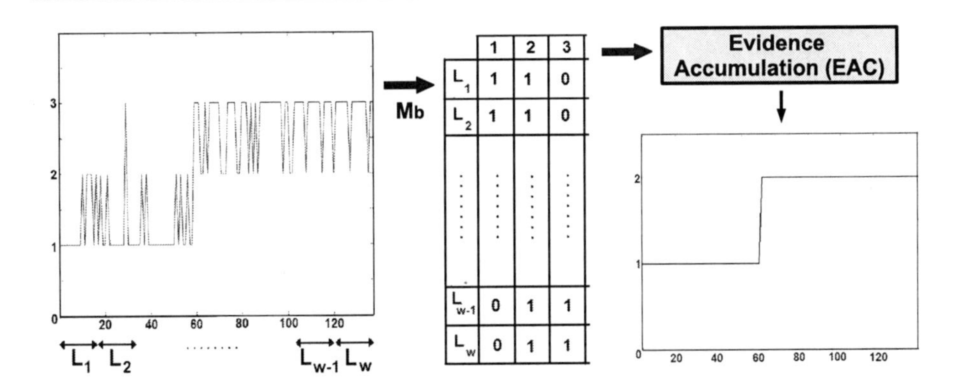

Fig. 4. State detection procedure

of those segments correspond to a unique combination of clusters (Fig. 3(b)) and **(3)** there are segments that correspond either to a single cluster or to a unique combination of clusters (Fig. 3(c)).

In order to identify states corresponding to distinct combinations of clusters, as per criteria 2 and 3, a meta-clustering procedure is used. Splitting the observation period (total duration of the acquired data) into adjacent windows (L_1, L_2, ..., L_w) each containing n samples, we define a cluster indicator $w \times k$ matrix, M_b. Each row of the matrix corresponds to a window, and each column to a cluster label, where the cluster combination for each window is expressed. Specifically, in each row, columns with the value 1 indicate the presence of the associated cluster in that window and so forth. This procedure is described in Algorithm 1, as well as in Fig.4.

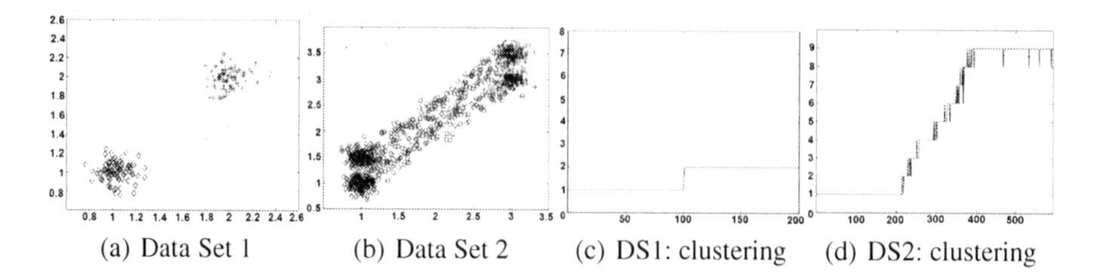

(a) Data Set 1 (b) Data Set 2 (c) DS1: clustering (d) DS2: clustering

Fig. 5. Original synthetic data sets: feature values change abruptly between the clusters of DS1 (Fig. 5(a)) and feature values change smoothly over time for the samples of DS2 (Fig. 5(b)). Temporal evolution of the clusters obtained after applying EAC over K-Means partitions: the samples of DS1 are correctly partitioned but samples of DS2 corresponding to the transition between the two main clusters are wrongly assigned.

3 Genetic Algorithm for Denoising of Data with Temporal Evolution

In this section we propose a genetic algorithm (GA) specifically designed to overcome the ambiguities induced by transient states present in signals characterized by temporal evolution. This GA is based on the assumption that, after removal from the original data set of the subset of samples that corresponds to transient states and performing EAC on the clustering ensemble based on this new reduced data set, a structure of separate states might emerge from the respective co-association matrix.

Fig. 5 illustrates the difficulty of clustering temporal data with smooth transitions. In these examples, data is represented by 2-D feature vectors. In data set 1 (DS1, see Fig. 5(a)), each state is modeled by a gaussian distribution, the transition between the two states (well separated mean values) occurring abruptly in time. Data set 2 (DS2, see Fig. 5(b)) illustrates a smooth evolution between states, each being composed by a mixture of two gaussians, where the mixtures of gaussians are quite distinct at the initial (left) and final (right) time periods; however the transitions between these occur smoothly. The clustering results obtained after applying Evidence Accumulation over K-means [3] produced partitions from both synthetic data sets are also depicted in Fig. 5. The partition found for the first data set represents truthfully the two states, (Fig. 5(c)); however the clustering method fails to label correctly the samples of the second data set due to smooth transitions between its groups of samples (Fig. 5(d)). It is not possible to assert, with confidence, where the first state ends and where the second state begins.

Fig. 6 illustrates the testing of the existence of well separated states, each corresponding to a single cluster, by denoising with a task specific genetic algorithm. Fig. 6(a) represents the co-associations matrix that corresponds to the original data set DS2; Fig. 6(b) illustrates the group of samples (marked pure black) to be removed in order to obtain separated quadrangular blocks, with no transient structures between them; finally Fig. 6(c) illustrates the co-association matrix obtained from the new data set. The temporal evolution of the clusters obtained for the original and denoised data sets are shown in Fig. 5(d) and Fig. 6(d), respectively. Two well separated states may be observed in the temporal evolution of the clusters obtained from the denoised data set

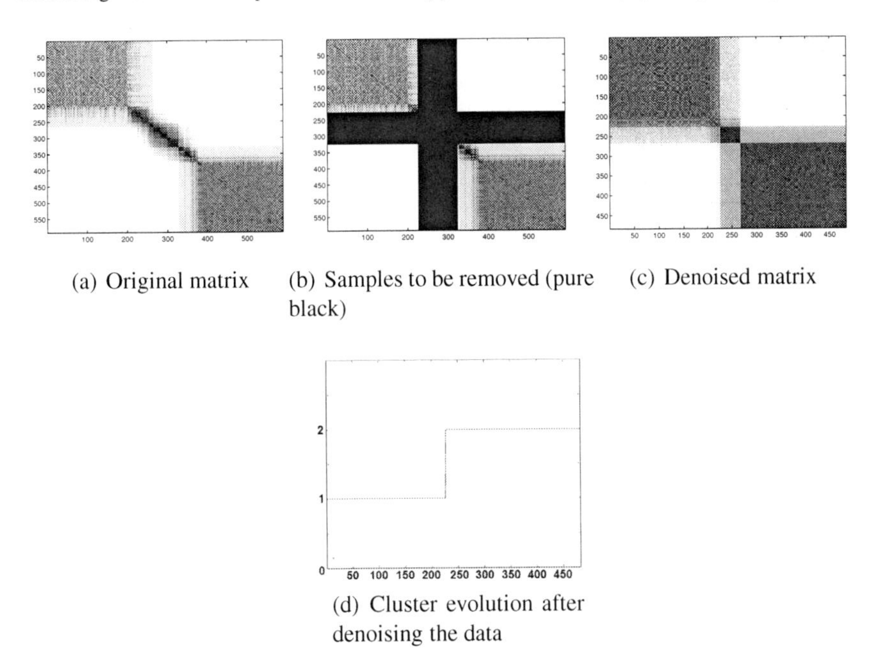

(a) Original matrix (b) Samples to be removed (pure black) (c) Denoised matrix

(d) Cluster evolution after denoising the data

Fig. 6. Example of the application of denoising

(see Fig. 6(d)). The intermediate groups of samples to be removed are determined by applying a task-specific genetic algorithm (GA). Several operators and procedures must be declared in order to define a particular GA [9]. These operators were defined for the denoising GA as follows. The algorithm itself is summarized by Algorithm 2.

Representation. Each individual is a set of samples, obtained after removal of one or more subsets of intermediate samples from the original data set. The first pattern of each removed subset is called minimum limit, (l_{min}), and the last pattern is referred to as maximum limit, (l_{max}).

Fitness Function. The evaluation of the fitness value of each individual is comprised of two stages, each concerning a partial fitness value function.

1. Determine if two or more of the H partitions that are associated to the individual m are equal: if partitions P_1 and P_2 are equal, then $I(P_1, P_2) = 1$, else $I(P_1, P_2) = 0$. The fitness value associated to consensus partitions equality is given by F_1 such that

$$F_1 = \frac{1}{\binom{H}{2}} \sum_{i=1}^{H-1} \sum_{j=i+1}^{H} I(P_i, P_j) \tag{2}$$

2. Determine, for each of the H partitions associated to the individual, the degree of cluster separability between temporal segments. We define the following two segments: segment (A) comprised of all the samples that occur before l_{min}, and

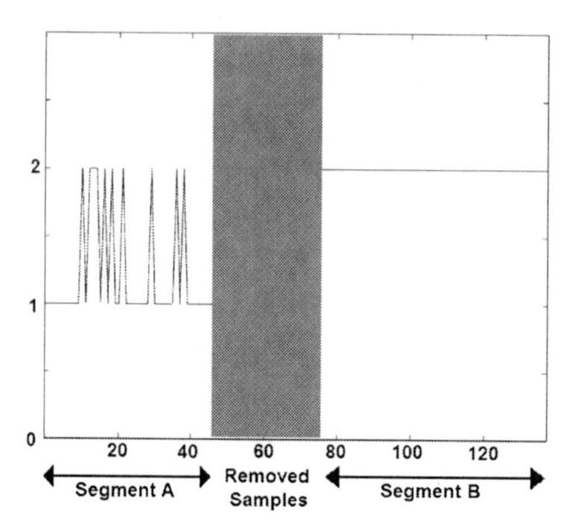

Fig. 7. Evaluation of clusters temporal separability. The subset of removed samples is marked gray.

segment (B) comprised of all the samples that occur after l_{max} (see Fig. 7). The subset of samples between l_{min} and l_{max} is the subset of removed samples from the original data set. For each segment, we determine the dominant cluster label. Samples with a different cluster label are considered outliers.

The fitness value associated to temporal separability is given by F_2:

$$F_2 = \frac{1}{H} \sum_{H} \frac{n_{samples} - n_{outliers}}{n_{samples}} \tag{3}$$

where $n_{samples}$ is the total number of samples associated to the evaluated individual. For the example shown in Fig. 7, the dominant cluster for segment A is cluster 1, with 9 outliers (that belong to cluster 2). The dominant cluster for segment B is cluster 2, with 0 outliers.

The final fitness value, F_{total}, for each individual is then

$$F_{total} = \alpha F_1 + (1 - \alpha) F_2 \tag{4}$$

where α is a weighting coefficient such that $\alpha \in [0, 1]$.

Selection. The selection of individuals for recombination is based on their fitness values by employing deterministic tournament selection (see [9]). Two individuals are selected for each selection step.

Recombination. Recombination of l_{min} and l_{max} of the selected individuals for generation of two new individuals with probability of occurrence p_r: the new l_{min} and l_{max} are chosen randomly as intermediate values of the selected individuals l_{min} and l_{max}, respectively.

Mutation. Modification of l_{min} and l_{max} of the new individuals by adding or subtracting to them a randomly chosen number, n_{mut}, with probability of occurrence p_m. Addition and subtraction have the same probability of occurrence.

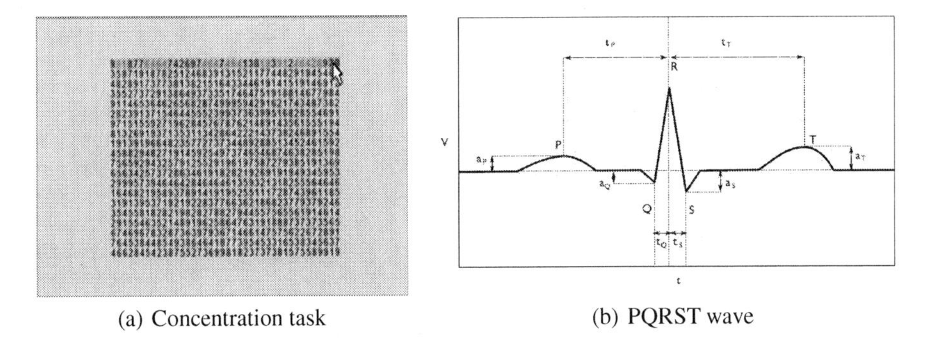

(a) Concentration task (b) PQRST wave

Fig. 8. Stress inducing task (examination and annotation of a matrix with 800 numbers) and typical morphology of an ECG wave

4 Application Domain: Detection of Stress from ECG Signals

The temporal series analyzed correspond to electrocardiography, or ECG, signals. These signals are part of a more vast experience of multi-modal acquisition of physiological signals - the HiMotion project [1]. The ECG signals were acquired from a group of 24 subjects performing a stress inducing cognitive task, illustrated in Fig. 8(a). This task is a concentration test that consists of the identification and annotation of pairs of numbers that add to 10, by examination of the lines of a matrix of 20 lines per 40 columns of numbers [1]. The population of subjects is comprised of 18 males and 9 females, being their mean ages 23.4 years. For each one of these subjects, a montage with two electrodes called V_2 bipolar single lead electrocardiogram was used to collect signals from the heart [1]. Given that this concentration task is stress inducing, the presented methodology is applied to the ECG signals in order to assess the existence of stress states.

4.1 ECG Processing and Feature Extraction

An ECG signal is a recording of the electrical activity of the heart that consists of sequences of heart beats. Each heart beat has a typical morphology which consists of five waves (P, Q, R, S, T), schematically represented in Fig. 8(b).

From the acquired time series corresponding to the ECG signal, signal processing techniques were applied for signal segmentation [1] and a mean wave form was calculated based on 10 consecutive heart beats, to remove some spurious noise. All the waves were aligned with respect to the R wave. The recorded signal for each subject is then summarized in a temporal sequence of 137 mean waves, each wave represented by a feature vector [1].

The representation of the ECG signals is based on the P, Q, S, T waves. The R wave is used for time alignment, setting the initial instant of the beat ($t_R = 0$). The following rules are used to locate the position of each of the P, Q, S, T waves and to extract the eight main features of each mean wave [1], as depicted in Fig. 8(b):

Algorithm 2. Genetic algorithm for denoising of data characterized by temporal evolution of its features

```
input  : Data to be clustered.
         Clustering algorithm: cAlgo;
         Number of partitions of each clustering ensemble: nParts;
         H distinct hierarchical extraction algorithms;
         Cluster extraction criterion;
         Fitness threshold: th;
         Number of generations: G.
         Number of individuals of each population: M.
output: Denoised representation of the data, solution.

obtain an initial population from the original data set, pop(g = 1), of M individuals,
randomly;
while g <= G do
    currentPop = pop(g);
    foreach m ∈ currentPop do
        cE ← clusteringEnsemble(m,cAlgo,nParts);
        coAssocs ← EAC(cE);
        for h ← 1 to H do
            m(h).partition ← extract(coAssocs,h);
        end
        fitValue ← fitness(m);
        if fit >= th then
            solution ← m;
            break;
        end
        solution ← m : fitValue = maxFitnessValue(currentPop);
    end
    g ← g + 1;
    m ← 0;
    while sizePop(pop(g + 1)) < M do
        parents ← select(currentPop);
        (recIndOne, recIndTwo) ← recombine(parents);
        (mutIndOne, mutIndTwo) ← mutate(recIndOne,recIndTwo);
        insert(pop(g + 1),mutIndOne,mutIndTwo);
        sizePop(pop(g + 1)) ← sizePop(pop(g + 1))+2;
    end
end
```

1. t_P - the first maximum before the R wave;
2. a_P - the amplitude of the P wave;
3. t_Q - the first minimum before the R wave;
4. a_Q - the amplitude of the Q wave;
5. t_S - the first minimum after the R wave;
6. a_S - the amplitude of the S wave;
7. t_T - the first maximum after the R wave;
8. a_T - the amplitude of the T wave.

Table 1. Algorithmic parameters

Description	Notation	Parameter Values
Clustering Algorithms	$cAlgo1$ and $cAlgo2$	$k \in [2, 6]$ $\sigma \in [0.3, 0.4, ..., 2.9, 3.0]$ $nParts = 140$ (for each $cAlgo$)
Number of individuals	M	20
Number of partitions associated to each individual	H	5
Number of generations	G	20
Minimum threshold of fitness value	th	0.95
Fitness coefficient	α	0.1
Probability of recombination	p_r	0.9
Probability of mutation	p_m	0.1
Range of mutation values	n_{mut}	$n_{mut} \in [0, 5]$
Window size	n	10 samples

Each mean wave is represented by a 53-dimensional feature vector: the aforementioned 8 features, plus the amplitudes of the signal at 45 points of the signal obtained by re-sampling of the mean wave [1] . Thus, for each of the 24 subjects, there is a group of 137 temporally sequenced samples or patterns, corresponding each sample to a vector of 53 features.

4.2 Algorithmic Parameters and Experiments

Table 1 synthesizes the algorithms and algorithmic parameters employed. Two spectral clustering algorithms were used to produce clustering ensembles for each of the 24 data sets. These algorithms were originally proposed by Ng et al, [6], and Shi et al [7], and referred to in Table 1 by $cAlgo1$ and $cAlgo2$, respectively. Each partition of the clustering ensemble is generated such that it corresponds to a combination of possible values of the spectral algorithms parameters (which consist on the number of clusters, k and a scaling parameter σ).

Five agglomerative hierarchical methods were used for the consensus partition extraction from the co-associations matrix thus generated, using the cluster lifetime criterion: Single Link (SL), Complete Link (CL), Average Link (AL), Ward's Link (WL) and Centroid's Link (CenL). Detailed descriptions and studies of these algorithms may be found for example in [3] or [10].

For the final step of the detection method described in Section 2.2, five different partitions are extracted with the cluster lifetime criterion (one for each of the hierarchical methods already mentioned), as well. The final state structure is chosen to be the one that the majority of the hierarchical methods finds.

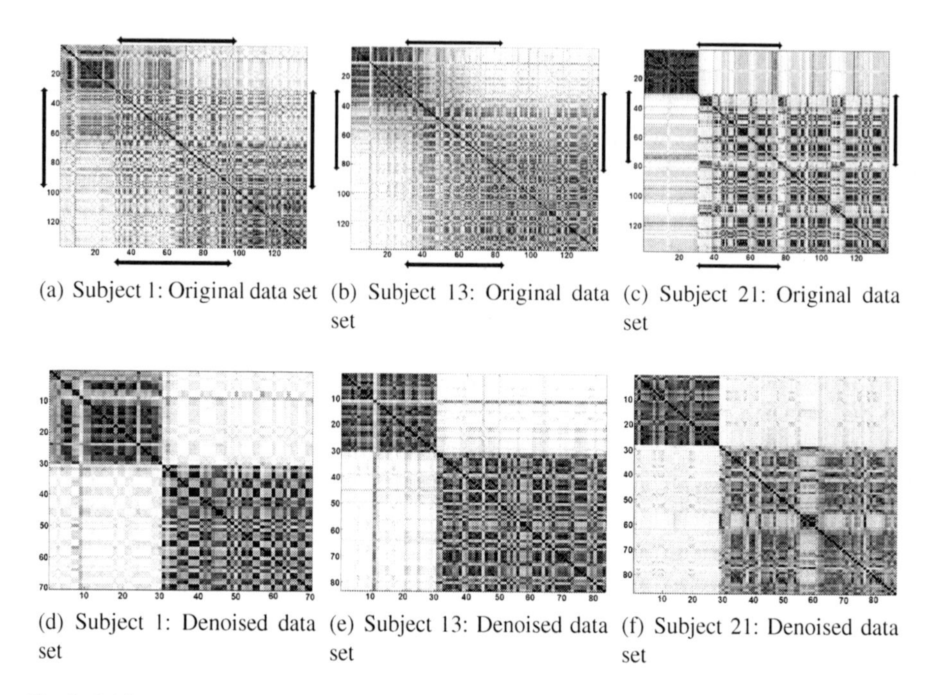

(a) Subject 1: Original data set (b) Subject 13: Original data set (c) Subject 21: Original data set

(d) Subject 1: Denoised data set (e) Subject 13: Denoised data set (f) Subject 21: Denoised data set

Fig. 9. Subjects 1, 13 and 21. The original data sets co-association matrices are depicted by Figs. 9(a),9(b) and 9(c), respectively. The denoised data sets co-association matrices are depicted by Figs. 9(d), 9(e) and 9(f), respectively. The samples removed from the original co-association matrices are within the area delimited by the arrows alongside the axes of Figs. 9(a) to 9(c).

4.3 Results and Discussion

Fig. 9 represents co-association matrices obtained for the original data set and for the denoised data set of subjects 1, 13 and 21, which show different levels of separability of the evolution into stress states.

By comparing the representations of both co-association matrices for the same subject, it is possible to observe that denoising of the original data sets will lead to the revelation of the structure of states in the ECG time series. Similarity relationships between contiguous samples are thus emphasized, which means that the clusters are separated such that a structure of differentiated states emerges, with no ambiguities in the observation and detection of these states.

This better separability of emotional states by the proposed GA-based method is further evaluated by observing the temporal evolution of clustering results produced from the learned similarities. Figs. 10(a) to 10(e) illustrates the temporal evolution of clusters obtained for the original data set of subject 6 (each partition corresponds to one of the five hierarchical methods used for combined partition extraction). By inspection of these 5 representations, we observe that different methods extract different partitions, in terms of number of clusters and samples assigned to each cluster. Though a structure of the data appears to be present, the transitions between clusters induce ambiguities in the observation of differentiated states.

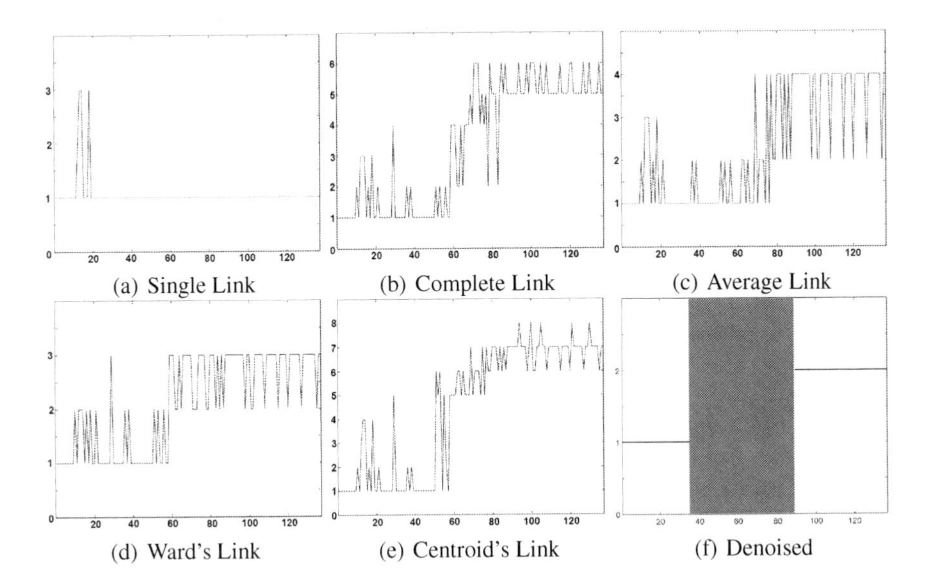

(a) Single Link (b) Complete Link (c) Average Link

(d) Ward's Link (e) Centroid's Link (f) Denoised

Fig. 10. Subject 6: original data set partition (Figs. 10(a) to 10(e)) and denoised data set partition (Fig. 10(f)), where the removed samples are marked gray. The x-axis represents the temporally ordered ECG samples and the y-axis the clusters to which they are assigned.

After applying the denoising GA the five methods extract the same partition of the data (depicted in Fig. 10(f)). This partition reveals two completely separated clusters each corresponding to a different emotional state. Thus, these results validate the observations of the original data set and it is possible to conclude that emotional states are observable in the ECG temporal series of subject 6.

Before denoising the data sets, the state detection method described (see Section 2.2) produces two or three states for 10 of the 24 subjects. The structures obtained have many transitions between clusters, which induces uncertainty in the observations of these results. Only in 3 of these 10 structures can we observe distinct state structures with no ambiguity.

For the remaining subjects, it is not clear how many states exist, nor which samples belong to each states (i.e., where does one state ends and the next begins). Figs. 11(a) to 11(c) depict structures found for 3 of such subjects: Fig. 11(a) refers to subject 22, Fig. 11(b) refers to subject 3, and Fig. 11(c) refers to subject 8. In all of these structures we observe several transitions between clusters.

After applying the denoising GA the results obtained by the state detection method reveal, with no ambiguities, the existence of distinct states for 18 subjects. We observe three typical structure types, depicted in Figs. 11(d) to 11(f): two distinct states, found for 5 subjects (of which subject 22 is an example, as shown in Fig. 11(d)), three distinct states, found for 11 subjects (of which subject 3 is an example, as shown in Fig. 11(e)), and four distinct states found for two subjects (of which subject 8 is an example, as shown in Fig. 11(f)). The duration of each state differs between subjects. The amount of samples removed by the denoising GA is also different for each individual, ranging from 31% to 59% of the total samples.

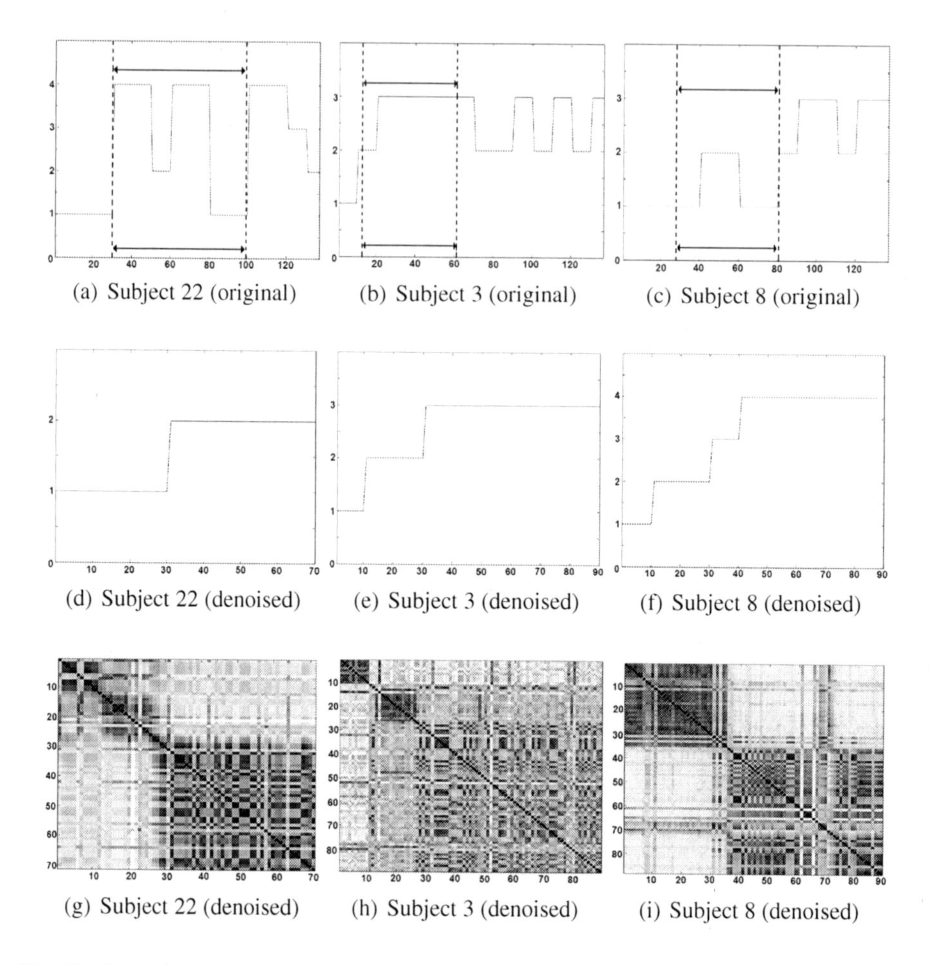

(a) Subject 22 (original) (b) Subject 3 (original) (c) Subject 8 (original)

(d) Subject 22 (denoised) (e) Subject 3 (denoised) (f) Subject 8 (denoised)

(g) Subject 22 (denoised) (h) Subject 3 (denoised) (i) Subject 8 (denoised)

Fig. 11. State detection: three types of structures are observable in the denoised data sets (Figs. 11(d) to 11(f)). These structures were detected in partitions extracted from the denoised co-association matrices represented in Figs. 11(g) to 11(i). The structures obtained with the detection method for the original data sets have more transitions between clusters, thus inducing uncertainty in the observation of differentiated states. The removed subsets of samples are delimited by dashed lines (Figs.11(a) to 11(c)).

5 Conclusions

In this work we proposed a methodology for the analysis of data characterized by temporal evolution, such as electrophysiological signals. This methodology is based on a clustering ensemble method, and on a genetic algorithm for assessment of the existence of differentiated states in time series. The presented results pertain to the application of the proposed techniques on ECG temporal series acquired during the performance of a cognitive task. These results validate our assumption that it is possible to infer the existence of differentiated emotional states in these signals by using the proposed methodology.

Ongoing work consists on a further extensive validation of this methodology in the herein presented application domain, as well as extrapolation to the automatic analysis of other time series, such as EEG data.

Acknowledgements. We acknowledge financial support from the FET programme within the EU FP7, under the SIMBAD project (contract 213250). This work was partially supported by the Portuguese Foundation for Science and Technology (FCT), Portuguese Ministry of Science and Technology, under grant PTDC/EIA CCO/1032230/ 2008.

References

1. Gamboa, H.: Multi-Modal Behavioral Biometrics Based on HCI and Electrophysiology. PhD Thesis, Instituto Superior Técnico (2008), http://www.lx.it.pt/~afred/pub/thesisHugoGamboa.pdf
2. Fred, A.L.N., Jain, A.K.: Combining multiple clusterings using evidence accumulation. IEEE Trans. Pattern Anal. Mach. Intell. 27, 835–850 (2005)
3. Jain, A.K., Dubes, R.C.: Algorithms for Clustering Data. Prentice-Hall, Inc., Englewood Cliffs (1988)
4. Jain, A.K., Murty, M.N., Flynn, P.J.: Data Clustering: A Review. ACM Computing Surveys 31(3), 264–323 (1999)
5. Xu, R., Wunsch, D.: Survey of Clustering Algorithms. IEEE Transactions on Neural Networks 16(3), 645–678 (2005)
6. Ng, A.Y., Jordan, M.I., Weiss, Y.: Clustering: Analysis and an Algorithm. In: Advances in Neural Information Processing Systems, vol. 14. MIT Press, Cambridge (2002)
7. Shi, J., Malik, J.: Normalized Cuts and Image Segmentation. IEEE Trans. Pattern Anal. Mach. Intell. 22(8), 888–905 (2000)
8. Lourenço, A., Fred, A.L.N.: Unveiling Intrinsic Similarity - Application to Temporal Analysis of ECG. Biosignals (2), 104–109 (2008); INSTICC - Institute for Systems and Technologies of Information, Control and Communication
9. Sumathi, S., Hamsapriya, T., Surekha, P.: Evolutionary Intelligence. Springer, Heidelberg (2008)
10. Manning, C.D., Raghavan, P., Schtze, H.: Introduction to Information Retrieval. Cambridge University Press, Cambridge (2008), http://nlp.stanford.edu/IR-book/html/htmledition/irbook.html

Part II
Agents

MASITS Methodology Supported Development of Agent Based Intelligent Tutoring System MIPITS

Egons Lavendelis and Janis Grundspenkis

Department of Systems Theory and Design, Riga Technical University
1 Kalku street, Riga, Latvia
egons.lavendelis@rtu.lv, janis.grundspenkis@cs.rtu.lv

Abstract. Many Intelligent Tutoring Systems (ITS) have been developed to add adaptivity and intelligence to e-learning systems. Intelligent agents are widely used in ITSs due to their characteristics such as modularity and facilitation of intelligent mechanism implementation. At the same time development of agent based ITS is complicated and methodological support is needed to enable industrial adoption of agent based ITSs. The paper describes a specific agent based ITS development methodology, named MASITS, and MIPITS system developed with the methodology. The system is created for the course "Fundamentals of Artificial Intelligence". It offers learning materials, provides practical problems and gives feedback to the learner about his/her solution evaluating his/her knowledge. The main focus of the system is on problem solving. The problems are adapted to the learner's knowledge level and preferences about difficulty, size and practicality of problems. The system offers three types of problems: tests, state space search problems and two person games algorithm problems.

1 Introduction

Nowadays, the availability of education, including lifelong learning, must be increased to successfully develop knowledge society. There is a need for easy available and individualised tutoring process that adapts to every learner. Different e-learning technologies are used to teach large numbers of students, facilitate availability of education and lifelong learning. Learning management systems like Blackboard (http://www.blackboard.com/) and Moodle (http://moodle.org/) are among the most popular ones. These systems are available from any place with an Internet connection and at any time. Thus learners can choose when and where to study the course.

Unfortunately, e-learning as well as traditional tutoring process which is done in the classroom is not effective if many learners with different knowledge levels and learning styles are taught simultaneously. E-learning systems mainly offer learning materials and different kinds of tests to evaluate learners' knowledge. The majority of them use the same materials and tests for all learners. Thus, traditional e-learning systems can not realize individualized tutoring by adapting to any specific characteristics of the learner. Moreover, e-learning systems usually are not capable to generate any learning material or test using domain knowledge. The teacher must create all learning materials and tests that are used in the course.

J. Filipe, A. Fred, and B. Sharp (Eds.): ICAART 2010, CCIS 129, pp. 119–132, 2011.

To eliminate the abovementioned drawbacks of e-learning systems, Intelligent Tutoring Systems (ITS) are developed. They simulate the teacher realizing individualized tutoring, using domain and pedagogical knowledge as well as knowledge about the learner. ITSs to a certain extent can adapt learning materials, generate tasks and problems from domain knowledge, evaluate learner's knowledge and provide informative feedback. So ITSs add adaptivity to above-mentioned benefits of e-learning systems [1]. During the last 40 years since the first ITS named SCHOLAR which taught geography [2], many ITSs have been developed. Well-known examples of ITSs are FLUTE [3], ABITS [4], Slide Tutor [5] and Ines [6].

ITSs mainly are built as modular systems consisting of four traditional modules: the tutoring module, the expert module, the student diagnosis module and the communication module. During the last decade many ITSs have been built using intelligent agents to implement these modules (for an overview see [7]). Well-known examples of agent based ITSs are ITS for enhancing e-learning [8], ABITS [4] and Ines [6]. Despite the fact that agents increase modularity of the system and facilitate implementation of intelligent mechanisms needed in ITSs [9], development of agent based ITSs is still complicated and appropriate development methodologies are needed to bring agent based ITSs into industry. Unfortunately, general purpose agent oriented software engineering (AOSE) methodologies [10] do not take into consideration all ITS characteristics and do not allow to plug in knowledge from ITS research [7].

The paper gives a brief overview of an agent based ITS development methodology MASITS and describes an ITS for the course "Fundamentals of Artificial Intelligence" (MIPITS). The MIPITS system offers learning materials and problems to a learner, evaluates learner's knowledge in each topic and provides feedback to a learner. The system adapts problems to the learner's knowledge level and preferences.

The remainder of the paper is organized as follows. A brief description of the MASITS methodology is given in the Section 2. The Section 3 contains general description of the developed system. The architecture of the system is given in the Section 4. The tutoring scenario implemented in the system is described in the Section 5. The Section 6 shows how to extend the system with new types of problems. The Section 7 concludes the paper and gives brief outline of the future work.

2 The MASITS Methodology

To facilitate adoption of agent-oriented software by software developers a large number of general purpose AOSE methodologies have been developed during the last years. The well known examples of AOSE methodologies are Gaia [11], Prometheus [12] and MaSE [13]. However general purpose AOSE methodologies are developed to support as many different agent systems as possible by providing high level techniques that are suitable for different agents. Thus these methodologies do not allow to adjust the development process to the characteristics of specific systems. On the other hand, ITSs are specific systems with the specific characteristics that limit effective usage of general purpose methodologies, for details see [9]. Additionally, ITSs have many important results gained in specific research that can facilitate the development process. The main results of ITS research are the following: identification of types of knowledge used in ITS, modular architecture, set of agents used to implement modules and tasks done by these agents as well as different agent architectures for ITS implementation [7], [9].

In the absence of general purpose methodologies and corresponding tools allowing to plug in domain specific knowledge and adjust the development process to the characteristics of specific systems, a specific agent based ITS development methodology, named MASITS methodology, and the corresponding tool have been developed [9], [14].

The MASITS methodology is a full life cycle methodology for agent based ITS development. The methodology includes the most important results of ITS research and reusable techniques of AOSE methodologies, which allow to integrate specific knowledge for the ITS development. Additionally, new techniques are introduced in steps where known techniques do not allow integration of ITS knowledge.

The methodology and tool support development of agent based ITSs using the open holonic architecture described in Section 4. The methodology is connected with the JADE platform (http://jade.tilab.com/) – the lower level design is done in concepts used in the JADE platform to enable direct code generation from the design.

The development process of the MASITS methodology consists of the following phases: analysis, design (divided into two stages: external and internal design of agents), implementation, testing, deployment and maintenance. The *analysis phase* consists of two steps. Hierarchy of goals is created during the goal modelling step. Use cases corresponding to the lower level goals are defined during the use case modelling step. The *external design of agents* answers the question what agents have to do. During this stage tasks are defined according to steps of use case scenarios and allocated to agents from the set of higher level agents of the architecture. Additionally, interactions among agents are designed in use case maps and the interaction diagram. Ontology is created to define predicates used in agent communications. The *internal design of agents* answers the question how agents achieve their behaviour specified during the external design of agents. Agents are designed in terms of events perceived, messages sent and received as well as behaviours carried out. The internal design of agents in the MASITS methodology is done according to the open holonic architecture for ITS development [15]. Choice of the architecture is justified in [9]. In the holonic architecture each agent can be implemented as a holon and consist of several smaller agents. For agents that are implemented as holons, design of the holon is done using techniques of both external and internal design of agents. Additionally, the MASITS methodology includes crosschecks among diagrams during the whole development process, to ensure that all elements from one diagram have corresponding elements in other diagrams [9].

Usage of open holonic architecture allows reusing some agents from previous projects during the *implementation phase*. Classes for agents that can not be reused and their behaviours as well as ontology classes are generated by the MASITS tool from the diagrams created during the design [16]. Agent and behaviour classes are manually completed after generation. *Testing phase* of the MASITS methodology is carried out in the following order, separate agents are tested first, holons as a whole are tested second and finally testing of the whole system is done. *Deployment* of the system is specified in the deployment diagram, which is used by the tool to generate configuration of the system. The MASITS methodology supports *maintenance phase* by

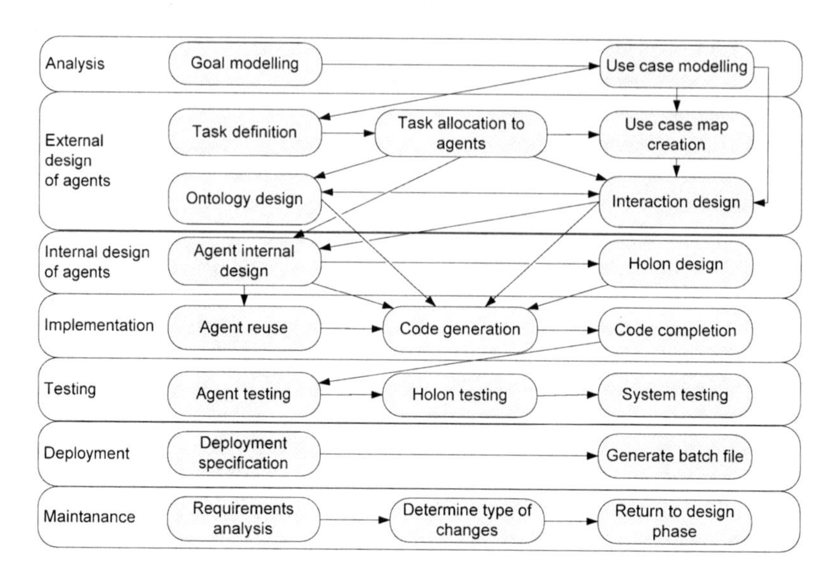

Fig. 1. Phases of the development process and steps done during each phase

providing mechanisms to implement some types of changes into the system. The way how to extend system with new tasks is shown in Section 6. For further details of the development process with the MASITS methodology and tool, see [9][14].

3 The MIPITS System

The MIPITS system is developed for the course "Fundamentals of Artificial Intelligence" simultaneously taught to more than 200 students at Riga Technical University. The course contains topics about different algorithms used in artificial intelligence like search algorithms and algorithms for two person games. Important part of learning such algorithms is practice. However, any guidance and feedback during the practice is limited due to the large number of students. Additionally, it is almost impossible to prepare unique problems and tasks for all students manually. The aim of the MIPITS system is to solve the issues of problem generation, limited guidance and feedback during the practice with different algorithms taught in the course. Moreover, students attending the course have very different knowledge levels and learning styles. Thus another goal of the system is to improve the tutoring process by adapting to the learner's knowledge level and learning style.

The MIPITS system is intended as an addition to the traditional classroom tutoring. Firstly, the learner attends lectures in the classroom. Later he/she has an opportunity to repeat the topics taught in the classroom and practice in the problem solving using the system. However, it is possible to use the system without attending the classes, because it covers all necessary activities to learn the basics of the corresponding topics. In each topic the MIPITS system offers the learning material, and the problem to be solved by the learner. In the MIPITS system the problem is any task, test or assignment used to evaluate the learner's knowledge. After the learner has finished the problem the system evaluates his/her solution and gives appropriate feedback.

The main focus of the MIPITS system is on problem solving. The system provides unique problems that are appropriate to the knowledge level and preferences of the individual learner. Initial version of the system is developed for first three modules of the course - „Introduction", „Uninformed Search" and „Informed Search" [17]. Thus, the system is capable to offer the corresponding types of problems:

- Different types of tests: single choice tests, multiple choice tests and tests, where a learner has to write the answer by him/herself. Figures and state spaces can be added to the question.
- Search algorithm problems, where a learner has to do a state space search using the specified algorithm and lists OPEN and CLOSED [17].
- Two person game problems, where a learner has to apply the MINIMAX algorithm or Alpha-Beta pruning to the given state space [17].

Other types of problems can be added to the system (for details see Section 6). When the learner requests a task the system finds the most appropriate problem to the learner's knowledge level and preferences among problems of all types that fit the topic and delivers it to the learner.

4 The Architecture of the MIPITS System

According to the MASITS methodology, the MIPITS system is developed using open holonic multi agent architecture for ITS development described in [15]. The architecture consists of the higher level agents that implement the traditional modules of ITSs. All higher level agents can be implemented as holons [18]. Each holon consists of one head agent and a number of body agents. The head of the holon communicates outside the holon and coordinates all body agents. Open and closed holons are distinguished. Open holons consist of the head and a certain type of body agents, however, the number and exact instances of body agents are not known during the design of the system and can be changed during the maintenance and runtime of the system so modifying the system's functionality. The body agents have to register their services at the directory facilitator agent. Heads of open holons use the directory facilitator agent to find actual body agents in the holon. Agent instances in closed holons are specified during the design and can not be changed afterwards.

Definition of specific agents used to implement the system is started during the external design of agents. Tasks are defined and allocated to the agents in this stage. During the internal design of agents a decision is made, whether to implement each higher level agent as a holon or as a single agent. The final architecture of the system is shown in Figure 2. Heads of open holons are denoted by gray colour. The developed system consists of the following higher level agents. The communication module is implemented as an *interface agent* that carries out all interactions with the learner. It is responsible for the following tasks: (1) Registration. (2) Log in. (3) Perceiving learner's requests and starting the processes in the system by forwarding learner's requests, actions and data to respective agents. (4) Giving all information to

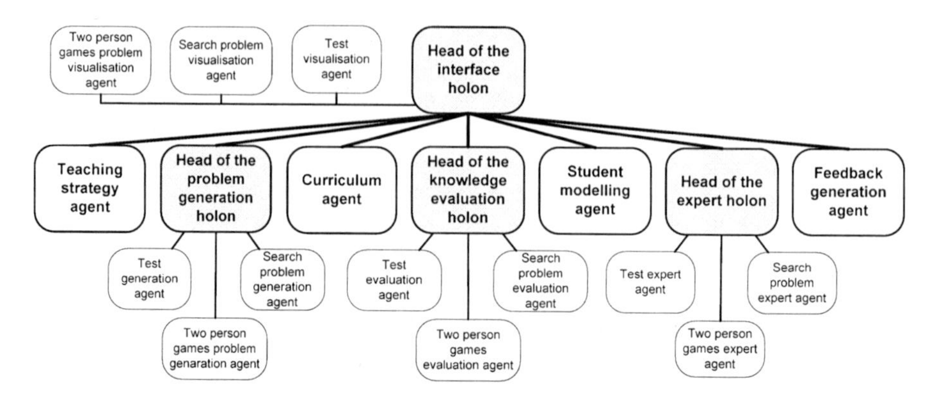

Fig. 2. Architecture of the MIPITS system

a learner, including learning materials, problems and feedback. The interface agent is the only agent interacting with a learner. Thus, it is the head of the higher level holon.

The tutoring module is implemented as the teaching strategy agent, the problem generation agent, the curriculum agent and the feedback generation agent. The *teaching strategy agent* is responsible for provision of the learning material in each topic. The *curriculum agent* is responsible for creation of the curriculum during the registration of a learner in the system. The *problem generation agent* is responsible for generation of all types of problems used in the system and adaptation of these problems to the knowledge level and preferences of the learner.

The expert module is implemented as the *expert agent,* which is responsible for solving all types of problems.

The student diagnosis module is implemented as the student modelling agent and the knowledge evaluation agent. The *student modelling agent* is responsible for creating, updating and providing the student model upon request of any other agent. The initial student model is created during the registration process. It is modified by reacting on the different actions reported by other agents. The student model contains:

- Personal data of a learner that are collected during the registration process.
- The learner's preferences that are collected during the registration process: the preferred (initial) difficulty of problems, the preferred practicality of problems and the preferred size of problems described below.
- The curriculum that is created for a learner during the registration process. Additionally, each topic has its status denoting what activities a learner has completed in the topic. The status has the following possible values: "initial", "started", "finished theoretical part", and "finished".
- All problems given to a learner and the results of all knowledge evaluations based on his/her solutions of the given problems.

The knowledge evaluation agent has to find the learner's mistakes in his/her solution by comparing it to the expert's solution. It must be able to evaluate solutions of all types of problems.

According to the MASITS methodology, to implement different types of problems and allow adding new problems, all higher level agents dealing with problems, namely the problem generation agent, the expert agent, the knowledge evaluation agent and the interface agent, are implemented as open holons. The problem generation holon consists of body agents that generate one type of problems: the test generation agent, the search problem generation agent and the two person games problem generation agent. Similarly, body agents of the expert holon are capable to solve problems of certain type. Knowledge evaluation body agents compare system's and learner's solutions of the given problem. Each interface body agent is capable to manage user interface of one type of problems. The heads of open holons are only capable to find the appropriate body agent and forward results received from them.

5 The Tutoring Scenario of the MIPITS System

The first activity a learner has to do is to *register in the system*, because a learner must be identified to adapt problems to his/her characteristics. For this purpose a learner fills a form containing his/her personal data and his/her preferences. After a learner has submitted the registration form, the interface agent collects and checks data from the form, inserts user data into database and sends data to the student modelling agent. The student modelling agent creates the initial student model based on learner's preferences and requests the curriculum agent to create the curriculum for a learner. After receiving the curriculum from the curriculum agent the student modelling agent completes the initial student model by adding the curriculum and sends it to the interface agent, who opens the main window with the curriculum and information about the first module. Interactions among agents are implemented using simple messages. Predicates from the domain ontology are used to express message contents. Messages sent during the registration process are shown in Figure 3.

Each time a registered learner logs into the system the learning process is restarted at the topic that was open when a learner quit the system last time. To do it, first, the interface agent validates learner's data and sends the data to the student modelling agent. Second, the student modelling agent reads the student model from the database and sends it to the user interface agent. Third, the interface agent requests the teaching strategy agent to provide the material in the current topic. Finally, when the interface agent receives the material, the main window of the system containing the curriculum and the learning material in current topic is opened. Interactions done during the login process are shown in Figure 4.

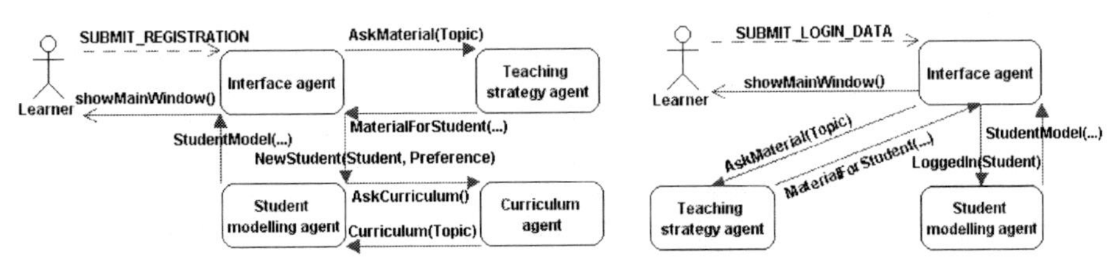

Fig. 3. Interactions done during the registration **Fig. 4.** Interactions done during the login

The curriculum of the course consists of modules that, in their turn, consist of topics. To teach a topic the MIPITS system performs the scenario consisting of three steps. When a learner starts a topic, the system starts the *theoretical step*. During this step a learner studies a theoretical material. After finishing it a learner requests a test. The system switches to the *problem solving step*. During this step a learner has to solve some problems in the topic. After finishing, a learner submits his/her solutions. The system moves to the *knowledge evaluation step*. As a result of this step a learner receives an evaluation of his/her knowledge in the current topic and constructive feedback about mistakes made in each problem. When the knowledge evaluation step is over, a learner can start a new topic. After finishing all topics of the module a learner has to pass the final test of the module that may contain problems from all topics included in this module. During the final testing of the module all actions of the problem solving and knowledge evaluation steps are done.

5.1 The Theoretical Step

The goal of the theoretical step is to hand out a learning material to a learner allowing him/her to repeat theory of the topic that has been given in the classroom. The step is carried out using the following scenario. When a learner chooses the topic to start learning, the interface agent requests the teaching strategy agent to generate a learning material in the chosen topic. The teaching strategy agent finds appropriate learning material and sends it to the interface agent. The interface agent shows a learning material in the user interface of the system. Additionally, the teaching strategy agent notifies the student modelling agent that a learning material in the current topic has been given to a learner. The student modelling agent modifies the student model by changing the status of the topic from "initial" to "started". Messages sent among agents during the theoretical step are shown in Figure 5.

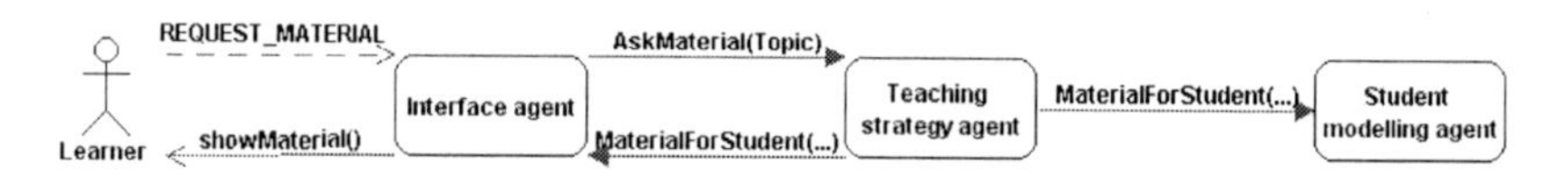

Fig. 5. Interactions done during the theoretical step

5.2 The Problem Solving Step

The goal of the problem solving step is to provide a learner an opportunity to practice in different types of problems. The knowledge evaluation step is based on learner's solutions in the problem solving step. The problem solving step starts when a learner submits that he has studied a material. The interface agent requests the problem generation agent to generate the problem in the current topic. The request is processed by the head of the problem generation holon, using the following algorithm (see Figure 6). Firstly, the head queries the student modelling agent to get full student model and the directory facilitator to find the body agents of the problem generation holon. If there are no problem generation body agents registered to the directory facilitator, the system error is generated. Otherwise, after receiving replies from the student modelling agent and the directory facilitator all body agents are queried to generate a problem in the current topic that is appropriate to learner's characteristics. Each problem generation body agent either generates the most appropriate problem to learner's characteristics

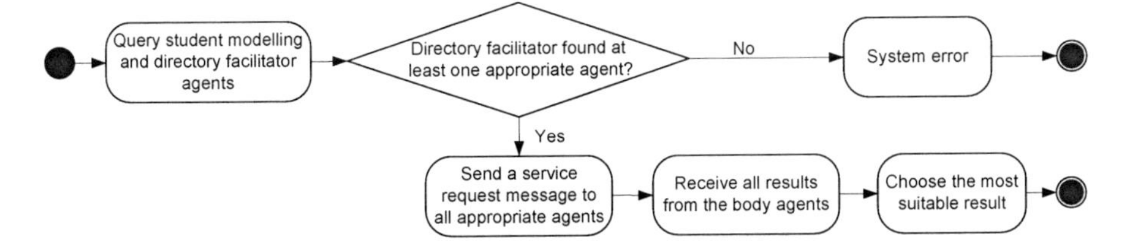

Fig. 6. Algorithm for the head of the problem generation holon

and sends it to the head of the holon or sends failure to the head of the holon if it can not generate a problem in the current topic, because the type of the problem does not match the topic.

After receiving problems from body agents the head chooses the most appropriate problem by using the following criteria, whose preferred values are calculated first:

- The degree of difficulty of the problem. The problem must match the preferred level of difficulty as close as possible, because it should not be too complex (unsolvable) nor too easy (not challenging) for a learner. During the registration process learner evaluates his/her knowledge level as an initial degree of difficulty in the scale from 1 (the lowest) to 5 (the highest). This is only a subjective his/her estimation that may be inaccurate. Thus, the system calculates the preferred degree of difficulty using the initial degree of difficulty and learner's results in previous knowledge evaluations. Moreover, the more problems a learner has solved, the more valuable is knowledge evaluation by the system and the less valuable is the value given by a learner. The preferred degree of difficulty is calculated as follows:

$$dif_{pref} = \frac{init*dif_{init} + max*l}{init+max}\text{, where} \quad (1)$$

init – coefficient denoting how much points for problem solving are equivalent to the initial degree of difficulty. Its value is determined empirically and is 50. For comparison, one test question is 2 to 4 points worth.
dif_{init} – initial degree of difficulty.
max – maximal number of points that can be scored for problems solved so far.
l – level of difficulty corresponding to results that a learner has achieved in problem solving. To calculate the level, firstly, learner's result is calculated in percents of maximal number of points that can be scored for the problems solved by him/her. Secondly, the level is determined using the following empirical function: 0-34% go in level 1, 35-49% go in level 2, 50-64% go in level 3, 65-80% go in level 4, results over 80% go in level 5.

- The size of the problem. During the registration learner may choose whether his/her knowledge evaluation will be carried out with small and concrete problems or large and time consuming problems.
- The practicality of the problem. During the registration learner may choose between more practical and more theoretical problems to match preferences of more practically and more theoretically oriented learners.

- Frequency of the type of problem. Different combinations of the learner's characteristics may lead to the situation, that only one type of problems is used and knowledge evaluation process becomes monotony. Thus, the frequency of the type of problems should be minimized. The frequency is 0 if a learner has not solved any problem yet, otherwise it is calculated by dividing the number of problems of certain type given to the learner by total number of problems given to him/her.

Each problem received from the problem generation agent contains the values of all criteria. So, after calculating the preferred values of criteria the difference between preferred and real values is minimized. The appropriateness is calculated as follows:

$$A = -(|dif_{pref} - dif_r| * c_d + |s_{pref} - s_r| * c_s + |pr_{pref} - pr_r| * c_p + f_t * c_f), \text{ where} \quad (2)$$

dif_{pref} – the preferred difficulty of the task;
dif_r – the real difficulty of the task;
c_d – the weight of the difficulty;
s_{pref} – the preferred size of the problem;
s_r – the real size of the problem;
c_s – the weight of the size;
pr_{pref} – the preferred practicality;
pr_r – the real practicality of the problem;
c_p – the weight of the practicality;
f_t – the frequency of problem's type;
c_f – the weight of the frequency.

Weights are determined empirically and are the following: c_d=2, c_s=3, c_p=3, c_f=6. With these weights all criteria have significant impact on the appropriateness.

After finding the problem with the highest appropriateness it is sent to three agents:

- To the interface agent, that is responsible for handing out the problem to a learner.
- To the expert agent, that has to find the correct solution of the problem.
- To the student modelling agent that changes the status of the topic from "started" to "finished theoretical part" in the student model.

Heads of the interface holon and the expert holon are not capable to accomplish the tasks that they are responsible for. Thus, they have to use body agents of their holons. After receiving the problem the heads of the holons use algorithm, similar to one shown in Figure 6. The head of the holon uses the directory facilitator to find the appropriate body agent. If such agent is found, the problem is forwarded to the body agent, otherwise system error is generated. The body agent does its job (respectively, passes the problem to a learner or solves it). The body agent of the expert holon sends the solution to the head of the holon that forwards it to the head of the knowledge evaluation holon, which saves the solution to use it during the knowledge evaluation. All messages sent among agents during this step are shown in Figure 7.

As a result of the problem solving step, the problem is handed out to a learner using the main window of the system (see Figure 8). The interface of the system is in Latvian, which is the language of the course. The window consists of two main parts: the curriculum denoted with 1 and the main panel denoted with 2. The main panel changes its contents depending on the step. It contains materials in the theoretical step and problems in problem solving and knowledge evaluation steps. The screenshot of the system shown in Figure 8 contains the state space search problem. The panel of the problem consists of three parts: the statement of the problem, the state space denoted with 3 and tools needed to do the search denoted with 4.

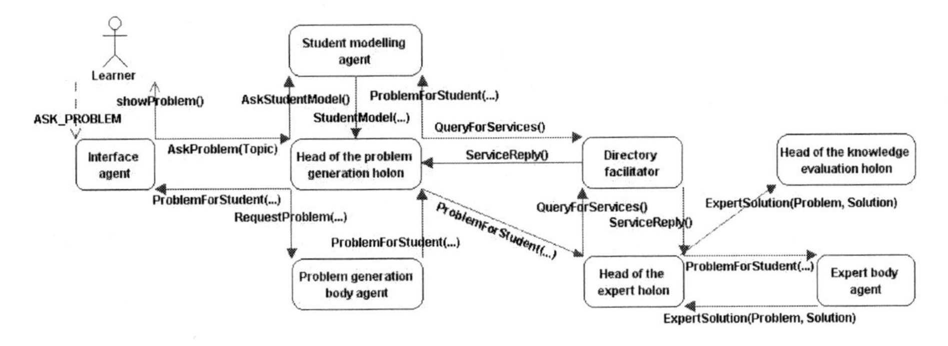

Fig. 7. Interactions done during the problem solving

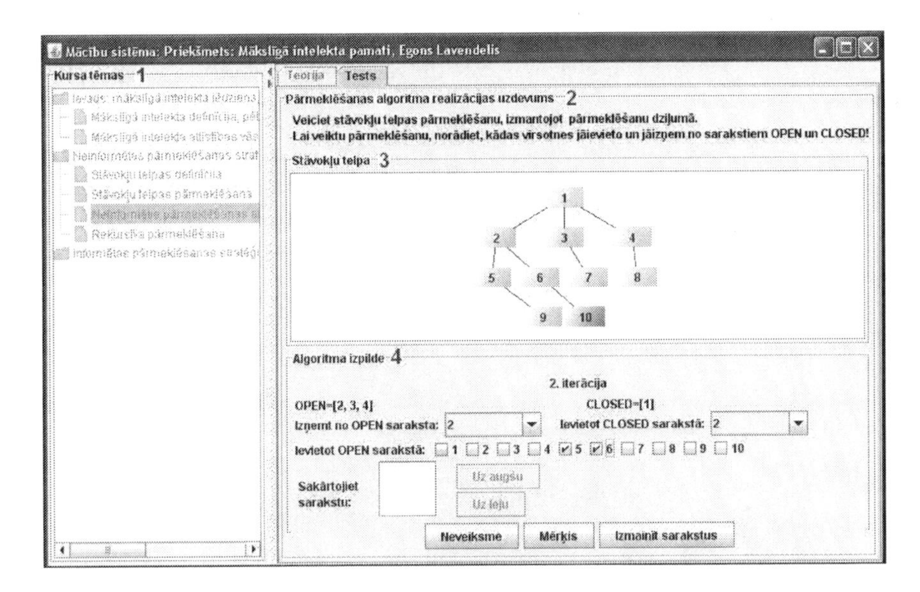

Fig. 8. The interface of the MIPITS system during the problem solving step

5.3 The Knowledge Evaluation Step

The goal of the knowledge evaluation step is to evaluate learner's solution created in the previous step and provide him/her the feedback about the solution. A learner starts the step by submitting his/her solution of the problem. The interface agent sends learner's solution to the knowledge evaluation agent. The head of the knowledge evaluation agent uses the same algorithm as the heads of the expert and interface holons. The body agent compares the system's and the learner's solutions finding the learner's mistakes and evaluating the solution. The head of the knowledge evaluation holon forwards the evaluation to the student modelling agent and the feedback agent. The student modelling agent records the knowledge evaluation in the student model and changes the status of the topic to "finished". The feedback agent creates the textual feedback about the result, like "You scored 19 points from 20! Great result!".

Additionally, it creates textual information about the learner's mistakes, like "You made a mistake determining the search goal during the last step of the algorithm". After the feedback is prepared it is sent to the interface agent, which passes it to a learner. Interactions among agents done in this step are shown in Figure 9.

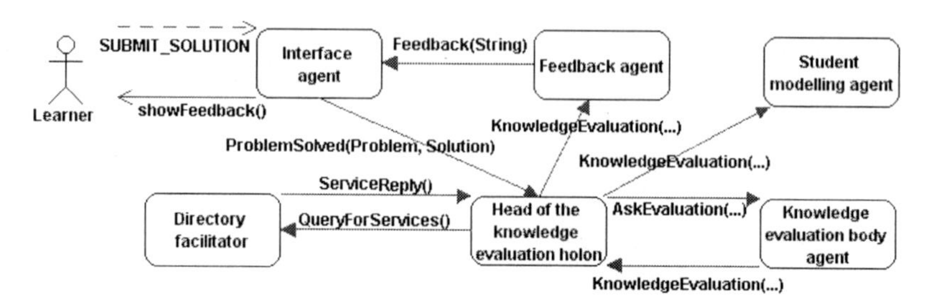

Fig. 9. Interactions done during the knowledge evaluation

6 Extending the System with New Types of Problems

The open architecture of the MIPITS system makes it extendable for teaching new topics of the course or even some other courses by including new types of problems and appropriate materials. To enable extension of the system, all steps of the tutoring scenario are implemented in the way that any new type of problems can be added to the system without modifying the code of already existing agents. It can be done by adding new agents to the open holons. A single body agent has to be added to all four open holons (heads of open holons are denoted with gray colour in Figure 2): the problem generation holon, the expert holon, the knowledge evaluation holon and the interface holon.

To illustrate the extendibility of the MIPITS system a topic about propositional logic and inference is added to the system. To practice in the basic concepts of inference, a problem about usage of the Modus Ponens is added to the system. The learner receives initial working memory with one fact and the knowledge base consisting of a number of implication rules. He/she then has to show how the rules are fired and the working memory changed.

To implement this task the following changes have been made in the system:

- Four new body agents are added. Modus Ponens problem generation agent is added to the problem generation holon, Modus Ponens visualisation agent is added to the interface holon, Modus Ponens knowledge evaluation agent is added to the knowledge evaluation holon and Modus Ponens expert agent is added to the expert holon. Newly added body agents register themselves to the directory facilitator agent in order the heads of the holons be able to find them.
- The ontology is extended with class of problem about Modus Ponens and its solution.
- Information about new type of problems is added to the corresponding table in the database to link up new type of problem with corresponding topics.

7 Conclusions and Future Work

The system is an example how an ITS for individualized tutoring can be created with the MASITS methodology and a tool. The system proves that the MASITS methodology and tool are suitable for development of open agent based ITS with holonic architecture. The MASITS methodology facilitates the development process by enabling usage of the main results of ITS research, like architecture, set of agents and tasks done by these agents. The usage of holonic agents allow to increase the modularity of the ITS, where agents are responsible for concrete and separate tasks. Moreover, the system can be modified by adding or removing types of problems used in the system without changing existing code.

An agent based ITS adapting the problems to the learner's knowledge level and preferences is proposed. The adaptation of the problems is done by minimizing the difference between the preferred and real values of problem's difficulty, practicality and size. Experiments with the system showed that learners received problems that matched their preferences closer than any problem that could be given to all learners.

There are two main directions of the future work in the MIPITS system. The first one is to add more types of problems corresponding to other topics. The second one is to use open holons to implement other types of openness, for example, usage of different types of learning materials is possible by implementing the teaching strategy agent as an open holon. Moreover, new types of adaptation (for different kinds of adaptation in ITS see [1]) can be implemented in the system. The main direction of the future work in the MASITS methodology is to create library of reusable agents and a basic reusable ITS ontology to facilitate reuse in ITS development.

Acknowledgements. This work has been supported by the European Social Fund within the project "Support for the implementation of doctoral studies at Riga Technical University".

References

1. Brusilovsky, P., Peylo, C.: Adaptive and intelligent Web-based educational systems. International Journal of Artificial Intelligence in Education 13(2-4), 159–172 (2003)
2. Carbonell, J.R.: AI in CAI: An Artificial Intelligence Approach to Computer-Assisted Instruction. IEEE Transactions on Man-Machine Systems 11(4), 190–202 (1970)
3. Devedzic, V., Debenham, J., Popovic, D.: Teaching Formal Languages by an Intelligent Tutoring System. Educational Technology & Society 3(2), 36–49 (2000)
4. Capuano, N., De Santo, M., Marsella, M., Molinara, M., Salerno, S.: A Multi-Agent Architecture for Intelligent Tutoring. In: Proceedings of the International Conference on Advances in Infrastructure for Electronic Business, Science, and Education on the Internet (SSGRR 2000), Rome, Italy (2000)
5. Crowley, R.S., Medvedeva, O.: An Intelligent Tutoring System for Visual Classification Problem Solving. Artificial Intelligence in Medicine 36(1), 85–117 (2005)
6. Hospers, M., Kroezen, E., Nijholt, A., op den Akker, R., Heylen, D.: An Agent-based Intelligent Tutoring System for Nurse Education. In: Nealon, J., Moreno, A. (eds.) Applications of Intelligent Agents in Health Care, pp. 141–157. Birkhauser Publishing Ltd., Basel (2003)

7. Grundspenkis, J., Anohina, A.: Agents in Intelligent Tutoring Systems: State of the Art. In: Scientific Proceedings of Riga Technical University Computer Science. Applied Computer Systems. 5th series, vol. 22, pp. 110–121 (2005)
8. Gascuena, J.M., Fernández-Caballero, A.: An Agent-based Intelligent Tutoring System for Enhancing E-learning/E-teaching. International Journal of Instructional Technology and Distance Learning 2(11), 11–24 (2005)
9. Lavendelis, E., Grundspenkis, J.: MASITS – A Multi-Agent Based Intelligent Tutoring System Development Methodology. In: Proceedings of IADIS International Conference Intelligent Systems and Agents 2009, Algarve, Portugal, June 21-23, pp. 116–124 (2009)
10. Henderson-Sellers, B., Giorgini, P.: Agent-Oriented Methodologies, p. 414. Idea Group Publishing, London (2005)
11. Zambonelli, F., Jennings, N.R., Wooldridge, M.: Multi-Agent Systems as Computational Organisations: The Gaia Methodology. In: Agent-Oriented Methodologies, pp. 136–171. Idea Group Publishing, London (2005)
12. Winikoff, M., Padgham, L.: The Prometheus Methodology. In: Methodologies and Software Engineering for Agent Systems. The Agent-Oriented Software Engineering Handbook, pp. 217–236 (2004)
13. DeLoach, S.: Analysis and Design Using MaSE and agentTool. In: Proceedings of the 12th Midwest Artificial Intelligence and Cognitive Science Conference, Oxford OH, March 31 - April 1, pp. 1–7 (2001)
14. Lavendelis, E., Grundspenkis, J.: MASITS - A Tool for Multi-Agent Based Intelligent Tutoring System Development. In: Proceedings of 7th International Conference on Practical Applications of Agents and Multi-Agent Systems (PAAMS 2009), Salamanca, Spain, March 25-27, pp. 490–500 (2009)
15. Lavendelis, E., Grundspenkis, J.: Open Holonic Multi-Agent Architecture for Intelligent Tutoring System Development. In: Proceedings of IADIS International Conference Intelligent Systems and Agents 2008, Amsterdam, The Netherlands, July 22 - 24, pp. 100–108 (2008)
16. Lavendelis, E., Grundspenkis, J.: Multi-Agent Based Intelligent Tutoring System Source Code Generation Using MASITS Tool. Scientific Journal of Riga Technical University (2010) (in Press)
17. Luger, G.F.: Artificial Intelligence: Structures and Strategies for Complex Problem Solving, p. 903. Addison-Wesley, Harlow (2005)
18. Fischer, K., Schillo, M., Siekmann, J.: Holonic Multiagent Systems: A Foundation for the Organisation of Multiagent Systems. In: Mařík, V., McFarlane, D.C., Valckenaers, P. (eds.) HoloMAS 2003. LNCS (LNAI), vol. 2744, pp. 71–80. Springer, Heidelberg (2003)

A 3D Indoor Pedestrian Simulator Using an Enhanced Floor Field Model

Chulmin Jun and Hyeyoung Kim

Department of Geoinformatics, University of Seoul, Seoul, Korea
{cmjun,mhw3n}@uos.ac.kr

Abstract. Many pedestrian simulation models for micro-scale spaces as building indoor areas have been proposed for the last decade and two models – social force model and floor field model – are getting attention. Among these, the CA-based floor field model is viewed more favorable for computer simulations than computationally complex social force model. However, previous floor field-based models have limitations in capturing the differences in dynamic values of different agents. In this study, we improved the floor field model in order for an agent to be able to exclude the influences of its own dynamic values by changing the data structure, and, also modified the initial dynamic value problem in order to fit more realistic environment. As the simulation data structure, we also proposed using a DBMS-based 3D modeling approach. In the simulations, we used real 3D building data stored in a spatial DBMS considering future integration with indoor localization sensors and real time applications. We illustrated the data construction processes and simulations using the proposed enhanced algorithms and DBMS approach.

Keywords: Pedestrian simulation, Floor field model, 3D model, Spatial DBMS.

1 Introduction

Many micro-scale pedestrian simulation models have been proposed for the last decade and applied to fire evacuation problems or building safety evaluation. Recent development in localization sensors such as RFID draws our attention to indoor spaces and real-time applications. In order for our pedestrian models to be applied to real world indoor applications, they need to use different data formats other than current experimental file formats. The data should include semantic and topological information of building 3D spaces. Also, to be able to communicate with the location sensors to capture the real pedestrian movement, the data should be stored in a DBMS. Once data are stored in a database, the simulation results can also be stored back in the DB for real time evacuation guidance.

In this paper we proposed a method to build a simplified 3D model which is suitable for pedestrian simulation. Instead of representing the complex details of indoor spaces, we used the floor surfaces focusing on the fact that pedestrian movements take place only on the surfaces.

As the simulation algorithm, we used the floor field model [4, 11] as our base model and revised the dynamic field strategy. Since the existing floor field-based

J. Filipe, A. Fred, and B. Sharp (Eds.): ICAART 2010, CCIS 129, pp. 133–146, 2011.

models have limitations in capturing the differences in dynamic values of different agents, we improved the algorithm in order for an agent to be able to exclude the influences of its own dynamic values. We illustrated the data construction processes and simulations using the proposed DBMS approach and modified algorithms.

2 Related Works

3D models currently used in the 3D GIS are actually 2.5 dimensional CAD-based data types focusing on visualization purpose in realistic way. They have limitations for analytical purposes in indoor space applications due to its lack of topological and semantic structure. As a solution to this, topological models along with using DBMSs for 3D objects have been recently investigated by some researchers [1, 18, 19, 21]. 3D models suggested by those are generally categorized as follows:

A. SOLID – FACE – EDGE – NODE
B. SOLID – FACE – NODE
C. SOLID – FACE

The three types are data models for defining 3D volumes not for interior spaces. CAD-based models have been used widely and there is a growing interest in using IFC(Industry Foundation Classes) format especially for modeling and developing building information systems. Although these formats offer flexibility in modeling indoor spaces with various data primitives, they are file-based formats and, thus, have limitations in being used in indoor information systems as mentioned earlier. On the other hand, CityGML which was adopted as a standard by OGC (Open Geospatial Consortium) is a 3D model that provides different levels of details ranging from region to interior spaces [13, 17]. CityGML is based on XML format for the storage of data and has capability of storing complex semantic information. However, it has not provided fully functional data base implementation. One of the reasons is attributed to the fact that current commercial DBMSs do not fully support topological structure of 3D objects yet.

Evacuation models have been studied in various fields such as network flow problems, traffic assignment problems, and are generally categorized into two; macroscopic and microscopic models [6]. Macroscopic models appear in network flow or traffic assignment problems and take optimization approach using node-link-based graphs as the data format. They consider pedestrians as a homogeneous group to be assigned to nodes or links for movements and do not take into account the individual interactions during the movement. On the other hand, microscopic models emphasize individual evacuees' movement and their responses to other evacuees and physical environment such as walls and obstacles. Microscopic models are mainly based on simulation and use fine-grained grid cells as the base format for simulation. They have been used by experts in different domains including architectural design for the analytical purposes of the structural implications on the human movement especially in emergency situations.

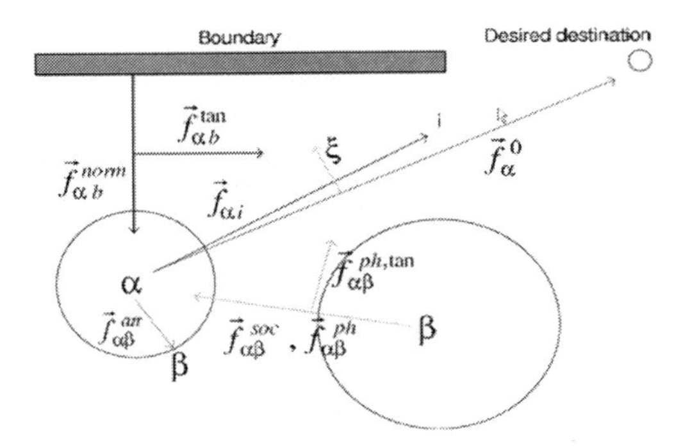

Fig. 1. Helbing's social force model

Different micro-simulation models have been proposed over the last decades [16] but two approaches are getting attention; social force model and floor field model [11]. A frequently cited model of former type is advanced by Helbing and collegues [7, 8] and is based on strong mathematical calculation acted on agents to determine its movement to destination. Helbing's model considers the effects of each agent upon all other agents and physical environment (Fig.1) leading to the computation of $O(n^2)$ complexity, which is unfavorable for computer-based simulation with many agents [9, 10].

In recent years there is a growing interest to use cellular automata as the base of micro-simulation [2, 12]. Kirchner and colleagues [11] have proposed CA-based floor field model, where two kinds of fields—static and dynamic—are introduced to translate Helbing's long-ranged interaction of agents into a local interaction. Although this model considers only local interactions, they showed that the resulting global phenomena share properties from the social force model such as lane formation, oscillations at bottlenecks, and fast-is-slower effects. The floor field model uses grid cells as the data structure and computes movement of an agent at each time step choosing the next destination among adjacent cells. This makes computer simulation more effective.

In this paper we focus on Kirchner's model as our base model. We will later describe the limitation of his dynamic field computation strategy and how we revised it.

3 A Simplified Indoor 3D Model

In our previous study [15] we had proposed a 2D-3D hybrid data model that can be used both in 2D-based semantic queries and 3D visualization. We used two separate models, 2D GIS layers and 3D models, and combined them using a database table as the linkage method.

Although the previous file-based approach was satisfactory in incorporating semantic and topological functionality into a 3D model, it has some drawbacks. First, two models are created separately and need additional table for linkage, which makes consistent maintenance difficult. Second, building a 3D model by separating compartments requires additional time and cost. Finally, such file-based models are not easy to store many buildings and, most importantly, they cannot be integrated with client/server applications such as sensor systems (i.e. RFID, UWB, thermal sensors).

To solve these problems, we proposed in this research a new approach that uses a DBMS instead of files. Because semantic information is now extracted from database tables and used for analyses and 2D/3D visualization, the new model does not require an additional table for linkage. This data model has a multi-layered structure based on 2D building floor plans as the previous file-based model. It retains 2D topology because building floor plans are converted into 2D GIS layers (shapefiles) and then are stored in a spatial database. Thus, it is possible to perform topology-based analyses and operations provided by the DBMS. Also, all records containing geometries can be visualized for 2D and 3D.

Indoor location-based application use locations and tracing information of pedestrians who move on the surface floors in the building. This means that it is possible to retrieve semantic data and perform analytical operations only using floor surfaces in such applications (i.e. indoor crowd simulation, indoor wayfinding). This is the reason that we choose to use building floor plans as the base data type. For the connection of floors, we also converted the stairs to a simple set of connected polygons and then stored in the DBMS. Fig. 2 illustrates the process for storing indoor objects in a database. This shows that we used only the bottom part of a room polyhedron.

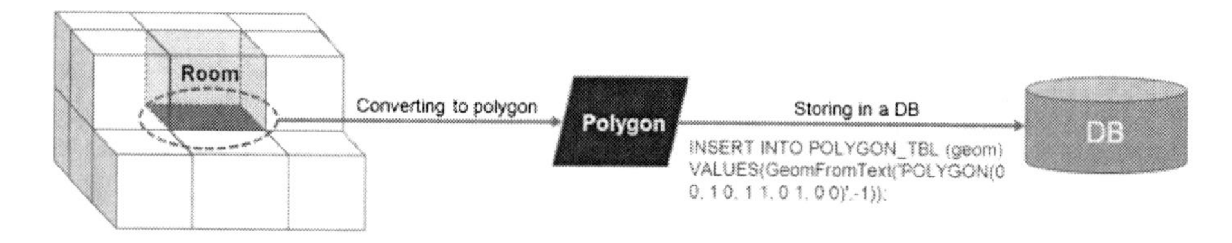

Fig. 2. An example of storing rooms floors in a spatial DB

This approach can well fit in DBMS-based applications due to less complex and simplified data construction process. Using a DBMS against file format gives many merits including data sharing, management, security, back-up and speed. It is also possible to integrate with sensor systems by storing the sensor information in the database. In this study, we used PostgreSQL/PostGIS for the DBMS. PostgreSQL is an open source object-relational database system, freely downloadable. To display indoor objects in 3D stored in the database, we used OpenGL library and it also interacts with the PostGIS database for the data retrieval and visualization (Fig. 3).

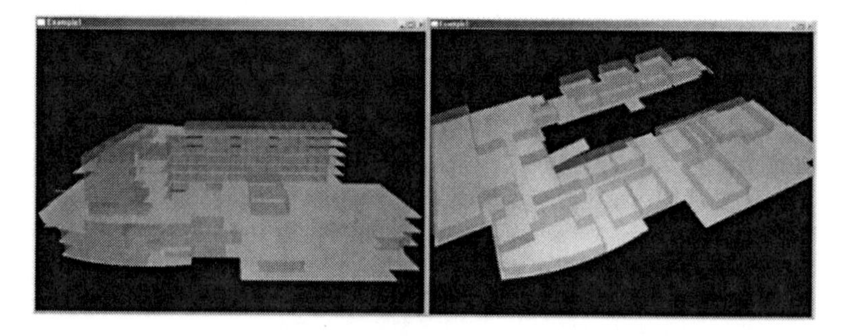

Fig. 3. 3D visualization using data from a spatial DBMS

4 Floor Field Model

We chose Kirchner model as our base model. His original model [11] and some variations [14] have demonstrated the ability to capture different pedestrian behaviors discussed in the previous section while being computationally efficient. First, we will describe the basic features of the floor field model and, then, describe how we improved the model.

4.1 Two Fields in Floor Field Model

Floor field model is basically a multi-agent simulation model. Here, each pedestrian is an agent who interacts with environments and other pedestrians. The group of such agents forms a multi-agent system (MAS). The agents in MAS have some important characteristics as follows [20].

- Autonomy: Agents are at least partially autonomous. An agent reacts to environment and other agents with autonomous manner.
- Local View: No agent has a full global view of the system. Each agent has no guidance to exits, instead, it moves only by local rules.
- Decentralization: Each agent in the system is equal and no agent controls others.

These characteristics of MAS in pedestrian models are frequently implemented using cellular automata (CA) and Kirchner model is also based on CA. CA theories are introduced in many related works, thus we will not introduce them here.

The basic data structure of Kirchner model is grid cells and each cell represents the position of an agent and contains two types of numeric values which the agent consults to move. These values are stored in two layers; static field and dynamic field. A cell in the static field indicates the shortest distance to an exit. An agent is in position to know the direction to the nearest exit by these values of its nearby cells. While the static field has fixed values computed by the physical distance, the dynamic field stores dynamically changing values indicating agents' virtual traces left as they move along their paths. As an ant use its pheromone for mating [3], the dynamic field is similarly modeled where an agent diffuses its influence and gradually diminishes it as it moves. Without having direct knowledge of where other agents are, it can follow other nearby agents by consulting dynamic values.

It is possible to simulate different pedestrian strategies by varying the degree to which an agent is sensitive to static or dynamic field. For example, we can model herding behaviors in panic situation by increasing sensitivity to the dynamic field. Such sensitivity factors are described in the following section.

4.2 Floor Field Rule

An agent in the floor field model consults the scores of its adjacent cells to move. A score represent the desirability or the attraction of the cell and the score of cell i is computed by the following formula [5]:

$$Score(i) = exp(k_d D_i) \times exp(k_s S_i) \times \xi_i \times \eta_i . \quad (1)$$

where,

> D_i : the dynamic field value in cell i
> S_i : the static field value in cell i
> k_d, k_s : scaling parameters governing the degree to which an agent is sensitive to dynamic field or static field respectively
> ξ_i : 0 for forbidden cells (e.g. walls, obstacles) and 1 otherwise
> η_i : Occupancy of agent in the cell. 0 if an agent is on the cell, and 1 otherwise.

Kirchner and his colleagues used probability *P(i)* , the normalized value of *score(i)* against all nine adjacent cells including itself. However, it turns out that using *score(i)* and *p(i)* has same effect since they are always proportional to each other in the adjacent nine cells.

The static field is first computed using a shortest distance algorithm such as the famous Dijkstra's algorithm. Then, all agents decide on their desired cells and they all move simultaneously.

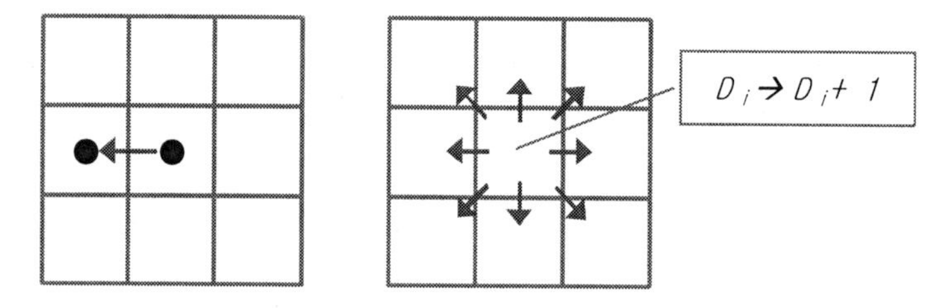

Fig. 4. Diffuse and decay of the dynamic value

After an agent has moved to one of its adjacent cells except its own, the dynamic value at the origin is increased by one: $D_i \rightarrow D_i + 1$ [4, 14]. Then a portion(α) of D_i is distributed equally to the adjacent cells (diffuse) and a portion(β) of D_i itself becomes diminished (Fig. 4). α and β is the input parameters to the model. This diffuse and decay process follows the analogy of ant pheromones which are left for a while and decayed gradually. Agents consult dynamic and static values at the same time. The scaling factors(k_d and k_s) are used to control the degree to which an agent react more to one of two fields. The ratio k_d / k_s may be interpreted as the degree of panic. The bigger the ratio, the more an agent tends to follow others.

5 Revised Dynamic Filed

The dynamic field is believed to be an effective translation of the long-ranged interaction of Helbing's model [7, 8] to local interaction. However, Kirchner model do not differentiate an agent's dynamic value with ones of others. The model simply adds the diffused and decayed values to the existing values.

It is reasonable that we consider that an agent should be able to avoid its own influence as an ant uses its pheromone. Kirchner's dynamic field does not cause a significant problem when an agent moves to one direction. However, as shown in Fig. 5,

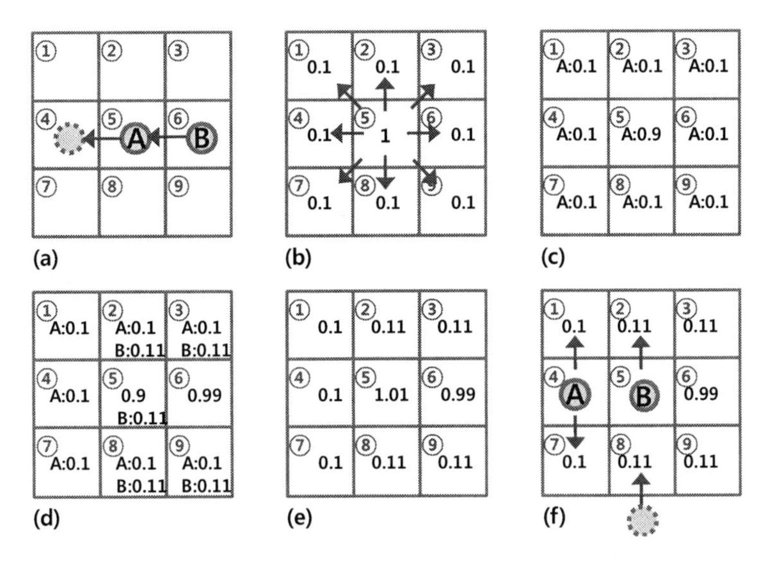

Fig. 5. Illustrating a problem of using the dynamic value of its own. ((a) Agents A and B move as in the figure. (b) D value increased to 1 in cell ⑤ and α portion of it diffuses to nearby cells (assuming no existing D values and α=β=0.1). (c) D value of ⑤ decays as much as β. (d) Diffuse & decay for Agent B. (e) The final D values. (f) Assuming agent B choose ② and another agent chooses ⑧, then, agent A has no choice but to choose ① or ⑦).

there may be cases when an agent has no choice but to get influenced by its own dynamic value if not much.

We modified the Kirchner's dynamic field such that an agent can exclude its own dynamic value when computing equation (1). To make it possible, the model should have a data structure that allows each cell to store a list of dynamic values of agents that have chance to leave their values to that cell. If we put the dynamic value of agent p as $d(p, k)$, then a set $D(k)$ having a list of dynamic values can be given by

$$D(k) = \{d(p, k) : p=1, 2, \ldots, n\}$$

Here, n is the number of agents that have the dynamic values that are greater than zero. We might easily presume that maintaining such set makes the model $O(n^2)$ complexity which are computationally unfavorable. However, $D(k)$ does not contain the entire agents' values and, instead, keep only those agents' values that pass k's nearby areas and keep non zero dynamic values. Thus, each cell keeps relatively small number of entries compared to the whole number of agents. For the implementation of the simulator, we used .NET C# language, and the data structure called Dictionary. The dictionary keeps a list of (*key, value*) pairs, where the *key* represent an agent while the *value* is its dynamic value.

If an agent p happens to leave any portion of its dynamic value to cell k more than once, $d(p, k)$ maintains only the maximum value among them. This makes sense if we imagine that the decaying scent get again maximized when an ant returns to that area.

$$d(p, k) = max\{d(p, k): p=1, 2, \ldots, m\}$$

Here, m is the number that an agent p leaves any portion of its dynamic values to cell *k*. Eventually, when consulting the score(*i*), agent q at cell *i* is able to exclude its own dynamic values in the adjacent cells and only takes the maximum one from each D(*k*) into account.

$$D(k)_q = max\{d(p, k): p \neq q\}$$

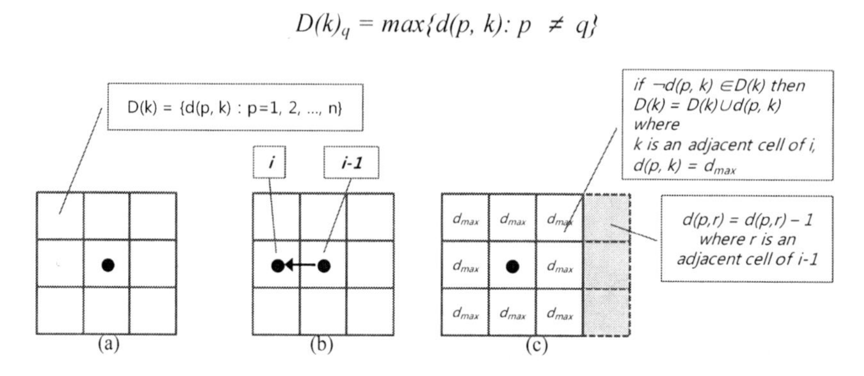

Fig. 6. The list of dynamic values at cell k(a), an agent's movement, and diffusion and decay(c)

We also modified the diffusion and decay strategy in our model. As shown in Fig. 6, right after an agent *p* moves, *d(p, k)* values of the adjacent cells of cell *i-1* is decreased by one, and then *d(p, k)*s of the adjacent cells of the current cell *i* are newly assigned the maximum dynamic value. Then, what is the maximum dynamic value? Let us first take an example before describing it.

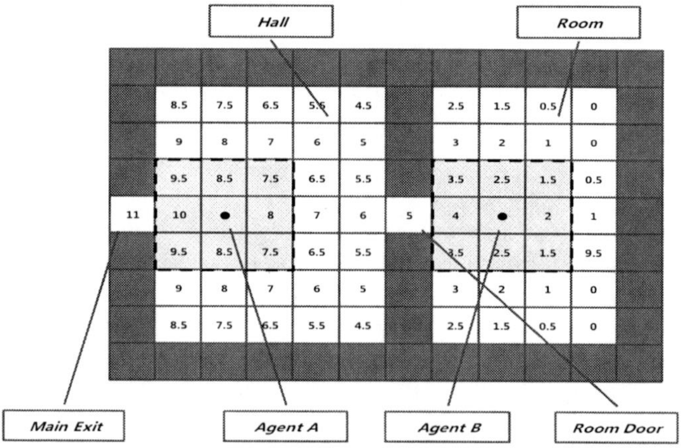

Fig. 7. The problem of initializing the dynamic values in a building with multiple compartments

Fig. 7 shows a building floor plan that has a main exit and a room inside with a door. We assume Agent A and B are located as in the figure. The numbers on the cells indicated the static field values computed from the main exit.

In Kirchner's model, the dynamic values are assigned regardless of the static values of the current location. In a simple rectangular space as those used by the author, such strategy may not cause much problem since the static values lead the agents to the exit eventually even though the dynamic values are much greater than its static counterparts at the current location. However, in using real building plans where multiple rooms are located inside, such strategy can cause a problem. If static values are gradually assigned from main exit(s), inner rooms can have very low values depending on the size of the building. Let's suppose Agent B reaches the room door. If there are multiple agents in the room and they happen to leave bigger dynamic values in the back of agent B than the static values in that area, then the agent can get stuck in the door because one or more empty adjacent cells in the back may be bigger than that of forward cells.

```
i = b                  // Set i to the beginning node
O=Φ                    // Set the open list to empty set
D(k)=Φ  for k∈N        // Set the Dynamic list for each node to empty set
P(i) = null            // Set the parent node of node i to null
s(i) = 0               // Set the score of the node i to 0
O = (i)                // Add the node i to the open list
While (i ≠ E)          // Iterate while node-i is not the destination node
{
    // Choose the maximum score node among the open list.
    Let  i ∈O  be a node for which
     s(i) = max{ s(i) : i ∈O } and s(i) > 0

    // If the agent- i has moved to a node other than itself
    if( i ≠ P(i) )
    {
        // For each node in the open list, if the node contains
        // the agent p's dynamic value, decrease it by one
        for each k ∈O
            if d(p, k) ∈ D(k) then d(p,k) = d(p,k) - 1

        O=Φ       // Reset the open list to empty set
        // For each of searchable adjacent nodes of i
        // (i.e. excluding those obstacles as walls and furniture and
        // including i and) set parent, and add to the open list
        // j: Adjacent nodes of i including i itself
        for each ( i, j ) ∈A(i)
        {
            // If not in the open list, add j to it
            if j∉O then O = O ∪(j)
            // Set the parent node of j to i
            P(j) = i
        }
        // Get the maximum static value   among those
        // in the open list nodes (t(k): static value in cell k)
        d_max = max{ t(k) : k ∈O }
        // For each node in the open list, if the dynamic list does
        // not contain the dynamic value of node k, add it to D(k)
        for each k ∈O
        {
            d(p, k) = d_max
            if d(p, k) ∉ D(k) then D(k) = D(k) ∪d(p, k)
        }
    }
}
```

Fig. 8. Pseudocode for an agent movement

To solve this problem, we changed the diffuse and decay strategy by letting an agent choose the maximum value of the adjacent static values as its initial dynamic value. This way, any agent inside the space can have the initial dynamic value which is proportional to the corresponding static values. The static and dynamic values are of different units; one is distance and the other is an abstract interpretation for attraction force. In order for a model to control the sensitivity to these two field values, two values should be comparable to each other at any cell. That is the reason we synchronizes the initial dynamic values with the static values every time an agent moves. Fig 8 is the pseudocode with the modified dynamic value computation.

6 Simulation

6.1 3D Data Model Construction

For the simulation, we constructed a 3D model of a real campus building following the proposed approach described in the previous section. The building has two main exits; the one in the front is wider than the side exit. We first simplified the CAD floor plans for the test purpose and they were converted to shapefiles, then stored into the PostGIS. The stairs were simplified and decomposed into several connected polygons and also stored in the DBMS. Once all data are stored, all connected floor surfaces now can be retrieved simply by SQL queries. Finally, the queried surface data were then converted to grid cell data for simulation. We set the cell size to 40cm × 40cm considering the human physical size.

We developed a simulator using the C# language and the OpenGL library. Fig. 9 shows the interface of the simulator and the 3D model used in the simulation. The simulator reads in the data from the PostGIS or the cellularized surface data. Once the data are read in, they can be visualized in 2D or 3D in OpenGL-based display module. In the simulator, we can input parameters such as k_s, k_d, d_{max}, time step, the number of agents, the number of iterations and the increments of the agents number in the iterations.

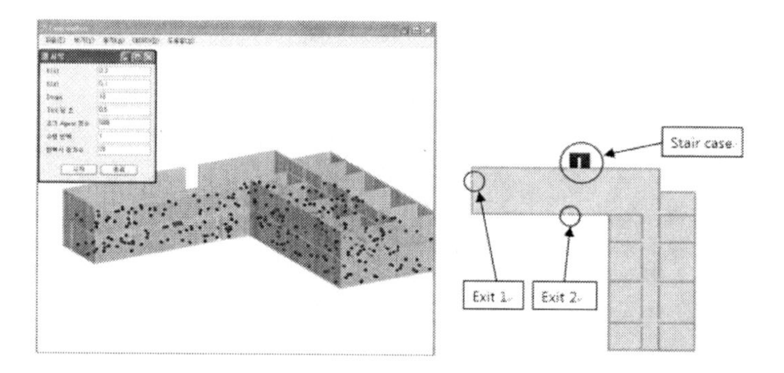

Fig. 9. The pedestrian simulator and the 3D model used in the test

6.2 Results

The simulator first constructs the static field computing the shortest distance from the two main exits to each cell in the building. We used varying numbers for the parameters. Fig. 10 shows the two extreme cases; $k_s = 0$ and $k_d = 0$. As can be easily guessed, $k_s = 0$ makes the agents wander around the space herding towards nearby agents without any clue of direction to exits. On the other hand, $k_d = 0$ causes the agents flow directly towards exits without any herding behaviors.

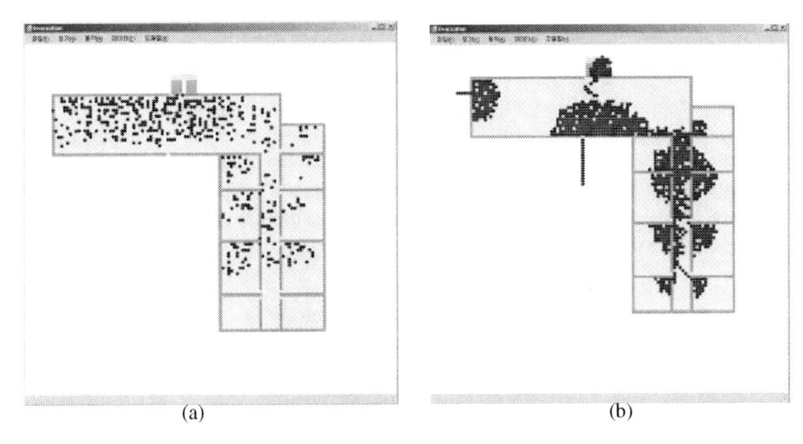

(a) (b)

Fig. 10. Snapshot of two extreme cases; $k_s = 0$(a) and $k_d = 0$(b)

Fig. 10 shows the effect of varying k_d values. While $k_d = 0$ correctly leads the agents to exits where the agents are belonged to based on their static values, $k_d > 0$ begins to show the herding behaviors. When a few agents happen to leave the group, others begin to follow them, leading to increasing the use of the side exit.

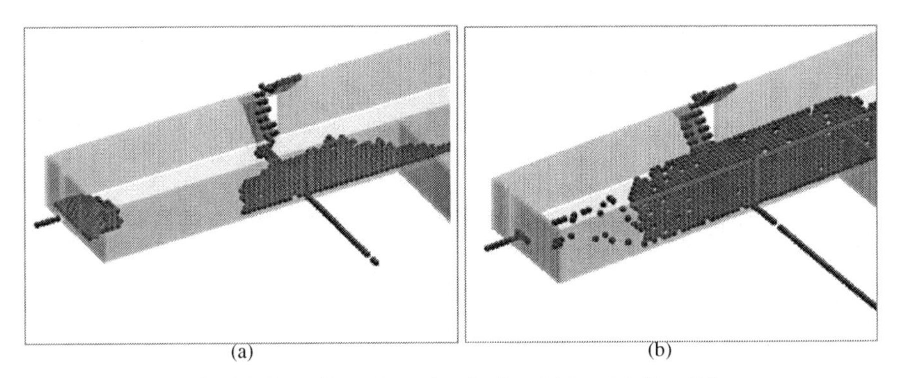

(a) (b)

Fig. 11. The effect of varying k_d; $k_d = 0$(a) and $k_d/k_s = 0.5$

We further investigated the effect of the dynamic term k_d using varying values. Table 1 shows the effect of k_d on the evacuation time and the use rate of the side exit (Exit 1). 2000 agents were used for the test. We observed that the use rate of the side exit gets increased in proportion to k_d. However, using k_d>0 slightly decreased the total number exited and then didn't change it significantly thereafter. This was because Exit 2, the main exit, is wider around twice as much as Exit 1. This indicates that using wider second exit can help decreasing the number exited.

Table 1. The effect of Varying k_d on evacuation time and use of the side exit

	k_d=0	k_d=0.05	k_d=0.1	k_d=0.25	k_d=0.5	k_d=1.0
Exit1	120	351	422	484	566	689
Exit2	1880	1649	1578	1516	1434	1311
Evactime	945	723	702	688	670	632

Another experiment was carried out to measure the time escaped with increasing agents and varying k_d/k_s. The results are provided in Fig. 12 showing the number of outgoing agents and the time taken with 6 sets of k_d/k_s. The (k_d, k_s) value pairs in the test were (0, 0.3), (1, 0.1), (0.25, 0.5), (0.1, 0.1), (0.05, 0.1), and (0.1, 1). The number of agents used were 500~5000. We observed that $k_d = 0$ made the curve almost linear increase while using different k_ds that are greater than 0 did not cause significant differences. However, the result shows leading people to alternative exit definitely decrease the overall escape time.

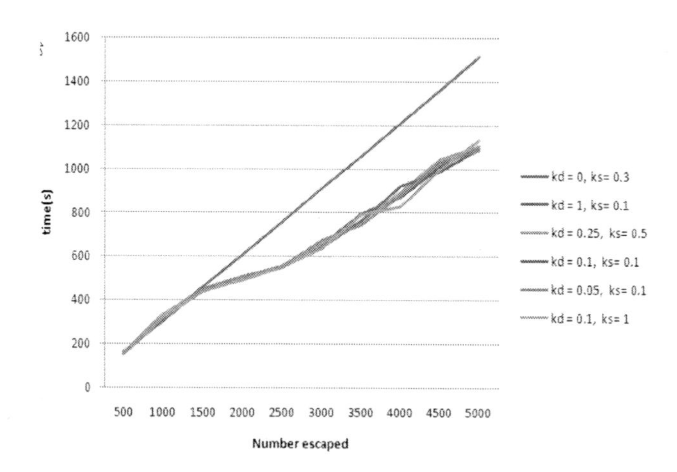

Fig. 12. Time taken for escape of varying number of agents with different sets of k_d/k_s

7 Concluding Remarks

In this study, we suggested a process to develop a 3D evacuation simulator instead of trying to improve the scientific investigation of crowd behaviors. In order to be able to integrate our system with real-time evacuation or rescuers' guidance, we suggested a less complex 3D indoor model focusing on the semantic information and navigation taking place on the floor surface. We also implemented the proposed model using a SDBMS and 3D visualization.

We also suggested a modified floor field pedestrian model using Kirchner's model. His model has demonstrated the ability to represent different pedestrian situations while maintaining basic MAS(multi-agent system) rules of autonomy and localization. However, his model is unable to capture the differences in dynamic values of different agents.

We have improved the floor field model in order for an agent to be able to exclude the influences of its own dynamic values by changing the data structure of dynamic field, which better conforms the analogy of ant pheromones. Also, by turning his constantly increasing and decreasing dynamic term D into dynamically changing term around agent's nearby static values, our model has shown the flexibility to more complex indoor configurations.

We currently keep improving the model by incorporating visibility effects and multiple velocities. Also, we focus on relating our model to real world applications. In this paper, we briefly introduced the use of spatial DBMS and 3D structures. However, with some refinements, we believe that our model can be adapted to real world 3D indoor applications equipped with indoor localization sensors. Then, we will be able to use the real distribution of indoor pedestrians captured by sensors instead of using randomly generated agents.

Acknowledgements. This research was supported by a grant (07KLSGC04) from Cutting-edge Urban Development - Korean Land Spatialization Research Project funded by Ministry of Land, Transport and Maritime Affairs.

References

1. Arens, C.A.: Maintaining reality: modelling 3D spatial objects in a GeoDBMS using a 3D primitive. M.Sc. Thesis, Delft University of Technology, The Netherlands (2003)
2. Blue, V.J., Adler, J.L.: Using cellular automata microsimulation to model pedestrian movements. In: Ceder, A. (ed.) Proceedings of the 14th International Symposium on Transportation and Traffic Theory, Jerusalem, Israel, pp. 235–254 (1999)
3. Bonabeau, E., Dorigo, M., Theraulaz, G.: Swarm intelligence: From natural to artificial systems. Oxford University Press, New York (1999)
4. Burstedde, C., Klauck, K., Schadschneider, A., Zittartz, J.: Simulation of pedestrian dynamics using a two-dimensional cellular automaton. Physica A 295, 507–525 (2001)
5. Colin, M., White, T.: Agent-Based Modelling of Force in Crowds. In: Davidsson, P., Logan, B., Takadama, K. (eds.) MABS 2004. LNCS (LNAI), vol. 3415, pp. 173–184. Springer, Heidelberg (2005)
6. Hamacher, H.W., Tjandra, S.A.: Mathematical modelling of evacuation problems- a state of art. In: Schreckenberg, M., Sharma, S. (eds.) Pedestrian and Evacuation Dynamics, pp. 227–266. Springer, Berlin (2001)
7. Helbing, D., Farkas, I., Molnár, P., Vicsek, T.: Simulation of pedestrian crowds in normal and evacuation situations. In: Schreckenberg, M., Sharma, S. (eds.) Pedestrian and Evacuation Dynamics, pp. 21–58. Springer, Berlin (2001)
8. Helbing, D., Molnár, P.: Self-organization phenomena in pedestrian crowds. In: Schweitzer, F. (ed.) Self-Organisation of Complex Structures: From Individual to Collective Dynamics. Gordon & Beach, London (1997)
9. Henein, C., White, T.: Agent-based modelling of forces in crowds. In: Davidsson, P., Logan, B., Takadama, K. (eds.) MABS 2004. LNCS (LNAI), vol. 3415, pp. 173–184. Springer, Heidelberg (2005)
10. Henein, C., White, T.: Macroscopic effects of microscopic forces between agents in crowd models. Physica A 373, 694–712 (2007)
11. Kirchner, A., Schadschneider, A.: Simulation of evacuation processes using a bionicsinspired cellular automaton model for pedestrian dynamics. Physica A 312, 260–276 (2002)
12. Klupfel, H., Konig, T., Wahle, J., Schreckenberg, M.: Microscopic simulation of evacuation processes on passenger ships. In: Proceedings of Fourth International Conference on Cellular Automata for Research and Industry, Karlsruhe, Germany (October 2002)
13. Kolbe, T.H.: Representing and exchanging 3D city models with CityGML. In: Lee, J., Zlatanova, S. (eds.) 3D Geo-information Sciences, pp. 15–31. Springer, Berlin (2008)

14. Nishinari, K., Kirchner, A., Namazi, A., Schadschneider, A.: Simulations of evacuation by an extended floor field CA model. In: Traffic and Granular Flow 2003, pp. 405–410. Spinger, Berlin (2005)
15. Park, I., Kim, H., Jun, C.: 2D-3D Hybrid Data Modeling for Fire Evacuation Simulation. In: ESRI International User Conference 2007, San Diego (2007), http://gis.esri.com/library/userconf/proc07/papers/papers/pap_1731.pdf
16. Schadschneider, A.: Cellular automaton approach to pedestrian dynamics - Theory. In: Schreckenberg, M., Sharma, S. (eds.) Pedestrian and Evacuation Dynamics, pp. 75–86. Springer, Berlin (2001)
17. Stadler, A., Kolbe, T.H.: Spatio-semantic coherence in the integration of 3D city models. In: Proceedings of 5th International ISPRS Symposium on Spatial Data Quality ISSDQ 2007 in Enschede (2007)
18. Stoter, J.E., van Oosterom, P.J.M.: Incorporating 3D geo-objects into a 2D geo-DBMS. In: ACSM-ASPRS 2002 (2002)
19. Stoter, J.E., Zlatanova, S.: Visualising and editing of 3D objects organised in a DBMS. In: Proceedings EUROSDR Workshop: Rendering and Visualisation, pp. 14–29 (2003)
20. Wooldridge, M.: An Introduction to MultiAgent Systems. John Wiley & Sons, Chichester (2002)
21. Zlatanova, S.: 3D GIS for urban development, PhD thesis, Institute for Computer Graphics and Vision, Graz University of Technology, Austria, ITC, the Netherlands (2000)

Toward a Self-adaptive Multi-Agent System to Control Dynamic Processes

Sylvain Videau, Carole Bernon, and Pierre Glize

Institut de Recherche Informatique de Toulouse
Toulouse III University, 118 Route de Narbonne
31062 Toulouse, cedex 9, France
{videau,bernon,glize}@irit.fr

Abstract. Bioprocesses are especially difficult to model due to their complexity and the lack of knowledge available to fully describe a microorganism and its behavior. Furthermore, controlling such complex systems means to deal with their non-linearity and their time-varying aspects.

A generic approach,relying on the use of an Adaptive Multi-Agent System (AMAS) is proposed to overcome these difficulties, and control the bioprocess. This gives it genericity and adaptability, allowing its application to a wide range of problems and a fast answer to dynamic modifications of the real system. The global control problem will be turned into a sum of local problems. Interactions between local agents, which solve their own inverse problem and act in a cooperative way thanks to an estimation of their own criticality, will enable the emergence of an adequate global function for solving the global problem while fulfilling the user's request.

This approach is then instantiated to an equation solving problem, and the related results are presented and discussed.

Keywords: Adaptive control, Multi-agent systems, Cooperation, Bioprocess, Criticality.

1 Introduction

Regulating a dynamic system is a complex task, especially when we consider a real-world application implying real-time constraints and limitations on computational power. Biology offers some of the best examples of such systems when bioprocesses have to be regulated.

Controlling a bioprocess is keeping a quasi-optimal environment in order to allow the growth of the expected microorganisms, while limiting and suppressing any product with toxic characteristics. However, this task is difficult, and this difficulty arises from, on the one hand, the bioprocess complexity, and, on the other hand, the amount of elements and interactions between them that are to be taken into account. Furthermore, controlling such a system implies dealing with uncertainty coming from lags in measures and delays in reactions.

Another point that has to be considered is the lack of *online* (which means obtained directly from the bioprocess) measures available. This limits the visible indicators of

J. Filipe, A. Fred, and B. Sharp (Eds.): ICAART 2010, CCIS 129, pp. 147–160, 2011.

the consequences of the action of the control, and leads the observer to rely on inferred data in order to describe the biological state of the system.

In this paper, we present a generic approach for controlling bioprocesses that uses an Adaptive Multi-Agent System (AMAS). Section 2 presents an overview of the existing methods of control before positioning our approach in section 3. This section also expounds what are AMAS and details the features of the agents composing the proposed one. Section 4 instantiates this system to an equation solving problem and gives some experimental results. Finally, the conclusions and perspectives that this work offers are discussed.

2 Bioprocess Control: A Brief Overview

Mathematically speaking, control theory is the subject of an extensive literature. Basically, two kinds of control systems may be considered: the first one is an open loop, meaning that there is no direct connection between the outputs of the controlled system and its inputs. The control being carried out without any feedback, it only depends on the model within the controller itself. The second kind of control is a closed loop, which is focused on the feedback, allowing the control system to make actions on the inputs by knowing the system's outputs. In those two cases, the function determining the control to apply is named the control law. The control system presented here is designed as a closed loop.

2.1 PID Control

Currently, the most widespread approach to control bioprocesses is the Proportional-Integral-Derivative (PID) controller. Controlling with such a tool means that three different functions will be applied to the received feedback, in order to select the adequate control. These functions are i) the proportional, which computes the current error multiplied by a "proportional constant", ii) the integral, which takes into account the duration and magnitude of the error, by summing their integral and multiplying by an "integral constant" and finally iii) the derivative, which estimates the rate of change of this error, allows to observe its variation, and multiplies it by a "derivative constant". These three functions are then summed.

However, several points need to be treated to make this approach adaptive enough to follow the bioprocess dynamics; the different constants appearing in the formulas have still to be defined and a way to allow them to be adjusted during the bioprocess has to be found. Such a modification may be done by using methods like Ziegler-Nichols [1] or Cohen-Coon [2].

This PID approach is quite generic, and can be applied to a wide range of control systems. However, its performances in non-linear systems are inconsistent. This drawback led to the hybridization of this method by adding mechanisms relying on fuzzy logic [3], or artificial neural networks [4].

2.2 Adaptive Control

The differences existing in the results coming from distinct runs of the same bioprocess led us to study the field of adaptive control; these differences are for example a noise addition or a delay in the chemical reactions that modify the system dynamics. This problem can be overcome by applying methods that dynamically modify the control law of the controller.

There are mainly three different categories of adaptive controller, the Model Identification Adaptive Control (MIAC), the Model Reference Adaptive Control (MRAC), and the Dual Control.

MIAC systems [5] use model identification mechanisms in order to enable the controller to create a model of the system it controls. This model can be created from scratch thanks to the observed data, or by using an already known basis. The identification mechanism updates the model using values coming from inputs and outputs of the controlled system.

MRAC systems, suggested by [6], employ a closed loop approach modifying the parameters of the controller thanks to one or several reference models. This time, the system does not create a model of the bioprocess, but it uses an existing model to update the control law by observing the difference between the predicted output and the measured ones. This adjustment is generally applied by the use of the *MIT rule* [7], which is a kind of gradient descent minimizing a performance criterion computed from the error measured.

The last system is called Dual Controller [8] and is especially useful for controlling an unknown system. In fact, such a control system uses two kinds of actions, the first one is a normal control, which aims at leading the system toward a certain value, while the other one is a probing action, which allows the controller to obtain more information on the controlled system by observing its reaction. The difference here is that the probing action is physically applied on the system, and not only predicted by the use of a model.

2.3 Intelligent Control

The last kind of controller is a subtype of the adaptive control, called intelligent control. It focuses on the use of methods coming from artificial intelligence to overcome problems linked to non-linearity and dynamic systems.

In the case of bioprocess control, the most used intelligent controller is the artificial neural network (ANN). Initially applied to bioprocesses to infer some non-measurable variables, it was then used to control such processes, or to improve already existing control methods by providing adaptation. Furthermore, ANN appear in pattern recognition control such as [9].

Unfortunately, the black box aspect of ANN is a limit to their use in the bioprocess control. And even if some works exist to reduce this aspect [10], it is to the detriment of their adaptability.

Among the Artificial Intelligence techniques used in intelligent control, we can also find expert systems [11] using knowledge databases to select the control needed, and fuzzy logic [3].

Bayesian controllers can be considered like intelligent controllers too, especially with the use of Kalman filters. This mathematical approach uses two distinct steps to estimate the state of the system. First, a prediction step enables to estimate the current state of the system using the estimation made in the previous state. Then, an update step improves this prediction with the help of observations made on the system.

2.4 Limitations

However, these approaches generally lack of reusability: the work required to apply them on a specific bioprocess is useless for applying them on another one. Indeed, the variables of mathematical models are specifically chosen to fit with a specific bioprocess, for example in the case of PID; and the learning set needed to train adaptive methods such as ANN are quite difficult to obtain on top of being meaningful only in a restricted range of variations of the bioprocess. This over-specification limits the predictive power of the controller when the bioprocess diverges from the expected scheme, and so, such a controller is unable to bring back the bioprocess into a desired state. Generally, black box models are poor at extrapolating and weak in accomodating lags [12]. The approach presented in this work, and its perspectives, aim at reducing the impact of such drawbacks, by offering a generic and adaptive control using a Multi-Agent System (MAS).

3 Control Multi-Agent System (CMAS)

Using a MAS to control and manage a process is an approach already experimented, for example in [13]. However, the complexity brought by a bioprocess implies the use of an adaptive architecture to organize the CMAS. As a result, the principles governing the MAS proposed for controlling a system in real-time come from the Adaptive Multi-Agent System (AMAS) theory [14]. This AMAS has to determine which control to apply on the bioprocess in order to drive the values of certain variables to reach a user-defined objective. This section begins by a description of these AMAS principles before giving an overview of the MAS and its positioning in the global control mechanism. The abilities and behavior of the agents composing this AMAS are then detailed before delineating the generic aspects of the proposed approach in order to instantiate it according to a specific problem.

3.1 The AMAS Approach

The functional adequacy theorem [14] ensures that the global function performed by any kind of system is the expected one if all its parts interact in a cooperative way. MAS are a recognized paradigm to deal with complex problems and the AMAS approach is focused on the cooperative behavior of the agents composing a MAS.

Here, cooperation is not only a mere resource or task sharing, but truly a behavioral guideline. This cooperation is considered in a proscriptive way, implying that agents have to avoid or solve any Non Cooperative Situation (or NCS) encountered. Therefore, an agent is considered as being cooperative if it verifies the following meta-rules:

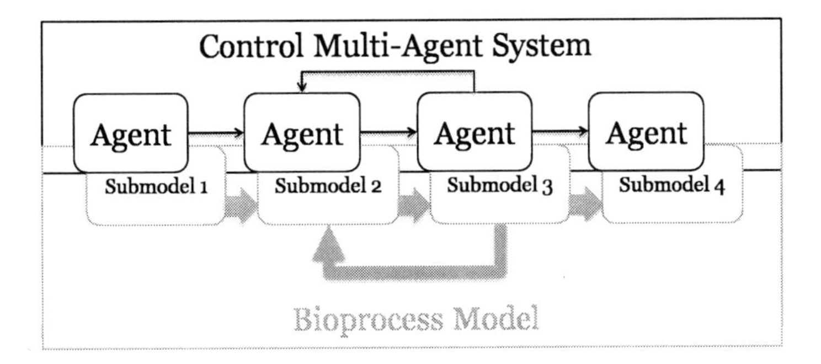

Fig. 1. Example of superposition of agents on the bioprocess model

- c_{per} : perceived signals are understood without ambiguity.
- c_{dec} : received information is useful for the agent's reasoning.
- c_{act} : reasoning leads to useful actions toward other agents.

When an agent detects a NCS ($\neg c_{per} \vee \neg c_{dec} \vee \neg c_{act}$), it has to act to come back to a cooperative state. One of the possible actions such an agent may take is to change its relationships with other ones (e.g., it does not understand signals coming from an agent and stops having relationships with it or make new ones for trying to find other agents for helping it) and therefore makes the structure of the global system self-organize. This self-organization led by cooperation changes the global function performed by the system that emerges from the interactions between agents. The MAS is thus able to react to changes coming from the environment and therefore becomes adaptive.

3.2 Structure of the CMAS

The AMAS described in this paper relies on the use of an existing model of the bioprocess it has to control. This model may be composed of any kind of different submodels (mathematical equations, ANN, MAS...) because this point only influences the instantiation of our agent detailed in section 3.2. Basically, a superposition of agent composing the CMAS on the bioprocess model must be done in order to create the structure of the CMAS. Figure 1 illustrates an example of such a superposition where one agent is associated with one submodel, but it would also be possible to describe one submodel with several agents. This choice is up to the system designer and offers an important flexibility, especially in the granularity of the created system.

When the general design of the CMAS is obtained, the description of the agents that are composing it is performed following the AMAS approach.

Generalities on CMAS Agents. Each of the agents composing the CMAS represents a variable, or a set of variables such as the quantities of different elements in a bioprocess, on which they have objectives of different criticality. This criticality symbolizes the priority of the objective, and agents can compute it thanks to the difference between their current value and the expected one. The main goal of an agent is to satisfy its most critical objective, which means to bring a variable toward a certain value.

Agents that compose the CMAS follow a common model, but they can be instantiated in different ways. This phenomenon underlines the fact that even if their ability are implemented differently, their behavior stay the same and so, the MAS is always composed of cooperative agents that are able to interact with one another.

Abilities of CMAS Agents. As stated by the AMAS approach, an agent has a strictly local view. From this local point of view, it computes its own objective which may be modified by the communication between agents. As a result, an agent must be able to evaluate the current objective that it has to achieve, and to update it according to the evolution of its *Criticality*. This ability includes the need to communicate with other agents, and to manage a set of received messages, by sorting them, or aggregating them according to the problem, in order to extract the current objective, which has the highest *Criticality*.

Our control system supposes the existence of a model of the bioprocess which has to be controlled. This model is used by the agents, which are able to extract certain abilities from it.

These abilities are the observation and the use of a local part of the bioprocess model. An agent is able to virtually inject some values (without any real control action) at the input points of the local model it observes in order to extract the corresponding output values. These observations enable this agent to have an idea of the variation direction that it has to apply in input for obtaining a desired output. Therefore, this mechanism grants an agent the abilities of its own direct problem solving, and gives it the tools needed to treat its own inverse problem, by determining which inputs it has to apply for achieving a certain output.

For example, let us consider an equation $y = f(x)$.

At time t, this equation is $y_t = f(x_t)$.

Then, at $t' = t + 1$, we obtain $y_{t'} = f(x_t + \Delta x)$, this Δx being a light variation of x.

Finally, by observing the sign of $y_{t'} - y_t$, the agent is able to find the modification Δx which moves y closer to its objective.

Thus, agents are able to deal with their inverse problem without needing a model that describes their inverse problem, such as the lagrangian.

Finally, each agent is able to compute its own objective, which is a set of values that the agent aims at. This computation depends on the communication between agents (described in part 3.2), and on the observation that an agent is making on its own local model. This objective can also be established by the user.

Behavior of CMAS Agents. The main mechanism guiding the behavior of the agents rests on a model in which this behavior is divided into two categories: the *Nominal* and the *Cooperative* one. The *Nominal* behavior describes the default behavior of an agent, the one used when this agent has no need to process one of the Non Cooperative Situations (NCS) described in 3.1, while the *Cooperative* behavior enables it to overcome the NCS met during the control. This *Cooperative* behavior is itself divided into three different behaviors. First, *Tuning* consists in trying to tweak parameters to avoid or solve a NCS. If this behavior fails to make an agent escape from this NCS, *Reorganization* takes place. This *Reorganization* aims at reconsidering the links established with other

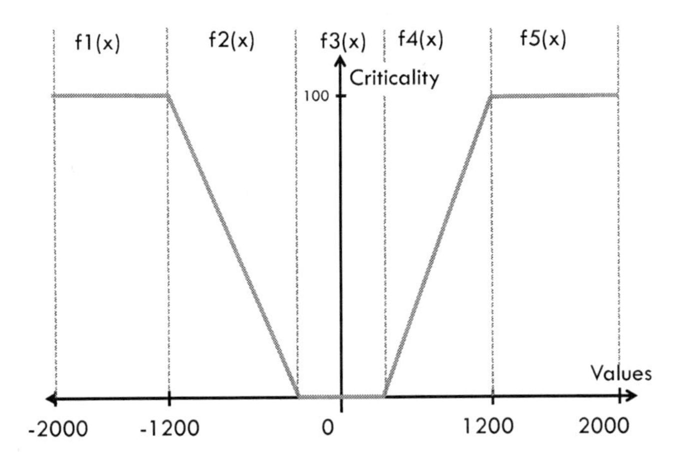

Fig. 2. Example of Criticality for an Agent

agents. Finally, *Evolution* enables the possibility for an agent to create another agent, or to self-destruct because it thinks itself as totally useless.

In the bioprocess control, the *Nominal* behavior has been instantiated in the following way. The agents communicate with one another in order to share their non-satisfaction degrees, and solve them if possible. This non-satisfaction degree is tied to a variable which value does not satisfy the objectives of the agent. This variable, used to make decisions, is called *Criticality*. *Criticality* expresses the degree of satisfaction of an agent, ranging from 0 to 100, with 100 being the most critical, and 0 representing a satisfied agent. Moreover, each agent possesses a set of equations allowing it to compute its own *Criticality* in relation to its current state. Figure 2 details this fact, presenting an example of an agent's *Criticality* obtained by the combination of five functions, for any value from -2000 to 2000. If the agent is the target of user-defined objectives, it then uses the difference between its current value and the expected one to set its *Criticality*, weighted to match the *Criticality* range.

Biologically speaking, the *Criticality* enables the modeling of some behaviors, like the knowledge of the user regarding the expected quantity of a specific element. For example, if the user knows that the temperature during the whole process must range from 36 to 37 degrees, he can define an appropriate *Criticality*, facilitating the evolution of this value toward the expected one.

After deciding what objective to follow, an agent can send fwo different kinds of messages:

– A *Request* message which expresses a non-satisfaction of the sending agent and asks for an action of control in order to change the value of the problematic value.
– An *Answer* message which notifies an applied control, or the observation of the modification of a value observed from the model by the agent.

As a consequence, even if two agents come from different instantiations, they share the same *Nominal* behavior which consists in sending requests asking for the modification of values that did not satisfy the agent, and acting if possible in order to answer those requests by carrying out a control action.

These two points are completed by the *Cooperative* behavior of *Tuning* stating that an agent receiving a request, to which it cannot answer positively, is able to modify it for conveying this modification towards the other agents linked with it. This modification is applied in order to make the request relevant to the receiving agent. It ensures that this request will be useful and comprehensible for these agents, by asking for modifications on variables that they know about. The inverse problem solving ability of an agent is used during this modification to decide which are the adjustments needed on the inputs to obtain the desired output.

The control can also be partially done when an agent is unable to make it completely; for example, if the agent is not permitted to make a modification of sufficient amplitude. In this case, it makes the maximum control possible and then, sends an answer to notify this modification to the other agents. Thus, if this objective is still the most critical for the agent source of the request, then a request related to the same objective will be sent again, and will finally be answered positively when another control will be possible. So, this behavior is still a *Tuning* behavior.

Eventually, in order to solve a specific control problem with this approach, our agents' abilities have to be instantiated according to the problem. For example, the methods used to observe and use models on which the agents are created must be defined depending on the kind of model used. The way the agents compute their current objective has also to be instantiated. To summarize, all the abilities described in section 3.2 may be implemented in different ways, without modifying the behavior of the agents. So, the MAS created for the control of bioprocesses can be composed of any number of different kinds of agents, provided that these agents possess the described abilities, instantiated to fulfill their role, and follow the same behavior.

In order to evaluate the control system that was developed, an instantiation to an equation solving problem was carried out.

4 Example of an Equation System

The goal of this example is to modify dynamically the values of some variables to fulfill some objectives that the user put on other variables. These objectives are threshold values that the variable must reach and the user can modify them during the simulation.

4.1 Description of Agents

Two different types of agents were instantiated: *Equation Agents* and *Variable Agents*.

Figure 3 explains how the MAS is generated from a mathematical equation. First, a *Variable Agent* is created for each variable appearing in the different equations. A *Variable Agent* is an agent that can make a control action by modifying the value of the mathematical variable associated with it. The model used by this agent is simply a model of the mathematical variable, namely the variable itself. Therefore, inverse and direct problem solvings are trivial for such an agent, the result being the value of the mathematical variable. Finally, a *Variable Agent* can be the target of user-defined objectives defining a value to achieve.

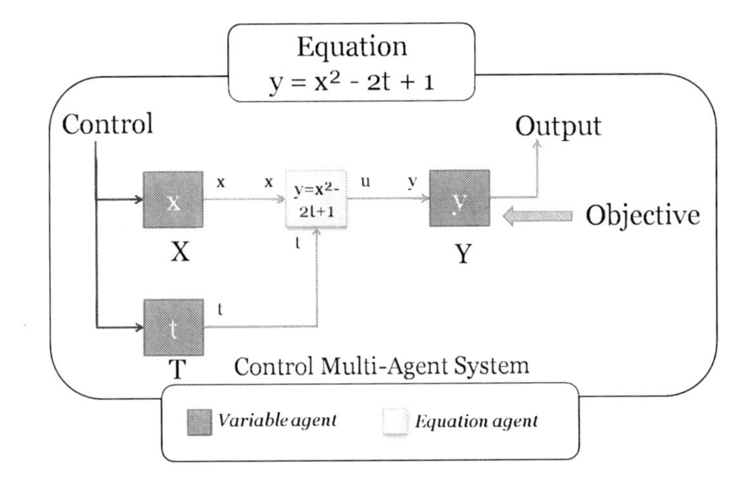

Fig. 3. Creation of the CMAS from an equation

After *Variable Agents* are created, an *Equation Agent* is added for each mathematical equation belonging to the system to solve. Each *Equation Agent* relies on the model of the mathematical equation it represents and this agent is able to know any output generated from a set of inputs thanks to its direct problem solving ability. Here, the inverse problem solving mechanism uses the same model, an *Equation Agent* tunes the inputs and, by observing the generated outputs, is able to estimate the modification needed to get closer to its objectives. However, while an *Equation Agent* is able to compute the amount and the direction of the modification to apply, it cannot make any control action. It has then to create requests and sends them to the corresponding *Variable Agents*.

When this step is done, an objective is allocated to a *Variable Agent* representing the output of the system, e.g., the agent Y on Figure 3. Agents X and T are able to apply control actions.

Finally, a last type of agent, named *Independent Variable Agents*, is derived from the *Variable Agent* one to which it adds a specific ability. Such an agent cannot make control actions but, instead, it represents a variable which value is modified over time, according to an inner function. That means that an *Independent Variable Agent* may receive requests but will never be able to fulfill them. On the other hand, at each simulation step, it will send an answer to notify the modification of its variable to other agents. The interest of this agent is to underline how the other *Variable Agents* act to make up for the drift brought by this uncontrolled modification.

The relationships between the agents composing the equation control system are as follows: an *Equation Agent* is linked, at its inputs, with every *Variable Agent* or *Independent Variable Agent* from the mathematical equation. Its outputs are connected with the *Variable Agent* representing the result variable. Communication between agents follows these links.

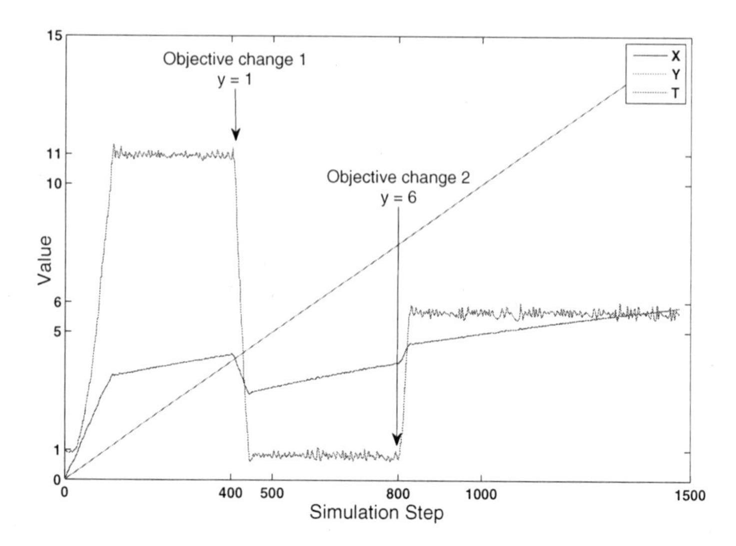

Fig. 4. Results from the control of an introductory example

4.2 Experimental Results

This section describes three examples of equation systems, highlighting different aspects of the presented approach. In these examples, the agents are constrained to update their value progressively, meaning that *Equation Agents* do not send the correct value that they have to reach to *Variable Agents*, but rather a modification step toward this value. This fact comes from the implementation of the inverse problem solving on *Equation Agents*, whose goal is to drive its inputs gradually toward the expected value, and not in a single jump. On top of that, when a *Variable agent* is created, it is named after the capital letter of the mathematical variable that it represents.

Controlling Single Polynomial Equation. This example consists of a single equation $y = x^2 - 2t + 1$, made up of a *Variable Agent* Y, which receives objectives from the user. Inputs are a *Variable Agent* X and an *Independent Variable Agent* T. During the process, the objective of agent Y is changed two times, depicted by arrows on Figure 4. Initially, its goal is 11, then the two changes occur at time 400 and time 800 when the objective is respectively set to 1 and 6.

Results presented in Figure 4 (on which time is expressed as simulation steps) highlight the reaction of the control performed by X, which compensates the uncontrolled evolution of T, while reacting to the objectives changes made on Y. The delay to reach the objective value when Y changes its goal comes from a maximum limit on the modification enforceable for each simulation step.

Controlling Multiple Polynomial Equations. This example, presented in Figure 5, is composed of 4 different equations, with a total of 10 variables whereof two of them are independent. Three objectives are defined at time 0 (10 for U, 5 for V and 14 for W) and remain static during the process. Those three variables are selected to receive

Table 1. "Multiple Polynomial" Equation Data

Equations	Variables	Independent Var.
$u = 0.2x + y + 0.3t$	u, x, y	t
$v = 0.8y + z$	v, y, z	
$w = 0.4x - a + o$	w, x, a	o
$x = d - 0.4e$	x, d, e	

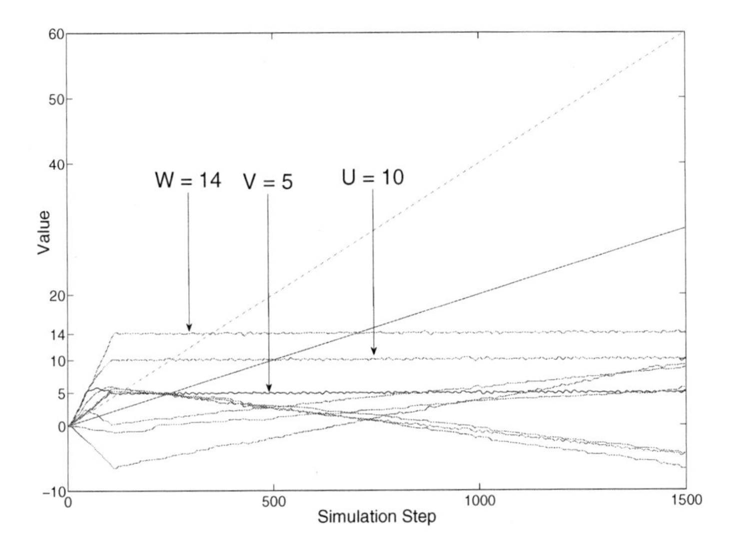

Fig. 5. Results from the control of the multiple polynomial example

objectives because they represent the outputs of the system, they are not used as an input for another equation. The full equations data are detailed in Table 1.

This example shows how multiple equations, that share variables, and can send antagonist objectives to them, are able to fulfill all the defined objectives. The values of the variables that are undergoing the greater changes are those of the non-shared variable. On top of that, the noise coming from the *Independent Variable Agents* is reduced thanks to the controls done by the *Variable Agents*.

Another interesting point comes from the adaptive feature of such a system. Indeed, several simulations were run to solve the presented problems, and we can observe that, given the few constraints on some variables, the system is able to find different balanced states. During each time step, all the agents composing the CMAS can act, but due to the stochastic order on which they behave, some delays may appear in the messages transmission. Therefore, some modifications occur before some others, implying a different dynamics. In those different cases, the CMAS is able to converge toward a stable state respecting the constraints. This fact can model a kind of management of noise coming from time delays, and highlights the robustness of the presented approach.

Table 2. "Interdependent" Equation Data

Equations	Variables
$u = 0.3x + 0.8y$	u, x, y
$v = 0.2u + 0.4z$	v, u, z
$x = 0.2v + 0.3m$	x, v, m

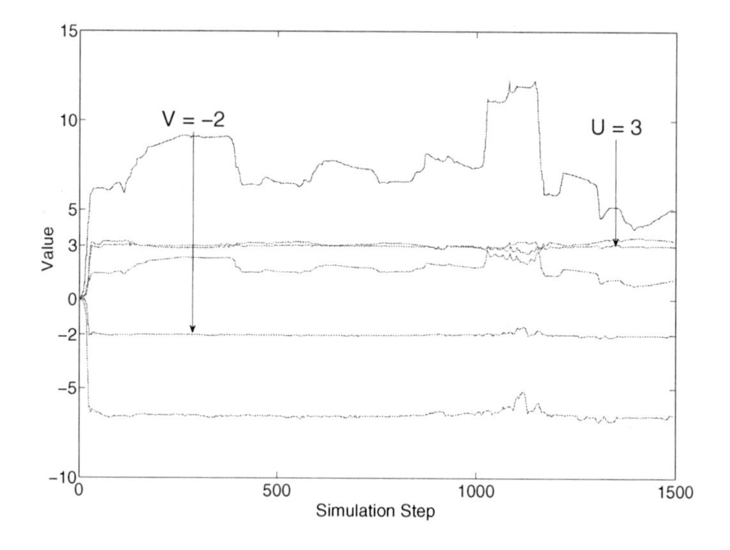

Fig. 6. Results from the control of the interdependent equations example

Controlling Interdependent Equations. The last example deals with the loops that appear in the controlled system, a common phenomenon in the equations used to describe bioprocesses. The results available on Figure 6 are made up of 3 equations, detailed in the Table 2, and possess 2 static objectives. The *Variable agent* V has to reach the value -2 while the *Variable agent* U aims toward 3.

This example underlines the message management ability of the *Variable Agents*. Indeed, an agent has to determine if a received request is still relevant. Here, we can have requests that are making a full loop and so, the agents must take this into account to avoid a divergence of results, by summing unnecessary requests.

Finally, it is noticeable that those three aspects, which are the dynamic change of objectives, independent variables and loops, presented here on separate examples, are managed in the same way when they are combined on the same equation system.

5 Conclusions and Perspectives

This paper focuses on the control of real-time, dynamic and non-linear problems, and presents a first step towards an adaptive control of bioprocesses. The approach given uses an AMAS made up of different types of generic cooperative agents. The behavior and abilities of these agents, as well as their relationships, were detailed. An instantiation of this generic approach was applied to an equation solving problem in order to prove the feasibility of this kind of control on different kinds of equations. The results obtained highlight the relevance of this approach, and its adaptability to a wide range of problems, especially the bioprocess control, a bioprocess being often modeled thanks to equations systems.

Currently, the application of the proposed approach to a full bioprocess model, with a true biological meaning, modeling the bioreactor physics and the evolution of microorganisms, is under development. This application will enable evaluating the performances of our control system, while validating its scalability, such a model being composed of hundred of equations.

On top of that, we are considering the time aspects, especially lags and delays coming from the scale diversity on which reactions occur. The design of a mechanism to manage those delays is in progress with a twofold aim. The first one is to measure the impact of different kinds of time constraints on the convergence of the system towards its objectives and the second one is to improve the robustness of this system while applied to strongly non-linear problems.

The final objective of this work is to combine this control system with another AMAS which dynamically models the bioprocess. Thus, the model needed by the agents belonging to the control system will itself be composed of agents, reducing the work of instantiation of the control agents. Therefore, the global control system will be viewed as a Model Identification Adaptive Control.

References

1. Ziegler, J.G., Nichols, N.B.: Optimum settings for automatic controllers. Transaction of the ASME 64, 759–768 (1942)
2. Cohen, G.H., Coon, G.A.: Theoretical consideration of retarded control. Journal of Dynamic Systems, Measurement, and Control, 827–834 (1953)
3. Visioli, A.: Tuning of pid controllers with fuzzy logic. IEE Proceedings - Control Theory and Applications 148, 1–8 (2001)
4. Scott, G.M., Shavlik, J.W., Ray, W.H.: Refining pid controllers using neural networks. Neural Computation 4, 746–757 (1992)
5. Astrom, K.J., Wittenmark, B.: Adaptive Control. Addison-Wesley Longman Publishing, Boston (1994)
6. Whitaker, H.P., Yamron, J., Kezer, A.: Design of model reference adaptive control systems for aircraft. Technical Report R-164, MIT (1958)
7. Kaufman, H., Bar-Kana, I., Sobel, K.: Direct adaptive control algorithms: theory and applications. Springer, NY (1994)
8. Feldbaum, A.: Dual control theory i-iv. Automation Remote Control 21,22, 874–880, 1033–1039, 1–12, 109–121 (1960-1961)
9. Megan, L., Cooper, D.J.: Neural network based adaptive control via temporal pattern recognition. The Canadian Journal of Chemical Engineering 70, 1208–1219 (1992)
10. Silva, R., Cruz, A., Hokka, C., Giordano, R.: A hybrid feedforward neural network model for the cephalosporin c production process. Brazilian Journal of Chemical Engineering 17, 587–597 (2000)

11. Dunal, A., Rodriguez, J., Carrasco, E., Roca, E., Lema, J.: Expert system for the on-line diagnosis of anaerobic wastewater treatment plants. Water Science and Technology 45, 195–200 (2002)
12. Alford, J.S.: Bioprocess control: Advances and challenges. Computers & Chemical Engineering 30, 1464–1475 (2006)
13. Taylor, J.H., Sayda, A.F.: Prototype design of a multi-agent system for integrated control and asset management of petroleum production facilities. In: Proc. American Control Conference, pp. 4350–4357 (2008)
14. Gleizes, M.P., Camps, V., Glize, P.: A Theory of Emergent Computation based on Cooperative Self-organization for Adaptive Artificial Systems. In: 4th European Congress of Systems Science, Valencia, Spain (1999)

The Provider Rating Agreement Pattern for Solving the Provider Selection Problem

José Javier Durán[1] and Carlos A. Iglesias[2,⋆]

[1] Centro para las Tecnologías Inteligentes de la Información y sus Aplicaciones (CETINIA)
Universidad Rey Juan Carlos, C/ Tulipan s/n, Mostoles, Spain
[2] Departmento de Ingenería de Sistemas Telemáticos, Universidad Politécnica de Madrid
Ciudad Universitaria s/n, Madrid, Spain
jjduran@ia.urjc.es, cif@gsi.dit.upm.es

Abstract. In this article, it is presented the provider selection problem, a typical SOC problem related to the initial phase of selecting a provider. The article proposes to model one of the solutions to this problem as the Provider Rating Pattern, taking advantage of agreement patterns notion. Agreement patterns can be defined as design patterns at the agreement level and aim at providing reusable patterns which assist developers in multidisciplinary areas, such as Agent Technology and Service Oriented Computing. This article details the Provider Rating pattern, that is shown as an example of how agreement patterns can provide a common vocabulary and collect best practices in the different phases of agreement fulfillment.

1 Introduction

The rapid development of Internet is enabling a whole new and innovative market of services providing a new experience to users [1], which is known as the vision of an Internet of Services.

Service Oriented Computing (SOC) is an emerging development paradigm which leads this Internet of Services. Nevertheless, there is not yet enough expertise as well as best practices for many of the challenges that a this dynamic and multiparty scenario brings.

In this context, this article proposes to identify and formalise patterns for reaching agreements, so-called *agreement patterns*, The objective of these patterns is to help developers with a huge catalogue of SOC good practices in common problems, how they are solved and a common vocabulary.

The rest of the article is organized as follows. First, section 2 presents the Provider Selection Problem, and the main approaches to solve it. Then, section 3 gives an overview of the notion of agreement patterns, their classification scheme an how agreement patterns are classified. Section 4 presents the Provider Rating pattern, describing its structure, usage and examples. Finally, section 5 draws out the main conclusions and future works of this work.

⋆ The second author has been partially supported by Germinus XXI (Grupo Gesfor) under the project RESULTA.

J. Filipe, A. Fred, and B. Sharp (Eds.): ICAART 2010, CCIS 129, pp. 161–172, 2011.

2 The Provider Selection Problem

Selection of the right parties to interact with is a fundamental problem in open and dynamic environments [2]. This problem is recurrent in SOC environments, where dynamic service selection [3] allows to combine available services to user needs. It is applied as a need to find and select an specific service for the purpose of the system, first needing to identify the service properties, which is called *quality-of-service* (QOS), and the resources that could be spent by that need, and after that, it is necessary to find a provider that offers that service, which in fact sometimes is not possible, and it is necessary to switch to the service provider that offers the more suitable service. Because of those properties, this problem is pillar for SOC.

This problem is present in different environments, which some of them are listed next:

- **Travel Agency [4].** An user requires an offer of a travel, which is formed of the flight and an hotel, selecting the one that suits it's preferences, like flight company or arrival time, and that has the lower cost.
- **E-Commerce [5].** Different providers offer a similar product, but with different qualities, and costs, and the user (or a mediator agent) should select one provider among them, based on similarity to its preferences, and cost restrictions.
- **WiFi Roaming [6].** The user must select among different access points, those that have a reliable security, and a good quality of signal.

Each of these domains represents a problem, in which confidence of bidders is necessary, and in which solutions taken are similar.

First, it is imperative to determine which roles are part of this problem, in fact which are the parts that interact between them to advert and negotiate service agreements.

- The main role is the consumer, or user, role, which suits with the system that needs the service to be implemented in, and that will be responsible to find, establish, and maintain different agreements with service providers to use a service.
- The other main role in this interaction is the service provider one, which is responsible of keep offering their services and helping to establish agreements to use those services.
- A secondary role of this interaction is performed by the marketplace, in which different providers are registered to help consumers to list them, and ask them about a service. The marketplace role could be implemented by the consumer, as a predefined list of providers, by a service provider offering that service, or throw a social collaborative network of provider's directory.

Also it is necessary to define the main aspects of this problem, in this case, what are the aspects of the interaction, which will grant a better quality to the solution adopted:

- **Providers Market-place.** It is a component that knows which providers are able to offer an specific service, in this case retrieving different offers as agreements. This component should be a predefined list of providers accessible to the user, another service that offers a provider search engine, or a collaboratively created directory of providers.

- **Service Offer Evaluation.** Each received offer is evaluated, so, they are comparable between them, using different techniques, that is more detailed in Section 2.1.
- **Agreement Negotiation.** Once one offer has been selected, it is done a negotiation process, as stated in Section 2.2.

2.1 Service Offer Evaluation

The main problem around service selection is how to select the most appropriate one, taking care of different factors [7]:

- **Quality of Service:** determined by similarity with required specifications, cost, and availability. It is necessary to remark that agreements' information could be difficult to understand by the system, because there should be different kinds of properties. The main properties are the functional ones, which could be measured easily, like final cost or minimal bandwidth offered. In counterpart, non-functional properties are more obscured, and difficult to treat, like the feedback from user.
- **Trust of the Provider [8]:** the service offered by a provider depends on the trust that is assigned to that provider. This information is totally non-functional, but in some cases should be treated like functional, e.g. mean times a services has failed, or what was the perceived quality of the service, for example, using a rating from 0 to 5.

Taking care of that aspects, there are several methods for rating a service. They can be distinguished the following approaches [7]:

- **Rating [2].** This technique use social collaboration to create a score based board, also known as "'Social service selection"', in which different user scores each provider, or also services of each provider, with a value used to create a feedback about user feel of QOS. This technique is useful when a service is obscure, for example non guaranteed inversions, but lacks of subjectivity of users, been manipulable by users. Also it is richer that other techniques, in the fact that there exists information about service feedback that could be used to enrich the selection process.
- **Ranking [3].** This other, instead, use heuristics to determine how the QOS of a service is near to the needed service, also known as "'Semantic service selection"', an treating semantic distance to required one as the evaluation metric, and *ranking* different offers to select the one that suits better. It is necessary to extract semantic information in which the service could be measured to understand how near it is to the required service.
- **Economic Service Selection.** Another technique to select a provider, simpler but the most appropriate in some cases, is to select those ones that are offering a service with a lower cost, in which only quantitative properties of the QOS are taken into account, for example time taken to treat a petition, or availability of the service in a determined time slot.

Selection of which technique to use is not trivial, for example, in a trusted network of providers it is better to select economic techniques, or ranking if the service definition is ambiguous, as it is expected that providers are trustworthy, but in no trusted networks, it is preferred to use rating ones. Also, it is possible to fuse different techniques, so the information of the service offer is richer, and in that way, the service will be selected wisely, but it is necessary to weight how each technique depends on the environment, to assure that the system is not highly restrictive with no trusted providers, and is cost-balanced.

An advantage of ranking techniques is that it is prepared to sustain any heuristic used to match offers with required service, been able to use anyone in a *trust-based component*. The advantage of abstracting this knowledge to a different component is that it is possible to fit the system requirements in this component, and reuse the rest of provider selection interaction.

2.2 Agreement Negotiation

Once a provider is chosen to establish an agreement with, it is done a negotiation process, in which consumer and provider offers each one an offer for that agreement, defining which valuables will be exchanged, in this case they are QOS and cost for service renting. For this process, there are different techniques [9], which assure that an agreement will be establish, but a problem of this negotiation is that it presents a Nash equilibrium problem, in which two parties have interest in conflict with each other.

Actually, this negotiation is performed after a provider is selected, but a more wide vision of the provider selection problem should be able to establish a negotiation process between the consumer and all the providers, been able to select the one that accepts first an offer from the consumer. This interaction should be treated as a bidding process, and presents more complexity that the provider selection problem, and would be suitable for a future agreement pattern.

3 Agreement Patterns

When working with a new technology, or beginning to work with one that previously existed, it is useful to have access to a collection of rules, examples, and descriptions, of the principal aspects of that technology. In software development there are programming languages as rules, source code examples, and software patterns as descriptions of good approaches taken to produce a specific piece of software. Those software patterns are intended to offer new developers a view of how to approach to a solution that is proven to have good properties, and also offering information that contrast that. Also, different works have proven that patterns are really usable as experience representation and distribution [10].

Agreement technologies [11] (AT) is a recent discipline which collects this multidisciplinary research and can be defined as *the technologies for the practical application of knowledge to the automated fulfilment of agreements*. Agreement technologies do not dictate the underlying technologies (objects, components, agents, services, ...), but are focused on the formalization of knowledge structures, protocols, algorithms and expertise that contribute to the establishing of agreements in an open dynamic environment.

Based on this previous research, **Agreement patterns** [12] are defined as *software patterns which helps software components coordination through the fulfilment of agreements*. Agreements patterns include all kind of agreements, both explicit ones (e.g. negotiation) and tacit ones (e.g. organization).

The main need of the pattern engineering is to define a pattern template that reflects the needs from the domain in which it will be used. In the case of agreement oriented services there is important to determine those ones:

- Participants, or Roles.
- Trigger
- Purpose

Taking those aspects in account, it should be possible to define the pattern template. In this case we propose the use of the canonical form (also called Alexandrian) for software patterns informal description[13], which it has been enrich with the elements listed before:

- **Name:** a meaningful name that provides a vocabulary for discussing.
- **Alias:** an alternative name to the pattern.
- **Participants:** who are the main participants in the interaction, and their roles.
- **Trigger:** why the interaction process begins, and with which interactions is related. It will describe when it is used.
- **Purpose:** what is the problem that it solves.
- **Problem:** a statement of the problem and the goals it wants to reach.
- **Context:** the preconditions under which the problem and its solution seem to recur.
- **Forces:** a description of the relevant forces and constraints and how they interact with one another and with the goals. Considerations to be taken into account to select a solution for a problem.
- **Solution:** static relationships and dynamic rules describing how to realize the desired outcome. It should be described using pseudo-code, class diagrams, reasoning diagrams, or any model that helps to understand the solution.
- **Examples:** one or more sample applications of the pattern which illustrate its application. Known occurrences of the pattern, which help in verifying that the pattern is a proven solution to a recurring problem.
- **Resulting Context:** the state or configuration of the system after the pattern has been applied.
- **Rationale:** a justification of the pattern, explaining how and why it works, and why it is "good".
- **Related Patterns:** compatible patterns which can be combined with the described pattern.

All this points forms a good batch of questions about the pattern itself, helping to determine the inner of the pattern, as assuring that it is a true pattern widely useful, and not an anti-pattern[14] of bad manners in software development, that don't offer and extensible and reusable interaction process for service oriented computing.

In order to classify agreement patterns, a classification scheme has been proposed [12], which has identified the following dimensions:

- **Duration.** Is the agreement established temporally, short term, or permanently, long term?
- **Normative context.** The pattern is strict as a established norm, or flexible?
- **Topic.** Which is the main purpose of the pattern? E.g.: Service offering, Service negotiation, or Service bidding.
- **Phase.** What moment of the agreement life-cycle it represents? E.g.: negotiation, conclusion or selection.
- **Decision Making.** How selection process are performed? E.g.: In provider selection it should be social-collaborative, but in agreement portability it should be rule based.

4 The Provider Rating Pattern

Using the provider selection problem as example of how the agreement patterns are applied, they are going to be defined a series of steps and models to be used as formalization of it. The purpose of those models is to define the interaction process that is part of this problem.

4.1 Problem Description

The pattern template previously showed in section 3 would be applied to the provider selection problem, which presents a good number of factors to consider it the basis of service oriented computing:

- It is use when it is required to create an adaptable system.
- Provider selection trust is required to assure system's assurance.
- It will dynamically establish agreements as required to manage different offers, and select the best that suits requirements, and cost factors.

The main purpose of the agreement is to unify all the information about different approaches taken in this scenario to be able to present it in a formal way, accessible by different developers, to take care of the pattern whenever a system requires its capabilities.

Those aspects assures the need for a formal description of how a solution is obtained, as presenting the problem enough complexities, and been widely used and generic.

4.2 Description of the Agreement Pattern

Based on the pattern structure described in section 3, the solution to the Provider Selection problem can be described as follows.

- **Name:** Provider Rating.
- **Alias:** Service selection
- **Duration.** Variable. Based on system and purpose of service.
- **Normative context.** Flexible.
- **Topic.** Service provider selection.

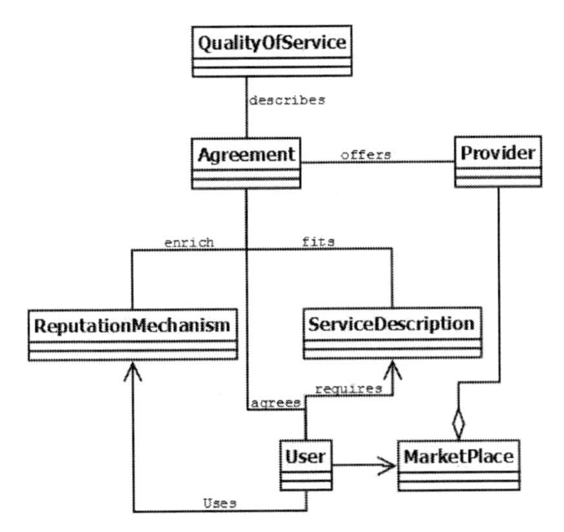

Fig. 1. GParticipant classes in the provider selection problem

- **Phase.** Agreement selection.
- **Decision Making.** Based on trust/reputation mechanism, mainly social-collaborative.
- **Participants.** *User*, that requires a service; *Service provider*, that offers an agreement for a service to be used by the *User*; and the *Market place*, which list the different *Service provider* that are offering services. Relations between roles are present in Figure 1.
- **Trigger.** A system requires a service, commonly with some restrictions or preferences on its non functional properties (QoS, price, ...), and there is more than one service provider that fits in that description.
- **Purpose.** Retrieve the best provider, and establish a service usage agreement with it.
- **Problem.** A user requires an agreement with a provider, that must offer a service with required properties, like quality of the service, cost or trust in that provider, and as a result, an agreement is done with the most appropriate provider.
- **Context.** The user has access to a market of offered services, and a trust system.
- **Forces.** Trust and reputation techniques to enrich bidders information.
- **Solution.** See Algorithm 1, for a pseudo-code description of the solution.
  1. The user asks bidders in a service providers' market, for a service offer, with an specific properties.
  2. Each provider offers a different proposed agreement, including non functional properties, such as costs or QOS information.
  3. The user enriches the information in each agreement with trust information, using a trust-based component, like a collaborative reputation system, self-experience, or heuristics for service matching.
  4. Agreements are evaluated, based on their non functional properties as well as based on the trust and reputation of the provider, using a specific evaluation function based on the system purpose, for example, a high security system will evaluate poorly any system without good trust information.

Algorithm 1. Pattern solution pseudo-code

```
Require: RequiredServiceDescription
Require: ReputationMechanism
Require: ProvidersMarketPlace
  OfferedAgreements = ProvidersMarketPlace.askFor( RequieredServiceDescription )
  while Agreement a in OfferedAgreements do
    ReputationMechanism.enrich( a )
    if RequiredServiceDescription.isBetter( a, best ) then
      best = a
    end if
  end while
  if best.assures( RequiredServiceDescription ) then
    ServiceAgreement = best.AgreeProposal( )
  end if
Ensure: ServiceAgreement.assures( RequiredServiceDescription )
```

5. If there is almost one provider with an acceptable evaluation, it is realized an agreement with it. A threshold must be defined to don't establish an agreement if all providers are offering invalid agreements. Once the target provider is chosen, it begins a negotiation process, in which the consumer tries to establish the agreement with highest utility.

- **Examples:** valuable throw an e-commerce scenario.
  - Broadband access negotiation [6], in which users selects the provider that fits with its needs, and use other users feedback to select the most appropriate.
  - Ad-Hoc service negotiation [15], in which a provider offers different services, in which the QOS changes, but fits better with the required service as an increase of cost instead. In this case the user tries to find equilibrium between service matching and cost assumed.
  - E-Commerce [5], in which several providers are offering the same good but with different important aspects, mainly shipping method and final cost, and the user measures the offering, with other users feedback, like comments in the provider web, and feedback about trustworthy of the provider.
- **Resulting Context.** The user establish an agreement, if an acceptable one is offered among the providers.
- **Rationale.** It defines the interaction basis in the search for an agreement when it is necessary to compare different offers, and enrich them with trust systems.
- **Related Patterns.** Agreement Portability.

4.3 Reasoning Cognitive Pattern

To help to understand how this problem could be driven, it is possible to divide it in different tasks, which should be threatened independently, except of how they are interconnected. This tasks interconnection is described in figure 2. The purpose of those tasks is as follows:

- **Estimate.** Enriches providers offers with trust information from the trust/reputation knowledge.

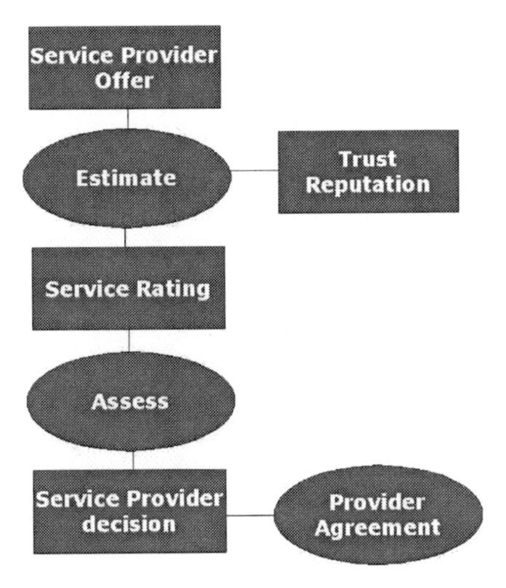

Fig. 2. Provider Rating reasoning diagram

- **Assess.** Selects a provider that fits the user requirements of the QOS. In this task it is measured how the service proposal is similar to the required service, and the trust information of the provider.

Those tasks treat different information from different knowledge sources:

- **Service Provider Offer.** This information is obtained from asking to the market place about service providers that fit a required need for a service. It should be described using an agreement definition language, like WS-Agreement.
- **Trust/Reputation.** This information should be obtained from a service of service providers reputation, the provider itself, or from a social-collaborative source. It should be a quantification of average number of service losses, a measure of principal properties, like average bandwidth in a WiFi access point; or feedback of other users, which requires to apply a new trust filtering to that information.
- **Provider Agreement.** This is the final product of the problem, in which the system, after selecting a service provider, creates an agreement, in which the consumer ask for access to the specified service. This agreement can be represented by an agreement language such as WS-Agreement.

5 Related Work and Conclusions

This article has presented *agreement patterns* as an instrument for modelling reusable solutions.

There are related works for defining design patterns in the areas of multi-agent systems and Service Oriented Computing (SOC).

Agent-Oriented Patterns have been defined for sharing multi-agent system development experiences. Oluyomi [16, 17] presents an agent pattern classification scheme based on two dimensions: stages of the agent-oriented software development and tasks in each stage of development. At each stage or level of development (analysis, multi-agent architecture, agent architecture, multi-agent implementation), the framework identifies the attributes of that level of abstraction, in order to classify these patterns. In addition, Oluyomi proposes to refine the canonical pattern form for defining an Agent-Oriented Pattern Template Structure, which adds more granularities depending on the pattern type (agent internal architecture structural, interactional or strategic patterns, etc.). Some of the patterns identified by Oluyomi, whose classification scheme includes other approaches, can be considered agreement patterns. The main differences between her classification and the one proposed in this article is that Oluyomi's classification is agent oriented, and it is hard to use if it is not implemented with agents (agent oriented development phase, agent architecture, etc.), while the one proposed here is independent of the technology to be used, although implementation examples can be presented with different technologies. In addition, agreements are not a key concept in Oluyomi's classification scheme as in our proposal. Future work will provide a mapping of the agreement related patterns classified by Oluyomi onto our classification scheme.

In the area of Service Oriented Architecture (SOA), *SOA patterns* have been defined [18–20]. For example, Erl [18] classifies patterns for architecture services, service compositions, service inventories and service oriented enterprise. Rotem-Gal-Oz [19] describes patterns for Message Exchange, Service Interaction, Service Composition, Structural, Security and Management. SOA patterns [20] provide high level architectural patterns, which do not detail yet agreement issues.

Inside the SOC community, the GRAAP Working Group (Grid Resource Allocation and Agreement Protocol WG) has defined the specification Web Services Agreement [21], which is particularly interesting for this research. The purpose of the specification is the definition of a Web Services protocol for establishing agreements defined in XML. The specification covers the specification of agreement schemas, agreement template schemas and a set of port types and operations for managing the agreement life cycle. This specification defines an agreement as *an agreement between a service consumer and a service provider specifies one or more service level objectives both as expressions of requirements of the service consumer and assurances by the service provider on the availability of resources and/or service qualities. An agreement defines a dynamically-established and dynamically-managed relationship between parties. The object of this relationship is the delivery of a service by one of the parties within the context of the agreement. The management of this delivery is achieved by agreeing on the respective roles, rights and obligations of the parties.* An agreement is characterized by its name, context and terms.

The OASIS Reference Architecture for SOA [22] is *an abstract realization of SOA, focusing on the elements and their relationships needed to enable SOA-based systems to be used, realized and owned.* The reference architecture defines three primary viewpoints: *business via services* that captures what SOA means for people using it to conduct business, *realizing service oriented architectures* deals with the requirements for constructing a SOA; and *owning service oriented architectures* addresses issues involved in owning and managing a SOA. The notion of agreement is included in several ways in the architecture, as an organizational concept (constitution) or as a formalization of a relationship (business agreement and contract).

These two initiatives, OASIS RA and WS-Agreement are compatible and complementary of our proposal, since they provide a modelling reference architecture as well as a language for describing the identified patterns boiling down to the implementation level. This integration will be include in future publications.

The pattern Provider Rating described within this article illustrates how agreement patterns can help in providing a common vocabulary as well as a collection of best practices for engineering agreement-based distributed applications.

Future works will validate this model with other problems, and represent a compendium of agreement based problems, and their solutions, which they purpose is to help developers to take each pattern an assemble a system capable of interact with service providers, without needing to know how the interaction must be done, instead knowing the required elements to be implemented.

References

1. Domingue, J.: Future internet service offer: An overview. John domingue on behalf of future internet services wg (2008)
2. Şensoy, M.: A Flexible Approach For Context-Aware Service Selectio In Agent-Mediated E-Commerce. PhD thesis, Boğaziçi University (2008)
3. Maximilien, E.M., Singh, M.P.: A framework and ontology for dynamic web services selection. IEEE Internet Computing 8(5), 84–93 (2004)
4. Billhardt, H., Hermoso, R., Ossowski, S., Centeno, R.: Trust-based service provider selection in open environments. In: Proceedings of the 2007 ACM Symposium on Applied Computing, SAC 2007, pp. 1375–1380. ACM, New York (2007)
5. Aydoğan, R.: Content-oriented composite service negotiation with complex preferences. In: Proceedings of the 7th International Joint Conference on Autonomous Agents and Multiagent Systems. International Foundation for Autonomous Agents and Multiagent Systems, AAMAS 2008, Richland, SC, pp. 1725–1726 (2008)
6. Merino, A.S., Matsunaga, Y., Shah, M., Suzuki, T., Katz, R.H.: Secure authentication system for public wlan roaming. Mob. Netw. Appl. 10(3), 355–370 (2005)
7. Singh, M.P., Huhns, M.N.: Service-oriented computing: Semantics, processes, agents. J. Wiley and Sons, Chichester (2005)
8. Yang, S.J.H., Hsieh, J.S.F., Lan, B.C.W., Chung, J.: Composition and evaluation of trustworthy web services. Int. J. Web Grid Serv. 2(1), 5–24 (2006)
9. Bromuri, S., Urovi, V., Morge, M., Stathis, K., Toni, F.: A multi-agent system for service discovery, selection and negotiation. In: Proceedings of The 8th International Conference on Autonomous Agents and Multiagent Systems, Richland, SC, International Foundation for Autonomous Agents and Multiagent Systems, AAMAS 2009, pp. 1395–1396 (2009)
10. Oluyomi, A., Karunasekera, S., Sterling, L.: Design of agent-oriented pattern templates. In: Proceedings of the Australian Software Engineering Conference, ASWEC 2006, Washington, DC, USA, pp. 113–121. IEEE Computer Society, Los Alamitos (2006)
11. Jennings, N.: Agreement technologies. In: IEEE / WIC / ACM International Conference on Intelligent Agent Technology, p. 17 (2005)
12. Iglesias, C.A., Garijo, M., Fernandez-Villamor, J.I., Durán, J.J.: Agreement patterns (2009)
13. Buschmann, F., Meunier, R., Rohnert, H., Sommerlad, P., Stal, M.: Pattern-oriented software architecture: a system of patterns. John Wiley & Sons, Inc., New York (1996)
14. Rising, L. (ed.): The patterns handbooks: techniques, strategies, and applications. Cambridge University Press, New York (1998)

15. Song, W.: Building dependable service-oriented application via dynamic reconfiguration and fault-tolerant reconfiguration collaboration protocol. PhD thesis, Tempe, AZ, USA (2008)
16. Oluyomi, A., Karunasekera, S., Sterling, L.: A comprehensive view of agent-oriented patterns. Autonomous Agents and Multi-Agent Systems 15(3), 337–377 (2007)
17. Oluyomi, A.O.: Patterns and Protocols for Agent-Oriented Software Development. PhD thesis, Faculty of Engineering. University of Melbourne, Australia (November 01, 2006)
18. Erl, T.: SOA Design Patterns. Prentice-Hall, Englewood Cliffs (2008)
19. Rotem Gal Oz, A.: SOA Patterns. Manning (2009)
20. Zdun, U., Hentrich, C., Aalst, W.M.P.V.D.: A survey of patterns for service oriented architectures. Int. J. Internet Protoc. Technol. 1(3), 132–143 (2006)
21. Andrieux, A., Czajkowski, K., Dan, A., Keahey, K., Ludwig, H., Nakata, T., Pruyne, J., Rofrano, J., Tuecke, S., Xu, M.: Web services agreement specification (WS-Agreement). Technical report, Grid Resource Allocation Agreement Protocol (GRAAP) Working Group (2007)
22. McCabe, F.G.: Reference architecture for service oriented architecture. Technical report, OASIS (April 2008)

Advanced Scheduling Techniques with the Pliant System for High-Level Grid Brokering

József Dániel Dombi and Attila Kertész

Institute of Informatics, University of Szeged, Árpád tér 2., 6720 Szeged, Hungary
MTA SZTAKI, Kende u. 13-17, 1111 Budapest, Hungary
dombijd@inf.u-szeged.hu, attila.kertesz@sztaki.hu

Abstract. Here we will present advanced scheduling techniques with a weighted fitness function for an adaptive grid meta-brokering using the Pliant system, in order to cope with the high uncertainty ruling current Grid systems. The algorithms are based on the Pliant concept, which is a specific part of fuzzy logic theory. This tool is also capable of creating functions, classifying sets and making decisions. We construct and show how well our new algorithms perform in a grid simulation environment. The results obtained demonstrate that these novel scheduling techniques produce better performance scores, hence the load of Grid systems can be more balanced.

Keywords: Pliant system, Sigmoid function, Grid computing, Meta-brokering.

1 Introduction

In 1998, a new computing infrastructure called the Grid was born when the bible for the Grid [12] was published by Ian Foster et. al. Since then, Grid Computing has become an independent field of research; current Grid Systems are intented for numerous worldwide projects, and production Grids serve various user communities all around the world. The emerging Web technologies have influenced Grid development, and the latest solutions from related research fields (e.g. autonomous computing, artificial intelligence and peer-to-peer technologies) also need to be taken into account in order to achieve better resource utilization and successfully transform the currently separate production Grids to the future Internet of Services [16].

As the management and advantageous utilization of highly dynamic, complex grid resources cannot be handled by the users themselves, various grid resource management tools have been developed and support different Grids. User requirements created certain properties that resource managers now provide. This development is continuing, and users still find it hard to distinguish brokers and to migrate their applications when they move to a different grid system. Scheduling in diverse and distributed environments requires sophisticated approaches because a high uncertainty is present at several stages of a Grid environment. The main contribution of this paper lies in an enhanced scheduling solution based on the *Pliant System* [7] that is applied to the resource management layer of grid middleware systems.

The Pliant system is very similar to a fuzzy system [6]. The difference between the two systems lies in the choice of operators. In fuzzy theory the membership function plays an important role, although the exact definition of this function is often not

J. Filipe, A. Fred, and B. Sharp (Eds.): ICAART 2010, CCIS 129, pp. 173–185, 2011.

clear. In Pliant systems we use a so-called distending function, which represents a soft inequality. In the Pliant system the various operators (conjunction, disjunction and aggregation) are closely related to each other. We also use unary operators (negation and modifiers), which are also related to the Pliant system. In consequence, the Pliant system involves only those operators whose relationship is clearly defined and the whole system is based on these identities.

In the next section we shall introduce the Pliant system, operators and functions. In Section 3 we will describe meta-brokering in Grids, while in Section 4 we discuss algorithms of an adaptive scheduling technique that seek to provide better scheduling in global Grids. Section 5 describes an evaluation of our proposed solution and finally, in the last section we will briefly summarize our results and draw some relevant conclusions.

2 Components of the Pliant System

In fuzzy logic theory [6] the membership function plays an important role. In pliant logic we introduce a distending function with a soft inequality. The *Pliant system* is a strict, monotonously increasing t-norm and t-conorm, and the following expression is valid for the generator function:

$$f_c(x) f_d(x) = 1, \tag{1}$$

where $f_c(x)$ and $f_d(x)$ are the generator functions of the conjunctive and disjunctive logical operators. Here we use the representation theorem of the associative equation [3]:

$$f_c(x_1, x_2, \cdots, x_n) = f_c^{-1}\left(\sum_{i=1}^{n} f_c(x_i)\right) \tag{2}$$

The generator function could be the *Dombi operator* [6]:

$$f(x) = \frac{1-x}{x}, f^{-1}(x) = \frac{1}{1+x} \tag{3}$$

Besides the above-mentioned logical operators in fuzzy theory, there is also another non-logical operator. The reason for this that for real world applications it is not enough to use just conjunctive or disjunction operators [17]. The rational form of an aggregation operator is:

$$a_{\nu,\nu_0}(x_1, \cdots, x_n) = \frac{1}{1 + \frac{1-\nu_0}{\nu_0} \frac{\nu}{1-\nu} \prod_{i=1}^{n} \frac{1-x_i}{x_i}}, \tag{4}$$

where ν is the neutral value and ν_0 is the threshold value of the corresponding negation.

The general form of the distending function is the following:

$$\delta_a^{(\lambda)}(x) = f^{-1}\left(e^{-\lambda(x-a)}\right), \lambda \in R, a \in R \tag{5}$$

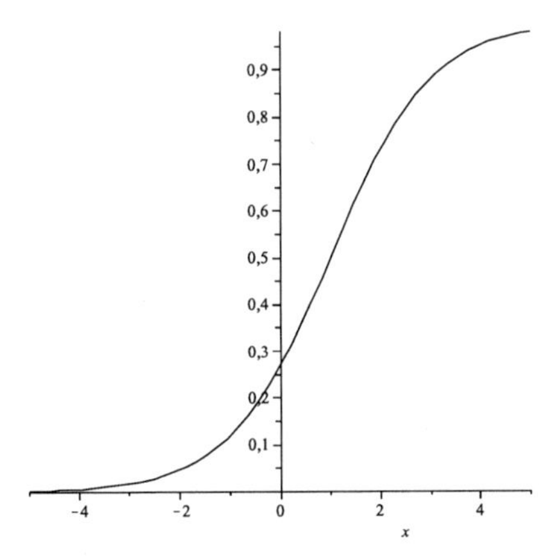

Fig. 1. Sigmoid function

Here f is the generator function of the logical connectives, λ is responsible for the sharpness and a denotes the threshold value. The semantic meaning of $\delta_a^{(\lambda)}$ is

$$truth(a <_\lambda x) = \delta_a^{(\lambda)}(x) \tag{6}$$

Important properties of the distending function are:

1. In the Pliant system f could be the generator function of the conjunctive operator or the disjunctive operator. The form of $\delta_a^{(\lambda)}(x)$ is the same in both cases.
2. In the Pliant concept the operators and membership are closely related.

2.1 Sigmoid Function

In the Dombi operator case, the distending function is the sigmoid function (see Figure 1):

$$\sigma_a^{(\lambda)}(x) = \frac{1}{1 + e^{-\lambda(x-a)}} \tag{7}$$

Here, it is clear that:

1

$$\sigma_a^{(\lambda)}(a) = \frac{1}{2} = \nu_0 \tag{8}$$

2

$$\left(\sigma_a^{(\lambda)}(a)\right)' = \frac{\lambda}{4} \tag{9}$$

3

$$\sigma_a^{(-\lambda)}(x) = n(\sigma_a^{(\lambda)}(x)) \text{ if } n(x) = 1 - x. \tag{10}$$

The sigmoid function naturally maps the values to the (0,1) interval.

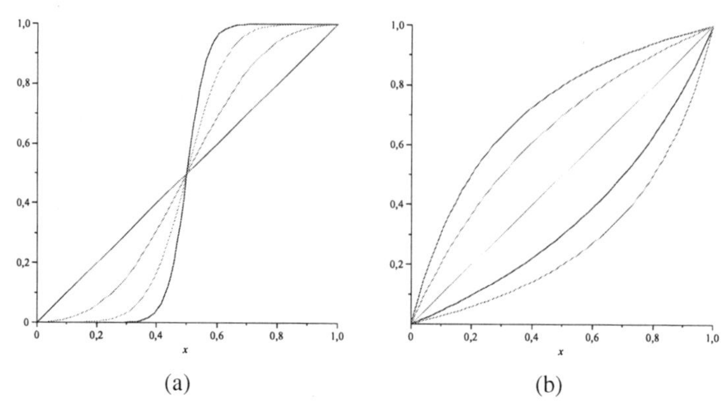

(a) (b)

Fig. 2. Modification operator with the parameter values (a) $\nu = 0.5, \lambda = 1, 2, 4, 8$ (b) $\lambda = 1, \nu = 0.1, 0.3, 0.5, 0.7, 0.8$

2.2 Kappa Function

In order to make a decision we need to define a unary transformation for the values produced by the pliant operator. Using pliant logic, the general form of the modification operators is:

$$\kappa_{\nu}^{\lambda} = \frac{1}{1 + \left(\frac{\nu}{1-\nu}\frac{1-x}{x}\right)^{\lambda}} \tag{11}$$

The behaviour of this function for different values of ν nand λ can be seen in figures 2.

3 Meta-Brokering in Grid Systems

Meta-brokering refers to a higher level of resource management, which utilizes an existing resource or service brokers to access various resources. In some generalized way, it acts as a mediator between users or higher level tools and environment-specific resource managers. The main tasks of this component are: to *gather* static and dynamic broker properties, and to *schedule* user requests to lower level brokers, i.e. match job descriptions to broker properties. Finally the job needs to be *forwarded* to the selected broker.

Figure 3 provides a schematic diagram of the Meta-Broker (MB) architecture [11], including the components needed to fulfil the above-mentioned tasks. Different brokers use different service or resource specification descriptions to understand the user request. These documents need to be written by the users to specify the different kinds of service-related requirements. For the resource utilization in Grids, OGF [1] developed a resource specification language standard called JSDL [4]. As JSDL is general enough to describe the jobs and services of different grids and brokers, this is the default description format of MB. The *Translator* component of the Meta-Broker is responsible for translating the resource specification defined by the user to the language of the appropriate resource broker that MB selects to use for a given request. These brokers have

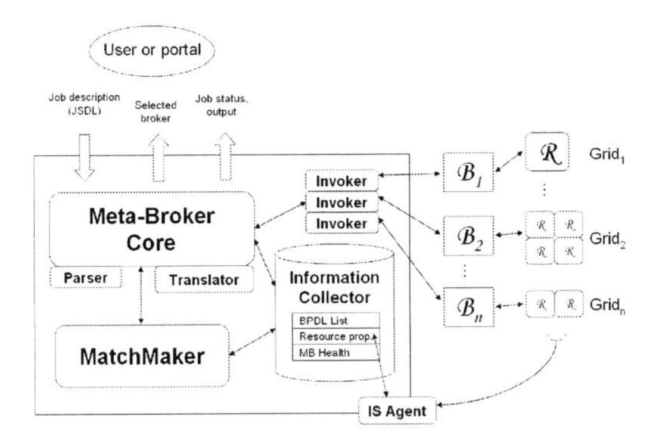

Fig. 3. Components of the Meta-Broker

various features for supporting different user needs, hence an extendable Broker Property Description Language (BPDL) [11] is required to express metadata about brokers and the services they provide. The *Information Collector* (IC) component of MB stores data about the accessible brokers and historical data about previous submissions. This information tells us whether the chosen broker is available, and/or how reliable its services are. During broker utilization the successful submissions and failures are tracked, and for these events a ranking is updated for each special attribute in the BPDL of the appropriate broker (these attributes are listed above). In this way, the BPDL documents represent and store the dynamic states of the brokers. In order to support load balancing, there is an *IS Agent* (IS stands for Information System) reporting to the IC, which regularly checks the load of the underlying resources of each linked broker, and stores this data. The matchmaking process consists of the following steps: The *MatchMaker* (MM) compares the received descriptions to the BPDL of the registered brokers. This selection determines a group of brokers that can provide the required service. Otherwise, the request is rejected. In the second phase the MM counts a rank for each of the remaining brokers. This rank is calculated from the broker properties that the IS Agent updates regularly, and from the service completion rate that is updated in the BPDL for each broker. When all the ranks have been counted, the list of the brokers is ordered by these ranks. Finally the first broker of the priority list is selected, and the *Invoker* component forwards the request to the broker.

Regarding related works, other approaches try to define common protocols and interfaces among scheduler instances enabling inter-grid usage. The meta-scheduling project in LA Grid [15] aims to support grid applications with resources located and managed in different domains. They define broker instances with a set of functional modules. Each broker instance collects resource information from its neighbors and saves the information in its resource repository. The resource information is distributed in the different grid domains and each instance will have a view of all resources. The Koala grid scheduler [9] was designed to work on DAS-2 interacting with Globus middleware services with the main features of data and processor co-allocation; lately it is being extended to support DAS-3 and Grid'5000. Their policy is to use a remote grid only if the local one is saturated. They use a so-called delegated matchmaking (DMM), where Koala instances delegate resource information in a peer-to-peer manner. Gridway introduces a Scheduling Architectures Taxonomy [14]. Its Multiple Meta-Scheduler Layers use Gridway instances to communicate and interact through grid gateways. These instances can access resources belonging to different administrative domains. They also pass user requests to another domain, when the current one is overloaded. Comparing these related approaches, we can state that all of them use a new method to expand current grid resource management boundaries. Meta-brokering appears in a sense that different domains are being examined as a whole, but they rather delegate resource information among domains, broker instances or gateways through their own, implementation-dependent interfaces. Their scheduling policies focus on resource selection by usually aggregated resource information sharing, while our approach targets broker selection based on broker properties and performances.

4 Scheduling Algorithms

In the previous sections we introduced the Pliant System and Grid Meta-Broker, and showed how the default matchmaking process is carried out. The main contribution of this paper is to *enhance* the scheduling part of this matchmaking process. To achieve this, we created a Decision Maker component based on functions of the *Pliant system*, and inserted it into the MatchMaker component of the Meta-Broker. The first part of the matchmaking is unchanged: the list of the available brokers is filtered according to the requirements of the actual job read from its JSDL. Then a list of the remaining brokers along with their performance data and the background grid load are sent to the Decision Maker in order to determine the most suitable broker for the actual job. The scheduling techniques and the scheduling process are described below.

The Decision Maker uses a random number generator, and we chose a JAVA solution that generates pseudorandom numbers. The JAVA random generator class uses a uniform distribution and 48-bit seed and the latter is modified by a linear congruential formula [13]. We also developed a unique random number generator which generates random numbers with a given distribution. We call this algorithm the generator function. In our case we defined a score value for each broker, and we created the distribution based on the score value. For example, the broker which has the highest score number has the biggest chance of being chosen.

To improve the scheduling performance of the Meta-Broker we need to send the job to the broker that best fits the requirements, and executes the job without failures with the shortest execution time. Every broker has *four properties* that the algorithm can rely on: a success counter, a failure counter, a load counter and the running jobs counter.

- The success counter gives the number of jobs which had finished without any errors.
- The failure counter shows the number of failed jobs.
- The load counter indicates the actual load of the grid behind the broker (in percentage terms).
- The running jobs counter shows the number of jobs sent to the broker which have not yet finished.

We developed *two* different kinds of decision algorithms that take into account the above-mentioned broker properties. These algorithms define a score number for each

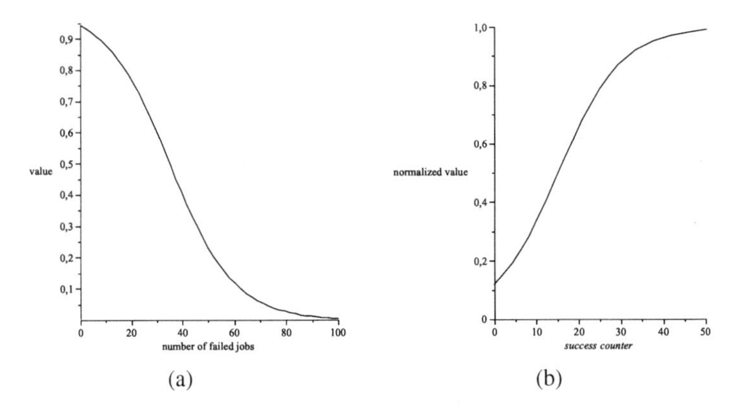

Fig. 4. Normalizing the (a) failed jobs counter / (b) the success counter using Sigmoid function

broker and use the generator function to select a broker. Both algorithms use the kappa function to determine the broker's score number.

Because the Pliant system is defined in the $[0, 1]$ interval, we need to *normalize* the input value. The two algorithms differ only in this step:

1. The first algorithm uses a linear transformation called Decision4.
2. The second algorithm uses the Sigmoid function to normalize the input values, which is called Decision5.

It is also important to *emphasize* that the closer the value is to one, the better the broker is, and if the value is close to zero, it means that the broker is not good. For example if the failure counter is high, both normalization algorithms should give a value close to zero because it is not a good thing if the broker has a lot of failed jobs (see in Figure 4). The opposite of this case is true for the success counter (see Figure 4).

In the next step we can modify the normalized property value by using the same Kappa function (see Figure 5). We can also define the expected value of the normalization via the ν and λ parameters.

To *calculate* the score value, we can make use of the conjunctive or aggregation operator. After running some tests we found that we get better results if we use the aggregation operator. In this step the result is always a real number lying in the $[0, 1]$ interval and then we multiply it by 100 to get the broker's score number.

When the Meta-Broker is running, the first two broker properties (the success and failure counters) are incremented via a feedback method that the simulator (or a user or portal in real world cases) calls after the job has finished. The third and fourth properties, the load value and the running jobs, are handled by the IS Agent of the Meta-Broker, queried from an information provider (Information System) of a Grid. During a simulation this data is saved to a database by the Broker entities of the simulator (described later and shown in Figure 6). This means that by the time we start the evaluation and before we receive feedback from finished jobs, the algorithms can only rely on the background load and running processes of the grids. To further enhance the scheduling we developed a *training process* that can be executed before the simulation in order to initialize the first and second properties. This process sends a small number of jobs

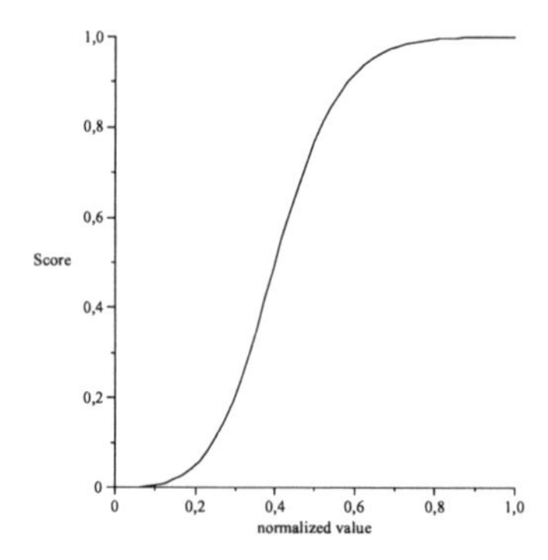

Fig. 5. Normalized parameter values using the Kappa function

with various properties to the brokers and sets the successful and failed jobs number at the BPDLs of the brokers. With this additional training method, we can expect shorter execution times because we will select more reliable brokers.

5 Results

5.1 Evaluation

In order to evaluate our proposed scheduling solution, we created a general *grid simulation environment*, where all the related grid resource management entities could be simulated and coordinated. The GridSim toolkit [5] is a fully extendable, widely used and accepted grid simulation tool. These are the main reasons why we chose this toolkit for our simulations. It can be used for evaluating VO-based resource allocation, workflow scheduling, and dynamic resource provisioning techniques in global grids. It supports the modeling and simulation of heterogeneous grid resources, users, applications, brokers and schedulers in a grid computing environment. It provides primitives for the creation of jobs (called gridlets), the mapping of these jobs to resources, and their management, thus resource schedulers can be simulated to study scheduling algorithms. GridSim provides a multilayered design architecture based on SimJava [8], a general purpose discrete-event simulation package implemented in Java. It is used for handling the interactions or events among GridSim components. Each component in GridSim communicates with another via message passing operations defined by SimJava.

Our general simulation architecture is shown in Figure 6. In the bottom right hand corner we can see the GridSim components used to simulate the grid system. The resources can be defined with different grid-types. These resources consist of more machines, for which workloads can be set. On top of this simulated grid infrastructure we can set up brokers. The Broker and Simulator entities were developed by us in order to simulate the meta-brokering process. Brokers are extended GridUser entities. Here

- they can be connected to one or more resources;
- different properties can be assigned to these brokers (agreement handling, co-allocation, advance reservation, etc.);
- some properties may be marked as unreliable;

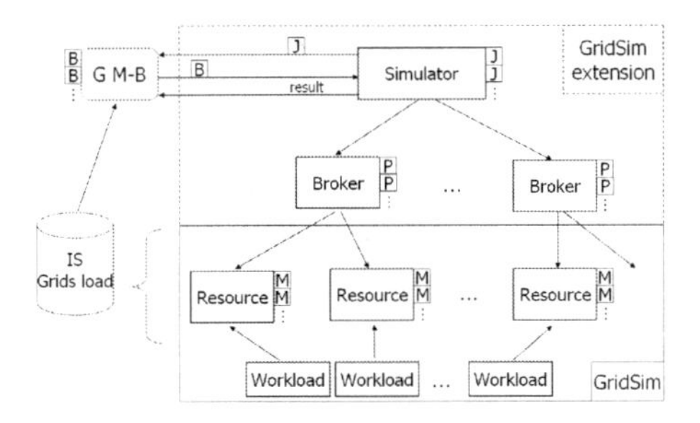

Fig. 6. Meta-Brokering simulation environment based on GridSim

- various scheduling policies can be defined (pre-defined ones: rnd – random resource selection, fcpu – resources having more free cpu time or fewer waiting jobs are selected, nfailed – resources having fewer machine failures are selected);
- in general resubmission is used when a job fails due to resource failure;
- after they report to the IS Grid load database by calling the feedback method of the Meta-Broker with the results of the job submissions (this database has a similar purpose to that of a grid Information System).

The Simulator is an extended GridSim entity. Hence

- it can generate a requested number of gridlets (jobs) with different properties, start and run times (length);
- it is connected to the created brokers and is able to submit jobs to them;
- the default job distribution is the random broker selection (where the middleware types should be taken into account);
- in the case of a job failure a different broker is selected for the actual job;
- it is also connected to the Grid Meta-Broker through its Web service interface and is able to call its matchmaking service for broker selection.

5.2 Evaluation Environment

Table 1 shows the *evaluation environment* used in our evaluation. The simulation setup was derived from real-life production grids: current grids and brokers support only a few special properties: here we used four. To determine the number of resources in our simulated grids we compared the sizes of current production grids (EGEE VOs, DAS3, NGS, Grid5000, OSG, etc.). In the evaluation we utilized 14 brokers. We submitted 1000 jobs to the system, and measured the makespan of all the jobs. Out of the 1000 jobs 100 had no special properties, while for the rest of the jobs four key properties were distributed in the following way: 300 jobs had property A, 300 had B, 200 had C and 100 had D. The second column above denotes the scheduling policies used by the brokers: fcpu means the jobs are scheduled to the resource with the highest free cpu time. The third column shows the capabilities/properties (like coallocation, checkpointing) of the brokers: here we used A, B, C and D in the simulations. The F subscript means unreliability, a broker having the kind of property that may fail to execute a job with the requested service with a probablity of 0.5. The fourth column contains the number

Table 1. Evaluation environment setup

Broker	Scheduling	Properties	Resources
1.	fcpu	A	6
2.	fcpu	A_F	8
3.	fcpu	A	12
4.	fcpu	B	10
5.	fcpu	B_F	10
6.	fcpu	B	12
7.	fcpu	B_F	12
8.	fcpu	C	4
9.	fcpu	C	4
10.	fcpu	$A_F D$	8
11.	fcpu	AD	10
12.	fcpu	AD_F	8
13.	fcpu	AB_F	6
14.	fcpu	ABC_F	10

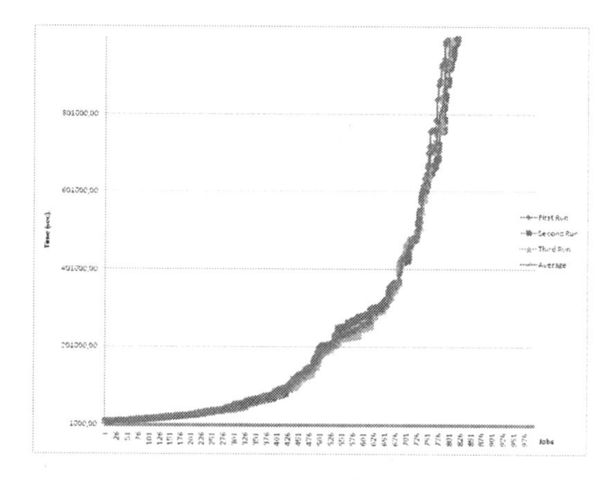

Fig. 7. Results of Decision 4 algorithm

of resources utilized by a broker. As a background workload, 50 jobs were submitted to each resource by the simulation workload entities during the evaluation timeframe. The SDSC BLUE workload logs were used for this purpose, taken from the Parallel Workloads Archive [2].

In order to test all the features of the algorithms, we submitted the jobs periodically: 1/3 of the jobs were submitted at the beginning then the simulator waited for 200 jobs to finish and update the performances of the brokers. After this phase the simulator again submitted 1/3 of all the jobs and waited for 200 more to finish. Lastly the remaining jobs (1/3 again) were submitted. In this way the broker performance results could be updated and monitored by the scheduling algorithms.

In the previous section we explained how the two algorithms called Decision4 and Decision5 (both based on the Pliant system) work. For the evaluation part we repeated each experiment *three times*. The measured simulation results of the Decision4

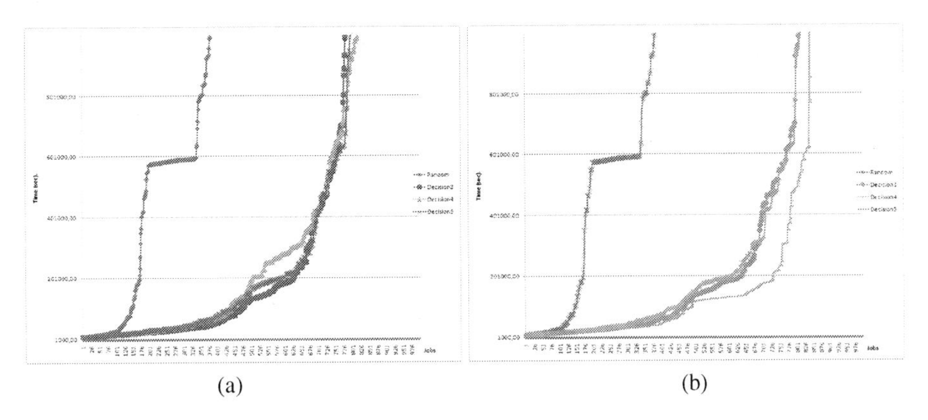

(a) (b)

Fig. 8. Simulation results for the three desicion algorithms (a) without training (b) with training compared with the random decision maker

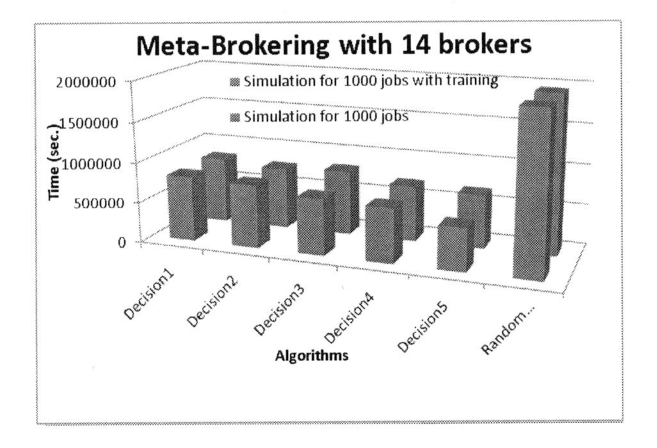

Fig. 9. Simulation in the main evaluation environment

algorithm can be seen in Figure 7. We noticed that the measured runtimes for the jobs were very close to each other. When comparing the various simulation types we always used the median: we counted the average runtime of the jobs in each of the three series and discarded the best and the worst simulations.

A comparison of the simulation results can be seen in Figure 8 above. In our previous paper [10] we used only random number generators to boost the Decision Maker, and proposed three algorithms called Decision1, Decision2 and Decision3. In that paper Decision3 gave the best results. This is why we will compare our new measurements with the results of this algorithm. We can see that for around 1/3 of the simulations, Decision3 provides better results, but the overall makespans are better for the new algorithms.

The simulation results for the algorithms with training can be seen in Figure 8. As we mentioned earlier, we used a training process to initiate the performance values of the brokers before job submissions. In this way, the decisions for the first round of jobs can be made better. Upon examining the results, Decision4 still performs about the same as Decision3, but Decision5 clearly *overperforms* the other two.

In Figure 9 above we provide a graphical summary of the various evaluation phases. The columns show the average values of each evaluation run with the same parameter values. The results clearly show that the more intelligence (more sophisticated methods) we put into the system, the better the performance. The *most advanced* version of our proposed meta-brokering solution is called the Decision Maker using the algorithm called *Decision5 with training*. Once the number of brokers and job properties are sufficiently high to set up this Grid Meta-Broker Service for inter-connecting several Grids, the new scheduling algorithms will be ready to serve thousands of users even under conditons of *high uncertainty*.

6 Conclusions

The Grid Meta-Broker itself is a standalone Web-Service that can serve both users and grid portals. The presented enhanced scheduling solution based on *Pliant functions* allows a higher level, interoperable brokering by utilizing existing resource brokers of different grid middleware. It gathers and utilizes meta-data about brokers from various grid systems to establish an adaptive meta-brokering service. We developed a new scheduling component for this Meta-Broker called Decision Maker that uses *Pliant functions* with a random generation in order to select a good performing broker for user jobs even under conditions of high uncertainty. We evaluated our algorithms in a grid simulation environment based on GridSim, and performed simulations with real workload samples. The evaluation results accord with our expected utilization gains: the enhanced scheduling provided by the revised Decision Maker results in a *more efficient* job execution.

Acknowledgements. The research leading to these results has received funding from the European Community's Seventh Framework Programme FP7/2007-2013 under grant agreement 215483 (S-Cube).

References

1. Open grid forum website (1999), `http://www.ogf.org`
2. Parallel workloads archive website (2009), `http://www.cs.huji.ac.il/labs/parallel/workload`
3. Aczl, J.: Lectures on Functional Equations and Applications. Academic Press, London (1966)
4. Anjomshoaa, A., Brisard, F., Drescher, M., Fellows, D., Ly, A., McGough, S., Pulsipher, D., Savva, A.: Job submission description language (jsdl) specification, version 1.0. Technical report (2005), `http://www.gridforum.org/documents/GFD.56.pdf`
5. Buyya, R., Murshed, M., Abramson, D.: Gridsim: A toolkit for the modeling and simulation of distributed resource management and scheduling for grid computing. Journal of Concurrency and Computation: Practice and Experience (CCPE), 1175–1220 (2002)
6. Dombi, J.: A general class of fuzzy operators, the de morgan class of fuzzy operators and fuzziness measures induced by fuzzy operators. Fuzzy Sets and Systems 8 (1982)
7. Dombi, J.: Pliant system. In: IEEE International Conference on Intelligent Engineering System Proceedings, Budapest, Hungary (1997)

8. Howell, F., Mcnab, R.: Simjava: A discrete event simulation library for java. In: Proc. of the International Conference on Web-Based Modeling and Simulation, pp. 51–56 (1998)
9. Iosup, A., Epema, D.H.J., Tannenbaum, T., Farrellee, M., Livny, M.: Inter-Operating Grids through Delegated MatchMaking. In: Proc. of the International Conference for High Performance Computing, Networking, Storage and Analysis (SC 2007), Reno, Nevada (2007)
10. Kertesz, A., Dombi, J.D., Dombi, J.: Adaptive scheduling solution for grid meta-brokering. Acta Cybernetica 19, 105–123 (2009)
11. Kertész, A., Kacsuk, P., GMBS: A New Middleware Service for Making Grids Interoperable. In: Future Generation Computer Systems, vol. 16, pp. 542–553 (2010)
12. Kesselman, C., Foster, I.: The Grid: Blueprint for a New Computing Infrastructure. Morgan Kaufmann Publishers, San Francisco (1998)
13. Knuth, D.E.: The art of computer programming. In: Seminumerical Algorithms, 3rd edn., vol. 2. Addison-Wesley Longman Publishing Co., Inc., Boston (1997)
14. Leal, K., Huedo, E., Llorente, I.M.: A decentralized model for scheduling independent tasks in Federated Grids. In: Future Generation Computer Systems, vol. 25(8), pp. 840–852 (2009)
15. Rodero, I., Guim, F., Corbalan, J., Fong, L.L., Liu, Y.G., Sadjadi, S.M.: Looking for an Evolution of Grid Scheduling: Meta-brokering. In: Proc. of Coregrid Workshop in Grid Middleware 2007, Dresden, Germany (2008)
16. Report, N.G.G.: Future for european grids: Grids and service oriented knowledge utilities – vision and research directions 2010 and beyond. Technical report (2006), `ftp://ftp.cordis.lu/pub/ist/docs/grids/ngg3_eg_final.pdf`
17. Zimmermann, H.: Fuzzy Set Theory and its applications. Kluwer Academic, Dordrecht (1991)

Predictive Learning from Demonstration

Erik A. Billing, Thomas Hellström, and Lars-Erik Janlert

Department of Computing Science, Umeå University, 901 87 Umeå, Sweden
{billing,thomash,lej}@cs.umu.se

Abstract. A model-free learning algorithm called Predictive Sequence Learning (PSL) is presented and evaluated in a robot Learning from Demonstration (LFD) setting. PSL is inspired by several functional models of the brain. It constructs sequences of predictable sensory-motor patterns, without relying on predefined higher-level concepts. The algorithm is demonstrated on a Khepera II robot in four different tasks. During training, PSL generates a hypothesis library from demonstrated data. The library is then used to control the robot by continually predicting the next action, based on the sequence of passed sensor and motor events. In this way, the robot reproduces the demonstrated behavior. PSL is able to successfully learn and repeat three elementary tasks, but is unable to repeat a fourth, composed behavior. The results indicate that PSL is suitable for learning problems up to a certain complexity, while higher level coordination is required for learning more complex behaviors.

1 Introduction

Recent years have witnessed an increased interest in computational mechanisms that will allow robots to *Learn from Demonstrations (LFD)*. With this approach, also referred to as *Imitation Learning*, the robot learns a behavior from a set of good examples, *demonstrations*. The field has identified a number of key problems, commonly formulated as *what to imitate*, *how to imitate*, *when to imitate* and *who to imitate* [3]. In the present work, we focus on the first question, referring to which aspects of the demonstration should be learned and repeated.

Inspiration is taken from several functional models of the brain and prediction is exploited as a way to learn state definitions. A novel learning algorithm, called *Predictive Sequence Learning (PSL)*, is here presented and evaluated. PSL is inspired by *S-Learning* [42, 43], which has previously been applied to robot learning problems as a model-free reinforcement learning algorithm [40, 41].

The paper is organized as follows. In Sect. 2 a theoretical background and biological motivation is given. Section 3 gives a detailed description of the proposed algorithm. Section 4 describes the experimental setup and results for evaluation of the algorithm. In Sect. 5, conclusions, limitations and future work are discussed.

2 Motivation

One common approach to identify what in a demonstration that is to be imitated is to exploit the variability in several demonstrations of the same behavior. Invariants among

J. Filipe, A. Fred, and B. Sharp (Eds.): ICAART 2010, CCIS 129, pp. 186–200, 2011.

the demonstrations are seen as the most relevant and selected as essential components of the task [3, 17]. Several methods for discovering invariants in demonstrations can be found in the LFD literature. One method presented by Billard et al. applies a time-delayed neural network for extraction of relevant features from a manipulation task [4, 5]. A more recent approach uses demonstrations to impose constraints in a dynamical system, e.g. [16, 25].

While this is a suitable method for many types of tasks, there are also applications where it is less obvious which aspects of a behavior should be invariant, or if the relevant aspects of that behavior is captured by the invariants. Since there is no universal method to determine whether two demonstrations should be seen as manifestations of the same behavior or two different behaviors [10], it is in most LFD applications up to the teacher to decide. However, the teacher's grouping of actions into behaviors may not be useful for the robot. In the well known imitation framework by Nehaniv and Dautenhahn [34], it is emphasized that the success of an imitation is observer dependent. The consequence of observer dependence when it comes to interpreting sequences of actions has been further illustrated with Pfeifer and Scheier's argument about the *frame of reference* [35, 36], and is also reflected in Simon's parable with the ant [45]. A longer discussion related to these issues can be found in [6].

Pfeifer and Scheier promotes the use of a *low level specification* [36], and specifically the *sensory-motor space* $I = U \times Y$, where U and Y denotes the *action space* and *observation space*, respectively. Representations created directly in I prevents the robot from having memory, which has obvious limitations. However, systems with no or very limited memory capabilities has still reached great success within the robotics community through the works by Rodney Brooks, e.g., [12–15], and the development of the *reactive* and *behavior based* control paradigms, e.g., [1]. By extending the definition of I such that it captures a certain amount of temporal structure, the memory limitation can be removed. Such a temporally extended sensory-motor space is denoted *history information space* $I^{\tau} = I_0 \times I_1 \times I_2 \times \ldots \times I_{\tau}$, where τ denotes the temporal extension of I *[10]*. With a large enough τ, I^{τ} can model any behavior. However, a large τ leads to an explosion of the number of possible states, and the robot has to generalize such that it can act even though the present state has not appeared during training.

In the present work, we present a learning method that is not based on finding invariants among several demonstrations of, what the teacher understands to be "the same behavior". Taking inspiration from recent models of the brain where prediction plays a central role, e.g. [22, 23, 27, 32], we approach the question of what to imitate by the use of prediction.

2.1 Functional Models of Cortex

During the last two decades a growing body of research has proposed computational models that aim to capture different aspects of human brain function, specifically the cortex. This research includes models of perception, e.g., Riesenhuber and Poggio's hierarchical model [38] which has inspired several more recent perceptual models [23, 32, 37], models of motor control [26, 42, 46–48] and learning [22]. In 2004, this field reached a larger audience with the release of Jeff Hawkins's book On Intelligence [28].

With the ambition to present a unified theory of the brain, the book describes cortex as a hierarchical memory system and promotes the idea of a *common cortical algorithm*. Hawkins's theory of cortical function, referred to as the *Memory-Prediction framework*, describes the brain as a prediction system. Intelligence is, in this view, more about applying memories in order to predict the future, than it is about computing a response to a stimulus.

A core issue related to the idea of a common cortical algorithm is what sort of bias the brain uses. One answer is that the body has a large number of reward systems. These systems are activated when we eat, laugh or make love, activities that through evolution have proved to be important for survival. However, these reward systems are not enough. The brain also needs to store the knowledge of how to activate these reward systems.

In this context, prediction appears to be critical for learning. The ability to predict the future allows the agent to foresee the consequences of its actions and in the long term how to reach a certain goal. However, prediction also plays an even more fundamental role by providing information about how well a certain model of the world correlates with reality.

This argument is supported not only by Hawkins's work, but by a large body of research investigating the computational aspects of the brain [8]. It has been proposed that the central nervous system (CNS) simulates aspects of the sensorimotor loop [29, 31, 33, 47]. This involves a modular view of the CNS, where each module implements one *forward model* and one *inverse model*. The forward model predicts the sensory consequences of a motor command, while the inverse model calculates the motor command that, in the current state, leads to the goal [46]. Each module works under a certain *context* or bias, i.e., assumptions about the world which are necessary for the module's actions to be successful. One purpose of the forward model is to create an estimate of how well the present situation corresponds to these assumptions. If the prediction error is low the situation is familiar. However, if the prediction error is high, the situation does not correspond to the module's context and actions produced by the inverse model may be inappropriate.

These findings have inspired recent research on robot perception and control. One example is the *rehearse, predict, observe, reinforce* decomposition proposed by [18, 20, 44] which adapts the view of perception and action as two aspects of a single process. Hierarchical representations following this decomposition have also been tested in an LFD setting [19] where the robot successfully learns sequences of actions from observation. In work parallel to this, we also investigates PSL as an algorithm for behavior recognition [11], exploring the possibilities to use PSL both as a forward and an inverse model. The present work should be seen as a further investigation of these theories applied to robots, with focus to learning with minimal bias.

2.2 Sequence Learning

PSL is inspired by *S-Learning*, a dynamic temporal difference (TD) algorithm presented by Rohrer and Hulet, [42, 43]. S-Learning builds sequences of passed events which may be used to predict future events, and in contrast to most other TD algorithms it can base its predictions on many previous states.

S-Learning can be seen as a variable order Markov model (VMM) and we have observed that it is very similar to the well known compression algorithm LZ78 [49]. This coincidence is not that surprising considering the close relationship between lossless compression and prediction [2]. In principle, any lossless compression algorithm could be used for prediction, and vice verse [21].

S-Learning was originally developed to capture the discrete episodic properties observed in many types of human motor behavior [39]. Inspiration is taken from the *Hierarchical Temporal Memory* algorithm [24], with focus on introducing as few assumptions into learning as possible. More recently, it has been applied as a model-free reinforcement learning algorithm for both simulated and physical robots [40, 41]. We have also evaluated S-Learning as an algorithm for behavior recognition [9]. However, to our knowledge it has never been used as a control algorithm for LFD.

The model-free design of S-Learning, together with its focus on sequential data and its connections to human motor control makes S-Learning very interesting for further investigation as a method for robot learning. With the ambition to increase the focus on prediction, and propose a model that automatically can detect when it is consistent with the world, PSL was designed.

3 Predictive Sequence Learning

PSL is trained on an *event sequence* $\eta = (e_1, e_2, \ldots, e_t)$, where each *event* e is a member of an alphabet $\sum$. η is defined up to the current time t from where the next event e_{t+1} is to be predicted.

PSL stores its knowledge as a set of hypotheses, known as a *hypothesis library* H. A *hypothesis* $h \in H$ expresses a dependence between an event sequence $X = (e_{t-n}, e_{t-n+1}, \ldots, e_t)$ and a target event $I = e_{t+1}$:

$$h : X \Rightarrow I \tag{1}$$

X_h is referred to as the *body* of h and I_h denotes the *head*. Each h is associated with a *confidence* c reflecting the conditional probability $P(I|X)$. For a given η, c is defined as $c(X \Rightarrow I) = s(X, I)/s(X)$, where the *support* $s(X)$ describes the proportion of transactions in η that contains X and (X, I) denotes the concatenation of X, and I. A transaction is defined as a sub-sequence of the same size as X. The length of h, denoted $|h|$, is defined as the number of elements in X_h. Hypotheses are also referred to as *states*, since a hypothesis of length $|h|$ corresponds to VMM state of order $|h|$.

3.1 Detailed Description of PSL

Let the library H be an empty set of hypotheses. During learning, described in Alg. 1, PSL tries to predict the future event e_{t+1}, based on the observed event sequence η.

If it fails to predict the future state, a new hypothesis h_{new} is created and added to H. h_{new} is one element longer than the longest matching hypothesis previously existing in H. In this way, PSL learns only when it fails to predict.

For example, consider the event sequence $\eta = ABCCABCCA$. Let $t = 1$. PSL will search for a hypothesis with a body matching A. Initially H is empty and consequently

Algorithm 1. Predictive Sequence Learning (PSL)

Require: an event sequence $\eta = (e_1, e_2, \ldots, e_n)$

1: $t \leftarrow 1$
2: $H \leftarrow \emptyset$
3: $M \leftarrow \left\{h \in H \mid X_h = \left(e_{t-|h|+1}, e_{t-|h|+2}, \ldots, e_t\right)\right\}$
4: **if** $M = \emptyset$ **then**
5:     **let** $h_{new} : (e_t) \Rightarrow e_{t+1}$
6:     **add** h_{new} to H
7:     **goto** 20
8: **end if**
9: $\hat{M} \leftarrow \{h \in M \mid |h| \geq |h'| \ for \ all \ h' \in M\}$
10: **let** $h_{max} \in \left\{h \in \hat{M} \mid c(h) \geq c(h') \ for \ all \ h' \in \hat{M}\right\}$
11: **if** $e_{t+1} \neq I_{h_{max}}$ **then**
12:     **let** h_c be the longest hypothesis $\{h \in M \mid I_h = e_{t+1}\}$
13:     **if** $h_c = null$ **then**
14:         **let** $h_{new} : (e_t) \Rightarrow e_{t+1}$
15:     **else**
16:         **let** $h_{new} : \left(e_{t-|h_c|}, e_{t-|h_c|+1}, \ldots, e_t\right) \Rightarrow e_{t+1}$
17:     **end if**
18:     **add** h_{new} to H
19: **end if**
20: update the confidence for h_{max} and $h_{correct}$ as described in Sect. 3
21: $t \leftarrow t + 1$
22: **if** $t < n$ **then**
23:     **goto** 2
24: **end if**

PSL will create a new hypothesis $(A) \Rightarrow B$ which is added to H. The same procedure will be executed at $t = 2$ and $t = 3$ so that $H = \{(A) \Rightarrow B; (B) \Rightarrow C; (C) \Rightarrow C\}$. At $t = 4$, PSL will find a matching hypothesis $h_{max} : (C) \Rightarrow C$ producing the wrong prediction C. Consequently, a new hypothesis $(C) \Rightarrow A$ is added to H. The predictions at $t = 5$ and $t = 6$ will be successful while $h : (C) \Rightarrow A$ will be selected at $t = 7$ and produce the wrong prediction. As a consequence, PSL will create a new hypothesis $h_{new} : (B, C) \Rightarrow C$. Source code from the implementation used in the present work is available online [7].

3.2 Making Predictions

After, or during, learning, PSL can be used to make predictions based on the sequence of passed events $\eta = (e_1, e_2, \ldots, e_t)$. Since PSL continuously makes predictions during learning, this procedure is very similar to the learning algorithm (Alg. 1). The prediction procedure is described in Alg. 2.

For prediction of a suite of future events, $\hat{e}_t$ can be added to η to create η'. Then repeat the procedure described in Alg. 2 using η' as event history.

Algorithm 2. Making predictions using PSL

Require: an event sequence $\eta = (e_1, e_2, \ldots, e_{t-1})$
Require: the trained library $H = \left(h_1, h_2, \ldots, h_{|H|}\right)$

1: $M \leftarrow \left\{h \in H \mid X_h = \left(e_{t-|h|}, e_{t-|h|+1}, \ldots, e_{t-1}\right)\right\}$
2: $\hat{M} \leftarrow \{h \in M \mid |h| \geq |h'| \ for\ all\ h' \in M\}$
3: **let** $h_{max} \in \left\{h \in \hat{M} \mid c(h) \geq c(h') \ for\ all\ h' \in \hat{M}\right\}$
4: **return** the prediction $\hat{e}_t = I_{h_{max}}$

3.3 Differences and Similarities between PSL and S-Learning

Like PSL, S-Learning is trained on an *event sequence* η. However, S-Learning does not produce hypotheses. Instead, knowledge is represented as *Sequences* ϕ, stored in a *sequence library* κ [43]. ϕ does not describe a relation between a body and a head, like hypotheses do. Instead, ϕ describes a plain sequence of elements $e \in \eta$. During learning, sequences are "grown" each time a matching pattern for that sequence appears in the training data. Common patterns in η produce long sequences in κ. When S-Learning is used to predict the next event, the beginning of each $\phi \in \kappa$ is matched to the end of η. The sequence producing the longest match is selected as a winner, and the end of the winning sequence is used to predict future events.

One problem with this approach, observed during our previous work with S-Learning [9], is that new, longer sequences, are created even though the existing sequence already has Markov property, meaning that it can predict the next element optimally. To prevent the model from getting unreasonably large, S-Learning implements a maximum sequence length m. As a result, κ becomes unnecessarily large, even when m is relatively low. More importantly, by setting the maximum sequence length m, a task-dependent modeling parameter is introduced, which may limit S-Learning's ability to model η.

PSL was designed to alleviate the problems with S-Learning. Since PSL learns only when it fails to predict, it is less prune to be overtrained and can employ an unlimited maximum sequence length without exploding the library size.

4 Evaluation

The PSL algorithm was tested on a Khepera II miniature robot [30]. In the first evaluation (Sect. 4.1), the performance of PSL on a playful LFD task is demonstrated. In a second experiment (Sect. 4.2), the prediction performance during training of PSL is compared to the performance of S-Learning, using recorded sensor and motor data from the robot. During both experiments, the robot is given limited sensing abilities using only its eight infrared proximity sensors mounted around its sides.

One important issue, promoted both by Rohrer et al. [40, 41] and ourselves [10], is the ability to learn even with limited prior knowledge of what is to be learned. Prior knowledge is information intentionally introduced into the system to support learning, often referred to as *ontological bias* or *design bias* [10]. Examples of common design biases are pre-defined state specifications, pre-processing of sensor data, the size of a neural network or the length of a temporal window. While design biases help in learning, they also limit the range of behaviors a robot can learn. A system implementing large amounts of design bias will to a larger extent base its decisions not on its own experience, but on knowledge of the programmer designing the learning algorithm, making it hard to determine what the system has actually learned.

In addition to design bias, there are many limitations and constraints introduced by other means, e.g., by the size and shape of the robot including its sensing and action capabilities, structure of the environment and performance limitations of the computer used. These kinds of limitations are referred to as *pragmatical bias* [10]. We generally try to limit the amount of ontological bias, while pragmatical bias should be exploited by the learning algorithm to find useful patterns.

In the present experiments, the robot has no previous knowledge about its surroundings or itself. The only obvious design bias is the thresholding of proximity sensors into three levels, *far*, *medium* and *close*, corresponding to distances of a few centimeters. This thresholding was introduced to decrease the size of the observation space Y, limiting the amount of training required. An *observation* $y \in Y$ is defined as the combination of the eight proximity sensors, producing a total of 3^8 possible observations.

An *action* $u \in U$ is defined as the combination of the speed commands sent to the two motors. The Khepera II robot has 256 possible speeds for each wheel, producing an action space U of 256^2 possible actions. However, only a small fraction of these were used during demonstration.

The event sequence is built up by alternating sensor and action events, $\eta = (u_1, y_1, u_2, y_2 \ldots, u_k, y_k)$. k is here used to denote the current stage, rather than the current position in η denoted by t. Even though events is categorized into observations and actions, PSL makes no distinction between these two types of events. From the perspective of the algorithm, all events $e_t \in \sum$ are discrete entities with no predefined relations, where $\sum = Y \cup U$.

In each stage k, PSL is used to predict the next event, given η. Since the last element of η is an observation, PSL will predict an action $u_k \in U$, leading to the observation $y_k \in Y$. u_k and y_k are appended to η, transforming stage k to $k+1$. This alternating use of observations and actions was adopted from S-Learning [42]. A stage frequency of 10 Hz was used, producing one observation and one action every 0.1 seconds.

4.1 Demonstration and Repetition

To evaluate the performance of PSL on an LFD problem, four tasks are defined and demonstrated using the Khepera II robot. *Task 1* involves the robot moving forward in a corridor approaching an object (cylindrical wood block). When the robot gets close to the object, it should stop and wait for the human teacher to "load" the object, i.e., place it upon the robot. After loading, the robot turns around and goes back along the corridor. *Task 2* involves general corridor driving, taking turns in the right way without hitting the walls and so on. *Task 3* constitutes the "unloading" procedure, where the robot stops in a corner and waits for the teacher to remove the object and place it to the right of the robot. Then the robot turns and pushes the cylinder straight forward for about 10 centimeters, backs away and turns to go for another object. *Task 4* is the combination of the three previous tasks. The sequence of actions expected by the robot

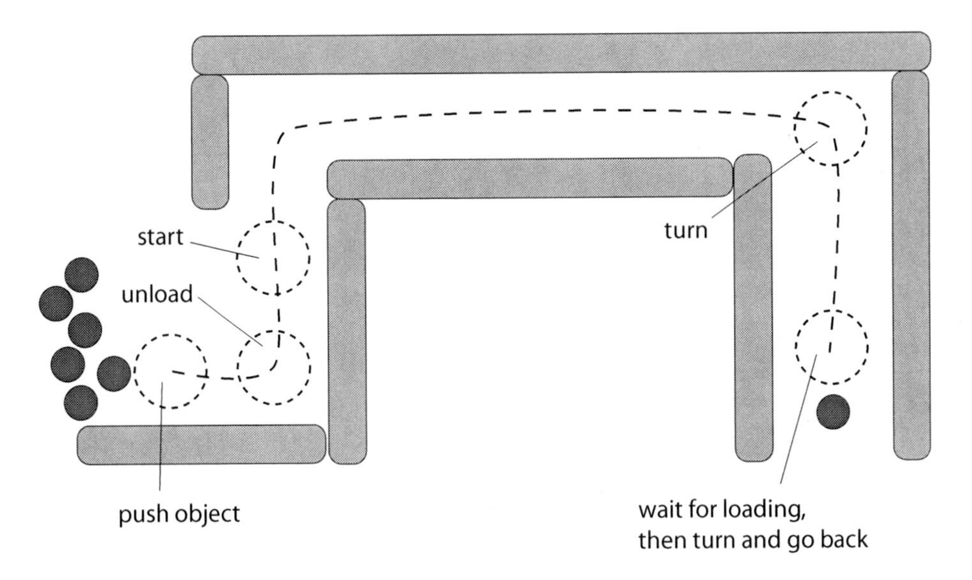

Fig. 1. Schematic overview of the composed behavior (*Task 4*). Light gray rectangles mark walls, dark gray circles mark the objects and dashed circles mark a number of key positions for the robot. See text for details.

is illustrated in Fig. 1. The robot starts by driving upwards in the figure, following the dashed line. until it reaches the object at the loading position. After loading, the robot turns around and follows the dashed line back until it reaches the unload position. When the cylinder has been unloaded (placed to the left of the robot), the robot turns and pushes the object. Finally, it backs away from the pile and awaits further instructions. The experimental setup can be seen in Fig. 2. Even though the setup was roughly the same in all experiments, the starting positions and exact placement of the walls varied between demonstration and repetition.

All tasks capture a certain amount of temporal structure. One example is the turning after loading the object in Task 1. Exactly the same pattern of sensor and motor data will appear before, as well as after, turning. However, two different sequences of actions is expected. Specifically, after the teacher has taken the cylinder to place it on the robot, only the sensors on the robot's sides are activated. The same sensor pattern appears directly after the robot has completed the 180 degree turn, before it starts to move back along the corridor. Furthermore, the teacher does not act instantly. After placing the object on the robot, one or two seconds passed before the teacher issued a turning command, making it more difficult for the learning algorithm to find the connection between the events. Even Task 2 which is often seen as a typical reactive behavior is, due to the heavy thresholding of sensor data, temporally demanding. Even longer temporal structures can be found in Task 3, where the robot must push the object and remember for how long the object is to be pushed. This distance was not controlled in any way, making different demonstrations of the same task containing slightly conflicting data.

After training, the robot was able to repeat Task 1, 2 and 3 successfully. For Task 1, seven demonstrations were used for a total of about 2.6 min. Task 2 was demonstrated

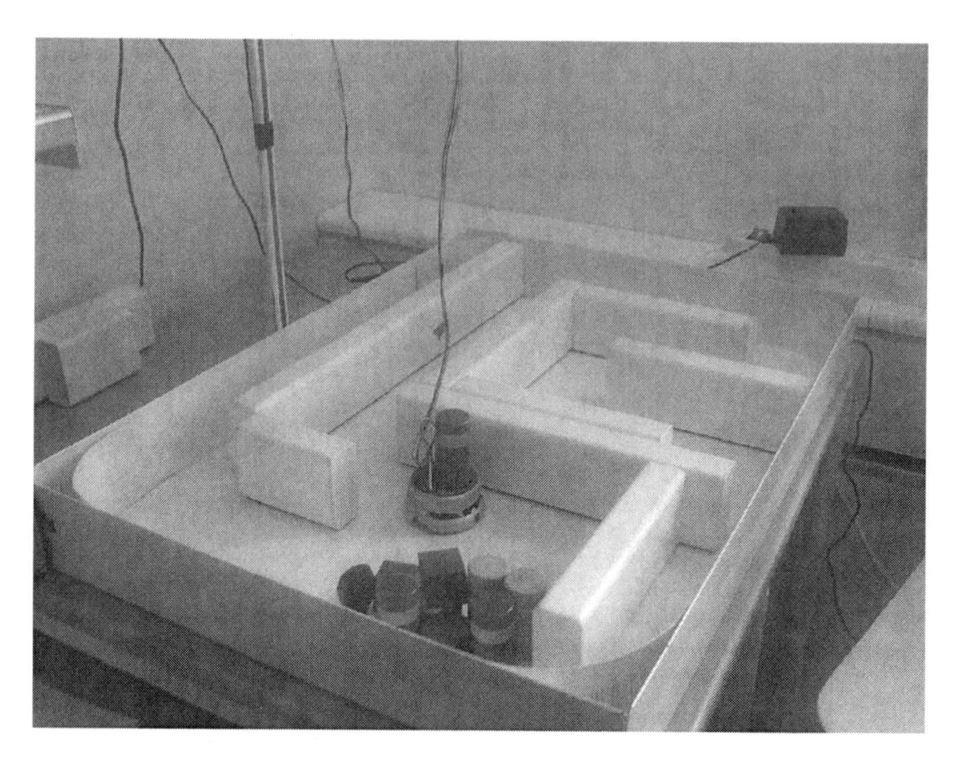

Fig. 2. Experimental setup

for about 8.7 min and Task 3 was demonstrated nine times, in total 4.6 min. The robot made occasional mistakes in all three tasks, reaching situations where it had no training data. In these situations it sometimes needed help to be able to complete the task. However, the number of mistakes clearly decreased with increased training, and mistakes made by the teacher during training often helped the robot to recover from mistakes during repetition.

For Task 4, the demonstrations from all three partial tasks were used, plus a single 2 min demonstration of the entire Task 4. Even after extensive training, resulting in almost 40 000 hypotheses in library, the robot was unable to repeat the complete behavior without frequent mistakes. Knowledge from the different sub-tasks was clearly interfering, causing the robot to stop and wait for unloading when it was supposed to turn, turning when it was supposed to follow the wall and so on. Detailed results for all four tasks can be found in Table 1.

PSL was trained until it could predict about 98% of the demonstrated data correctly. It would be possible to train it until it reproduces all events correctly, but this takes time and initial experiments showed that it did not affect the imitation performance significantly.

4.2 Comparison between S-Learning and PSL

In Sect. 3.3, a number of motivations for the design of PSL were given, in relation to S-Learning. One such motivation was the ability to learn and increase the model size only

Table 1. Detailed statistics on the four evaluation tasks. Training events is the number of sensor and motor events in demonstrated data. Lib. size is the number of hypotheses in library after training. Avg. $|h|$ is the average hypothesis length after training.

Task	Training events	Library size	Avg. $\lvert h\rvert$
Task 1	3102	4049	9.81
Task 2	10419	30517	16
Task 3	5518	8797	11
Task 4	26476	38029	15

when necessary. S-Learning always learns and creates new sequences for all common events, while PSL only learns when prediction fails. However, it should be pointed out that even though S-Learning never stops to learn unless an explicit limit on sequence length is introduced, it quickly reduces the rate at which new sequences are created in domains where it already has extensive knowledge.

To evaluate the effect of these differences between PSL and S-Learning, prediction performance and library size were measured during training in three test cases. *Case 1* contained a demonstration of the loading procedure (*Task 1*) used in the LFD evaluation, Sect. 4.1. During the demonstration, the procedure was repeated seven times for a total of about 150 seconds (3000 sensor and motor events). *Case 2* encapsulated the whole composed behavior (*Task 4*) used in LFD evaluation. The behavior was demonstrated once for 120 seconds (2400 events). *Case 3* constituted 200 seconds of synthetic data, describing a 0.1 Hz sinus wave discretized with a temporal resolution of 20 Hz and an amplitude resolution of 0.1 (resulting in 20 discrete levels). The 4000 elements long data sequence created a clean repetitive pattern with minor fluctuations due to sampling variations.

In addition to PSL and S-Learning, a first order Markov model (*1MM*) was included in the tests. The Markov model can obviously not learn the pattern in any of the three test cases perfectly, since there is no direct mapping $e_t \Rightarrow e_{t+1}$ for many events. Hence, the performance of 1MM should be seen only as reference results.

The results from the three test cases can be seen in Fig. 3. The upper part of each plot show accumulated training error over the demonstration while lower parts show model growth (number of hypotheses in library). Since the Markov model does not have a library, the number of edges in the Markov graph is shown, which best corresponds to sequences or hypotheses in S-Learning and PSL, respectively.

5 Description

A novel robot learning algorithm called *Predictive Sequence Learning (PSL)* is presented and evaluated in an LFD setting. PSL is both parameter-free and model-free in the sense that no ontological information about the robot or conditions in the world is pre-defined in the system. Instead, PSL creates a state space (hypothesis library) in order to predict the demonstrated data optimally. This state space can thereafter be used to control the robot such that it repeats the demonstrated behavior.

In contrast to many other LFD algorithms, PSL does not build representations from invariants among several demonstrations that a human teacher considers to be

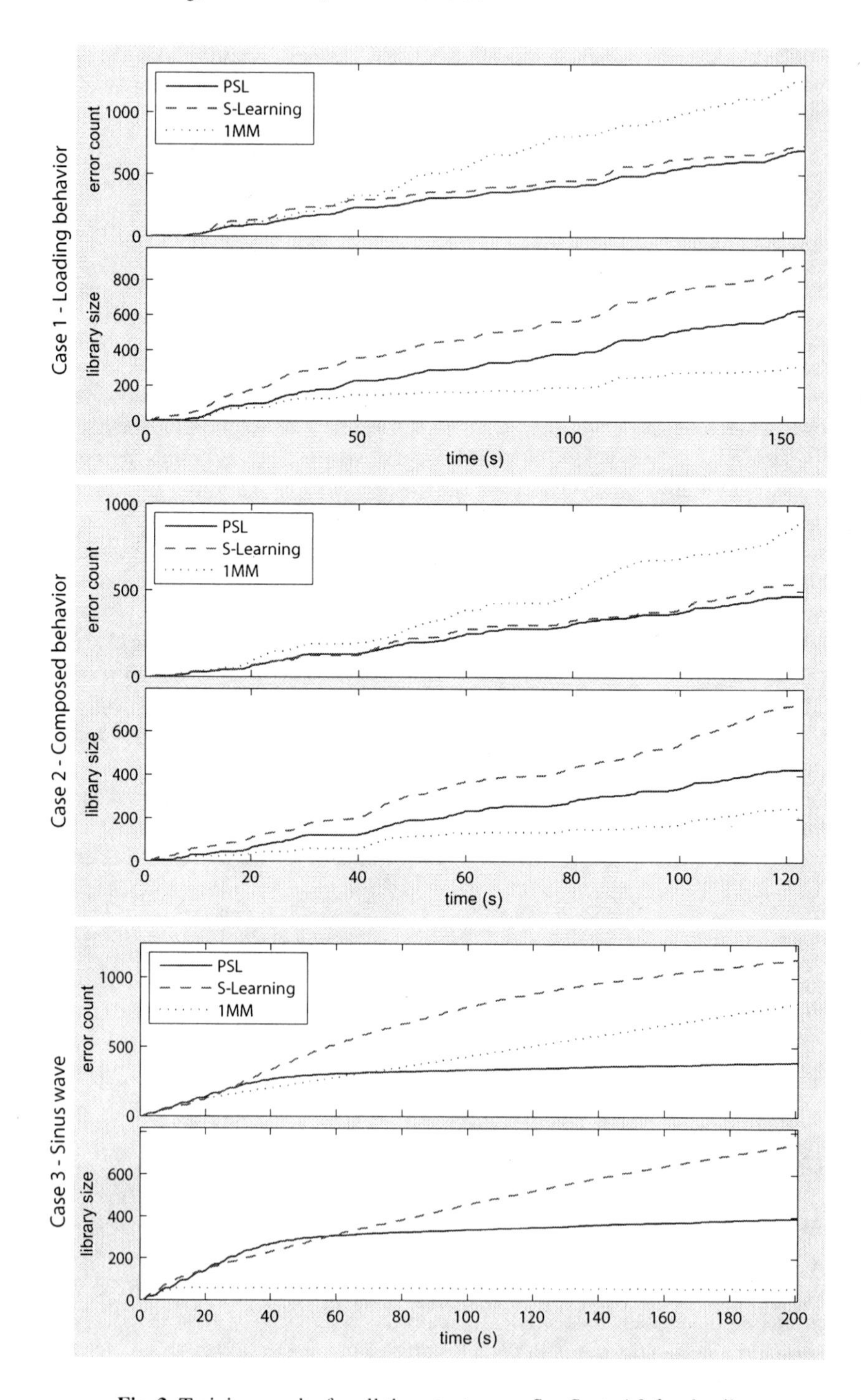

Fig. 3. Training results for all three test cases. See Sect. 4.2 for details.

"the same behavior". All knowledge, from one or several demonstrations, is stored as hypotheses in the library. PSL treats inconsistencies in these demonstrations by generating longer hypotheses that will allow it to make the correct predictions. In this way, the ambiguous definitions of *behavior* is avoided and control is seen purely as a prediction problem.

In the prediction performance comparison, PSL produces significantly smaller libraries than S-Learning on all three data sets. The difference is particularly large in Case 3 (Fig. 3), where both algorithms learn to predict the data almost perfectly. In this situation, S-Learning continues to create new sequences, while PSL does not.

In Case 3, PSL also shows the clearly fastest learning rates (least accumulated errors). The reason can be found in that PSL learns on each event where it fails to predict, while S-Learning learns based on sequence length. When the model grows, S-Learning decreases its learning rate even though the performance is still low. In contrast, the learning rate of PSL is always proportional to performance, which can also be seen in the plots for all three test cases (Fig. 3). However, even though PSL commits less accumulated errors than S-Learning in all three tests, the performance difference in Case 1 and 2 is small and how these results generalize to other kinds of data is still an open question.

In the demonstration-repetition evaluation, tasks 1, 2 and 3 were repeated correctly. Even though the robot made occasional mistakes, the imitation performance clearly increased with more demonstrations. However, in Task 4, which was a combination of the three first tasks, an opposite pattern could be observed. Despite the fact that PSL was still able to predict demonstrated data almost perfectly, knowledge from the three elementary tasks clearly interfered. The reason for this interference is that Task 4 requires much longer temporal dynamics than any of the elementary tasks did when learned separately.

One example of how this knowledge interference is manifested is the turning versus unloading. When the robot approaches the position marked as *turn* in Fig. 1, coming from the left and is supposed to take a right turn, it no longer sees the right wall behind it. Consequently, the situation looks identical to that of unloading. When the robot is to unload, it goes downward in Fig. 1 (position *unload*) but instead of turning it must wait for the cylinder to be placed to its right side. To make the right prediction, PSL has to base its decision on information relatively far back in the event history. Even though PSL has no problem to build a sufficiently large model from training data, the large temporal window produces a combinatorial explosion and the chance of the right patterns reappearing during repetition is small. As a result, PSL decreases the temporal window (i.e., uses shorter hypotheses), and the two situations become inseparable.

5.1 Conclusions and Future Work

The results show that the proposed algorithm is feasible for LFD problems up to a certain complexity. PSL implements very few assumptions of what is to be learned and is therefore likely to be applicable to a wide range of problems.

However, PSL also shows clear limitations when the learning problem increases and longer temporal dynamics is required. PSL is subject to combinatorial explosion and the amount of required training data increases exponentially with problem complexity.

In these situations, some higher-level coordination is clearly necessary. One possible solution is to place PSL as a module in a hierarchical system. PSL learns both to predict sensor data as a response to action (forward model) and to select actions based on the current state (inverse model). In the present work, PSL is viewed purely as a controller and the forward model is consequently not considered. However, in work parallel to this, we show that PSL can also be used as an algorithm for behavior recognition [11], i.e., as a predictor of sensor values. A big advantage of using PSL for both control and behavior recognition is that the forward and inverse computations are in fact based on the same model, i.e., the PSL library. This approach has several theoretical connections to the view of human perception and control as two heavily intertwined processes, as discussed in Section 2.1.

The present work should be seen as one step towards a hierarchical control architecture that can learn and coordinate itself, based on the PSL algorithm. The model-free design of PSL introduces very few assumptions into learning, and should constitute a good basis for many types of learning and control problems. Integrating PSL as both forward and inverse model to achieve a two-layer modular control system, is the next step in this process and will be part of our future work.

Acknowledgements

We would like to thank Brandon Rohrer at Sandia National Laboratories and Christian Balkenius at Lund University for valuable input to this work.

References

1. Arkin, R.C.: Behaviour-Based Robotics. MIT Press, Cambridge (1998)
2. Begleiter, R., Yona, G.: On prediction using variable order markov models. Journal of Artificial Intelligence Research 22, 385–421 (2004)
3. Billard, A., Calinon, S., Dillmann, R., Schaal, S.: Robot programming by demonstration. In: Siciliano, B., Khatib, O. (eds.) Handbook of Robotics. Springer, Heidelberg (2008)
4. Billard, A., Epars, Y., Cheng, G., Schaal, S.: Discovering imitation strategies through categorization of multi-dimensional data. In: Proceedings of IEEE/RSJ International Conference on Intelligent Robots and Systems, vol. 3, pp. 2398–2403 (2003)
5. Billard, A., Mataric, M.J.: Learning human arm movements by imitation: Evaluation of a biologically inspired connectionist architecture. Robotics and Autonomous Systems 37(2-3), 145–160 (2001)
6. Billing, E.A.: Representing behavior - distributed theories in a context of robotics. Technical report, UMINF 0725, Department of Computing Science, Ume University (2007)
7. Billing, E.A.: Cognition reversed (2009), `http://www.cognitionreversed.com`
8. Billing, E.A.: Cognition Reversed - Robot Learning from Demonstration. PhD thesis, Ume University, Department of Computing Science, Ume, Sweden (December 2009)
9. Billing, E.A., Hellström, T.: Behavior recognition for segmentation of demonstrated tasks. In: IEEE SMC International Conference on Distributed Human-Machine Systems, Athens, Greece, pp. 228–234 (March 2008)
10. Billing, E.A., Hellström, T.: A formalism for learning from demonstration. Paladyn: Journal of Behavioral Robotics 1(1), 1–13 (2010)

11. Billing, E.A., Hellström, T., Janlert, L.E.: Behavior recognition for learning from demonstration. In: Proceedings of IEEE International Conference on Robotics and Automation, Anchorage, Alaska (May 2010)
12. Brooks, R.A.: A robust layered control system for a mobile robot. IEEE Journal of Robotics and Automation RA-2 1, 14–23 (1986)
13. Brooks, R.A.: Elephants don't play chess. Robotics and Autonomous Systems 6, 3–15 (1990)
14. Brooks, R.A.: Intelligence without reason. In: Proceedings of 1991 Int. Joint Conf. on Artificial Intelligence, pp. 569–595 (1991)
15. Brooks, R.A.: New approaches to robotics. Science 253(13), 1227–1232 (1991)
16. Calinon, S., Guenter, F., Billard, A.: On learning, representing and generalizing a task in a humanoid robot. IEEE Transactions on Systems, Man and Cybernetics, Part B. Special Issue on Robot Learning by Observation, Demonstration and Imitation 37(2), 286–298 (2007)
17. Delson, N., West, H.: Robot programming by human demonstration: The use of human inconsistency in improving 3D robot trajectories. In: Proceedings of the IEEE/RSJ/GI International Conference on Intelligent Robots and Systems 1994. Advanced Robotic Systems and the Real World, IROS 1994, Munich, Germany, vol. 2, pp. 1248–1255 (September 1994)
18. Demiris, J., Hayes, G.R.: Imitation as a dual-route process featuring predictive and learning components: a biologically plausible computational model. In: Imitation in Animals and Artifacts, pp. 327–361. MIT Press, Cambridge (2002)
19. Demiris, Y., Johnson, M.: Distributed, predictive perception of actions: a biologically inspired robotics architecture for imitation and learning. Connection Science 15(4), 231–243 (2003)
20. Demiris, Y., Simmons, G.: Perceiving the unusual: Temporal properties of hierarchical motor representations for action perception. Neural Networks 19(3), 272–284 (2006)
21. Feder, M., Merhav, N.: Relations between entropy and error probability. IEEE Transactions on Information Theory 40(1), 259–266 (1994)
22. Friston, K.J.: Learning and inference in the brain. Neural Networks: The Official Journal of the International Neural Network Society 16(9), 1325–1352 (2003) PMID: 14622888
23. George, D.: How the Brain might work: A Hierarchical and Temporal Model for Learning and Recognition. PhD thesis, Stanford University, Department of Electrical Engineering (2008)
24. George, D., Hawkins, J.: A hierarchical bayesian model of invariant pattern recognition in the visual cortex. In: Proceedings of IEEE International Joint Conference on Neural Networks (IJCNN 2005), vol. 3, pp. 1812–1817 (2005)
25. Guenter, F., Hersch, M., Calinon, S., Billard, A.: Reinforcement learning for imitating constrained reaching movements. RSJ Advanced Robotics, Special Issue on Imitative Robots 21(13), 1521–1544 (2007)
26. Haruno, M., Wolpert, D.M., Kawato, M.: Hierarchical MOSAIC for movement generation. In: International Congress Series, vol. 1250, pp. 575–590. Elsevier Science B.V., Amsterdam (2003)
27. Haruno, M., Wolpert, D.M., Kawato, M.M.: MOSAIC model for sensorimotor learning and control. Neural Comput. 13(10), 2201–2220 (2001)
28. Hawkins, J., Blakeslee, S.: On Intelligence. Times Books (2002)
29. Jordan, M., Rumelhart, D.: Forward models: Supervised learning with a distal teacher. Cognitive Science: A Multidisciplinary Journal 16(3), 307–354 (1992)
30. K-Team. Khepera robot (2007), `http://www.k-team.com`
31. Kawato, M., Furukawa, K., Suzuki, R.: A hierarchical neural-network model for control and learning of voluntary movement. Biological Cybernetics 57(3), 169–185 (1987) PMID: 3676355
32. Lee, T.S., Mumford, D.: Hierarchical bayesian inference in the visual cortex. J. Opt. Soc. Am. A Opt. Image Sci. Vis. 20(7), 1434–1448 (2003)

33. Miall, R.C., Wolpert, D.M.: Forward models for physiological motor control. Neural Netw. 9(8), 1265–1279 (1996)
34. Nehaniv, C.L., Dautenhahn, K.: Of hummingbirds and helicopters: An algebraic framework for interdisciplinary studies of imitation and its applications. In: Demiris, J., Birk, A. (eds.) Learning Robots: An Interdisciplinary Approach, vol. 24, pp. 136–161. World Scientific Press, Singapore (2000)
35. Pfeifer, R., Scheier, C.: Sensory-motor coordination: the metaphor and beyond. Robotics and Autonomous Systems 20(2), 157–178 (1997)
36. Pfeifer, R., Scheier, C.: Understanding Intelligence. MIT Press, Cambridge (2001)
37. Poggio, T., Bizzi, E.: Generalization in vision and motor control. Nature 431(7010), 768–774 (2004)
38. Riesenhuber, M., Poggio, T.: Hierarchical models of object recognition in cortex. Nature Neuroscience 2(11), 1019–1025 (1999) PMID: 10526343
39. Rohrer, B.: S-Learning: a biomimetic algorithm for learning, memory, and control in robots. In: CNE apos;07. 3rd International IEEE/EMBS Conference on Natural Engineering, Kohala Coast, Hawaii, pp. 148–151 (2007)
40. Rohrer, B.: S-learning: A model-free, case-based algorithm for robot learning and control. In: Eighth International Conference on Case-Based Reasoning, Seattle Washington (2009)
41. Rohrer, B., Bernard, M., Morrow, J.D., Rothganger, F., Xavier, P.: Model-free learning and control in a mobile robot. In: Fifth International Conference on Natural Computation, Tianjin, China (2009)
42. Rohrer, B., Hulet, S.: BECCA - a brain emulating cognition and control architecture. Technical report, Cybernetic Systems Integration Department, Univeristy of Sandria National Laboratories, Alberquerque, NM, USA (2006)
43. Rohrer, B., Hulet, S.: A learning and control approach based on the human neuromotor system. In: Proceedings of Biomedical Robotics and Biomechatronics, BioRob (2006)
44. Schaal, S., Ijspeert, A., Billard, A.: Computational approaches to motor learning by imitation. Philosophical Transactions of the Royal Society B: Biological Sciences 358(1431), 537–547 (2003) PMC1693137
45. Simon, H.A.: The Sciences of the Artificial. MIT Press, Cambridge (1969)
46. Wolpert, D.M.: A unifying computational framework for motor control and social interaction. Phil. Trans. R. Soc. Lond. B(358), 593–602 (2003)
47. Wolpert, D.M., Flanagan, J.R.: Motor prediction. Current Biology: CB 11(18), 729–732 (2001)
48. Wolpert, D.M., Ghahramani, Z.: Computational principles of movement neuroscience. Nature Neuroscience 3, 1212–1217 (2000)
49. Ziv, J., Lempel, A.: Compression of individual sequences via variable-rate coding. IEEE Transactions on Information Theory 24(5), 530–536 (1978)

ISReal: A Platform for Intelligent Simulated Realities

Stefan Nesbigall, Stefan Warwas, Patrick Kapahnke, René Schubotz,
Matthias Klusch, Klaus Fischer, and Philipp Slusallek

German Research Center for Artificial Intelligence
Stuhlsatzenhausweg 3, 66123 Saarbrücken, Germany
{stefan.nesbigall,stefan.warwas,patrick.kapahnke,
rene.schubotz,matthias.klusch,klaus.fischer,
philipp.slusallek}@dfki.de
www.dfki.de

Abstract. Steadily increasing computational power leverages the use of more advanced techniques from artificial intelligence, computer graphics, and areas that have not been directly associated to virtual worlds, like Semantic Web technology. The development of highly realistic simulated realities is a non-trivial task and a cross-discipline endeavour. The ISReal (Intelligent Simulated Realities) platform uses these technologies to simulate virtual environments that are enriched with a high-level semantic description and populated by intelligent agents using Semantic Web technologies to perceive, understand, and interact with their environment. In this paper, we present the basic architecture of the ISReal platform and show the user interaction in an agent assisted learning scenario.

Keywords: Simulated reality, Multiagent system, Semantic web.

1 Introduction

Realistic virtual worlds are increasingly used for training, decision making, entertainment, and many other purposes, which require the convincing modeling and animation of virtual characters as well as the faithful behavior of devices in their environment. Technological development during the recent years mainly focused on the graphical aspect. The behavior of avatars has often been implemented with more or less powerful scripting engines. Instead, the intelligent agent paradigm offers a clean, intuitive, and powerful way of modeling the behavior of intelligent entities in virtual worlds.

Intelligent agents are represented in real-time virtual worlds through avatars (their virtual bodies) and can interact with their environment through sensors and actuators. Sensors cause perceptions that update the agent's beliefs. The agent can reason about its beliefs and plan its actions in order to achieve a given goal. Virtual environments are especially demanding since they are usually dynamic, non-deterministic. To enable agents to interact with their environment a semantic description of this environment is necessary.

3D computer graphic description languages (e.g. X3D, COLLADA) are used in order to describe virtual environments in so called *scene graphs*. These languages specify the objects in the virtual environment by defining their shape, position, orientation,

J. Filipe, A. Fred, and B. Sharp (Eds.): ICAART 2010, CCIS 129, pp. 201–213, 2011.

appearance, etc. The X3D/VRML[1] standard (ISO/IEC 19775) has become a unifying (at least conceptual) base, but it lacks when it comes to the high-level descriptions like they are required by intelligent agents. Semantic annotations can be used to link the X3D objects to their semantic description given in a formal semantic description language (e.g. OWL[2]). Furthermore, the functional behavior in the virtual environment can be specified as a semantic service by its input, output, precondition, and effect (IOPE). The formal specification of the virtual world enables the use of Semantic Web technology (reasoning, matchmaking, service composition planning, etc.).

In this paper we introduce the *Intelligent Simulated Reality* (ISReal) platform, which can be used to deploy semantically enabled virtual worlds that are inhabited by intelligent agents. Intelligent agents perceive the semantic annotations of geometric objects and use, beside traditional BDI planning, Semantic Web technology for reasoning and service composition. The platform is based on standards such as X3D for the scene graph, OWL for semantics, and OWL-S for service descriptions.

In the remainder of this paper we provide a detailed overview of the ISReal platform. Section 2 presents the architecture of the whole platform with a focus on the architecture of the agents that can be deployed. In Section 3 we use an agent assisted learning scenario, in which an user has to operate a virtual machine, to demonstrate the interaction of the different components. The user in this scenario can explore the world and interact with agents, e.g. assign tasks like *"Open the door!"* or *"Show me what to do in order to get the machine running again."* Finally, Section 4 points out related work and Section 5 concludes the paper.

2 ISReal System Architecture

The ISReal platform provides an open and extensible framework for real-time simulated realities. The kind of scenarios that can be deployed cover a wide range of applications such as demonstrators, decision support systems, and virtual training environments. For this purpose, the ISReal architecture has to meet requirements such as (i) scalability in the size and complexity of the simulated worlds, (ii) modularity and exchangeability of the different simulation components, (iii) extensibility (customizability) regarding the supported domains and application scenarios, and (iv) allow a highly realistic simulation of the world regarding geometry, physical properties, and behavior of the entities in a scene. To make our platform as modular as possible, we base our work on standards where possible (e.g. OWL, X3D). The central conceptual building block of our platform is the *semantic world model*, which encompasses the semantically annotated scene graph. It unifies the geometrical shape, semantical properties, and functionality of objects in virtual worlds. The semantic properties are defined with ontologies and the functionality of the objects is described and implemented as semantic services. Services play a key role in our platform. We distinguish between *object services* which are offered by semantic objects in the virtual world, and *platform services* which are offered by the ISReal platform. Both, semantic property and service descriptions, are referenced by the semantic world model.

[1] Web3D: `http://www.web3d.org/x3d/specifications`

[2] W3C: `http://www.w3.org/TR/owl-semantics/`

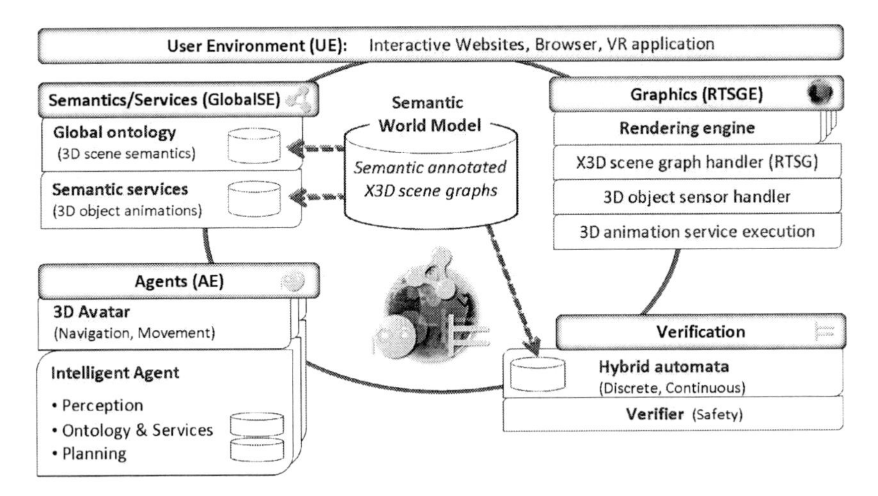

Fig. 1. Top-level view of the ISReal architecture

With reference to Figure 1, the ISReal platform consists of several components. The *Real-Time Scene Graph Environment* (RTSGE) manages (i) the geometric representation of the objects in a 3D environment (scene graph), (ii) their semantic meta data, together building the semantic world model, (iii) their physical properties, and (iv) their animations. The *Global Semantic Environment* (GlobalSE) maintains the global semantic high-level representation of the scene and provides implementations of the semantic services to interact with the RTSGE. Based on these two core components of the ISReal platform, we can connect further modules. The *Agent Environment* (AE) is responsible for the realistic behavior of intelligent entities in the virtual world. It provides the agents with (i) perceptions from the RTSGE, (ii) a semantic knowledge base, (iii) semantic reasoning, and (iv) enables them to invoke object services. A further module is the user environment (UE), which provides the interface for user interactions. The verification component is not covered by this paper.

In the remainder of this section we provide a detailed overview of the different components of the ISReal platform. We focus on the internal architecture of the agents that can be deployed on the ISReal platform. For this purpose, we first introduce in Section 2.1 the RTSGE and in Section 2.2 the GlobalSE. Both components build the core of the ISReal platform and manage the semantic world model. Section 2.3 is the main part and focuses on the agent architecture. The UE is not covered in this section, but we provide some more information in the example in Section 3.

2.1 Real-Time Scene Graph Environment

The RTSGE maintains the geometrical representation and offers an interface for external components to manipulate the virtual environment. We base the scene graph on the X3D standard and use the X3D *Scene Access Interface*[3] (SAI) standard as interface

[3] Web3D: `http://www.web3d.org/x3d/specifications`

for accessing it. The pure geometric description is enriched by semantic annotations, which is attached directly to the geometric object it belongs to (through X3D metadata objects). The RTSGE communicates with a variety of different services, all of which have a semantic description stored in the GlobalSE (see Section 2.2) and operate directly on the scene graph through SAI.

Semantic objects are a central concept of the ISReal platform. They are parts of the semantic world model. Every semantic object is annotated by (i) the URI that refers to the individual representing a high-level description for this object (ABox instance), (ii) the URIs that refer to the most specific concepts this 3D object belongs to (TBox concepts), (iii) the URIs to a set of semantic services (*object services*), describing the useability and/or behaviour of the object for a user or an agent. Beside the semantic properties, a semantic object also encompasses (i) geometric information, (ii) animations, and (iii) physical properties like the condition of its surface. These properties are maintained by the RTSGE. Every object interaction an user or agent can perform in the 3D environment is described as an object service in OWL-S[4]. These services are described with IOPE and enable the user or an agent to use Semantic Web technologies to retrieve, plan, or select interaction tasks. The RTSGE has been realized by the *Real-Time Scene Graph* [20].

2.2 Global Semantic Environment

The GlobalSE maintains the high-level description of the scene graph and the semantic services to interact with the 3D environment. It consists of (i) an *Ontology Management System* (OMS) to a) govern the semantic description of the world, b) allow queries in SPARQL[5], a query language to primarily query RDF graphs, and c) semantic reasoning, and (ii) a service registry to maintain semantic services for the interaction with the objects in the 3D environment (object services). With the initialization of the ISReal system, we assume, that the high-level description specified in OWL2-DL ontologies is consistent and accurate to the low-level description specified in the X3D scene graph. The GlobalSE provides an interface to maintain and query the OMS, register/add and read out object services. Figure 2 depicts the core components of the ISReal platform and their interfaces.

The OMS is initialized with OWL2-DL ontologies (global knowledge base KB) that are internally stored. It provides maintenance possibilities to add, update, and remove statements from the internal store and if the system shuts down write the store back into an ontology file. To read out the store the OMS provides following types of queries:

a) Object Reasoning. For queries about the concrete objects in the 3D environment, i.e. ABox knowledge retrieval, SPARQL is used. For details we refer to the SPARQL W3C recommendation.

b) Concept Reasoning. To answer questions about the concepts (terminological knowledge) in the 3D environment T-Box reasoning is provided by the OMS in form of a set

[4] W3C: `http://www.w3.org/Submission/OWL-S/`

[5] W3C: `http://www.w3.org/TR/rdf-sparql-query`

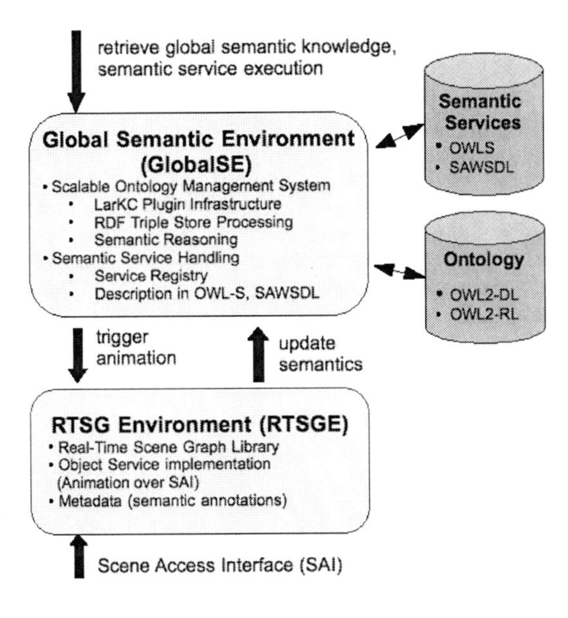

Fig. 2. ISReal core components

of fixed methods to answer tasks like global consistency ($KB \models false?$) or class consistency ($C \equiv \perp ?$) checks.

c) Relational Reasoning. To find non-obvious relations between different objects, i.e. a set of entities $\{e_1, \ldots, e_n\}$, the OMS can find the smallest tree of the RDF graph representing the KB, such that it contains all the entities $\{e_1, \ldots, e_n\}$ [9]. We provide an example for this kind of query in Section 3.

The implementation uses the LarKC architecture [6]. The OMS is implemented as a *LarKC Decider* consisting of different *LarKC Reasoner* plug-ins. As processing plug-ins the OWLIM triple store system [11] and Pellet[6] are used.

2.3 Agent Environment

Intelligent behavior of entities in the ISReal platform is modeled by agents. The agent architecture depends mainly on (i) the properties of the virtual worlds that can be deployed on the ISReal platform, and (ii) the end-user requirements. According to environment properties defined by [21], the environments considered by the ISReal platform are (i) inaccessible, (ii) non-deterministic, (iii) dynamic, and (iv) continuous. The inaccessible property is caused by the fact that an agent only perceives that part of the world which is currently covered by its sensors. Moreover, non-determinism and dynamism are owed to the fact that there is usually more than one agent in the world and the actions performed by these agents can interfere with each other. Furthermore, the number of states that can be reached is not finite, which causes the environment to be

[6] `http://clarkparsia.com/pellet`

continuous. Finally, the agents are acting in a real-time environment, meaning that they have to react in a timely manner. These properties have direct influence on the agent architecture. The end-user of the virtual worlds deployed on the ISReal platform has different possibilities to interact with the agents. He can ask the agent to perform a certain action, assign some declarative goal to it, or can query the agent's local knowledge base.

The ISReal agent architecture is based on the *Belief, Desire, Intension* (BDI) architecture [18] which is well suited for dynamic and real-time environments. Figure 3 depicts an overview of the ISReal agent architecture. We distinguish between the agent core, which encompasses the core functionality provided by an agent execution platform, and the *Local Semantic Environment* (LocalSE), which extends the core with (i) an OWL-based KB (referred to as KB_a) maintained by an OMS, (ii) a service registry S_a, and (iii) a *Service Composition Planner* (SCP). An agent directly controls its avatar (virtual body of an agent) which is situated in the virtual world. The interface of the agent to its virtual environment is realized by sensors and actuators. Sensors generate perceptions that contain information about semantic objects. Actuators are realized as semantic services that are offered by the agent itself or by the semantic objects the agent can interact with. The perception, information, and behavior components are introduced in the following subsections.

Perception Component. Sensors provide an agent with perceptions from its current environment. The perceptions in the ISReal platform are caused by the RTSGE which manages the X3D-based scene graph. Since the X3D standard does not specify the kind of sensor that is required by ISReal agents, it is necessary to extend RTSGE with the required functionality. This functionality is given by a sensor node, that defines a field of view, a resolution and a frequency. During runtime from the sensor position the field of view is scanned with regards to the given resolution in every frequency circle. This is done by ray shooting. The object hits are then checked for annotated data, to determine if they are semantic objects. Agents connected to the ISReal platform perceive only semantic objects (see Section 2.1). A perception event contains following information: (i) the object's ID in the scene graph, (ii) the object's individual URI, (iii), the object's concept URI, (iv) the URIs to the object services, and (v) the URIs to the context rules assigned to the object. The perception handling of the agent is done in the following order: (i) receive the perception event coming from the environment, (ii) use the individual URI to get the corresponding ontological facts from the GlobalSE, and (iii) add these facts to the KB_a of the agent's LocalSE (see Figure 3). Additionally, add the URIs of the object services to the respective lists S_a in the LocalSE.

Information Component. To enable agents to process OWL-based information we extended the agent core with the LocalSE. Equivalent to the GlobalSE the LocalSE consists of an OMS that can be queried in the same three ways. Initializing an agent (and with it its LocalSE) the LocalSE has only a partial description of the world representing the knowledge KB_a the agent has in this scenario. The main difficulty is to integrate the LocalSE in a transparent way, so that the agent's internal mechanisms do not break. The information component provides a transparent layer between the agent

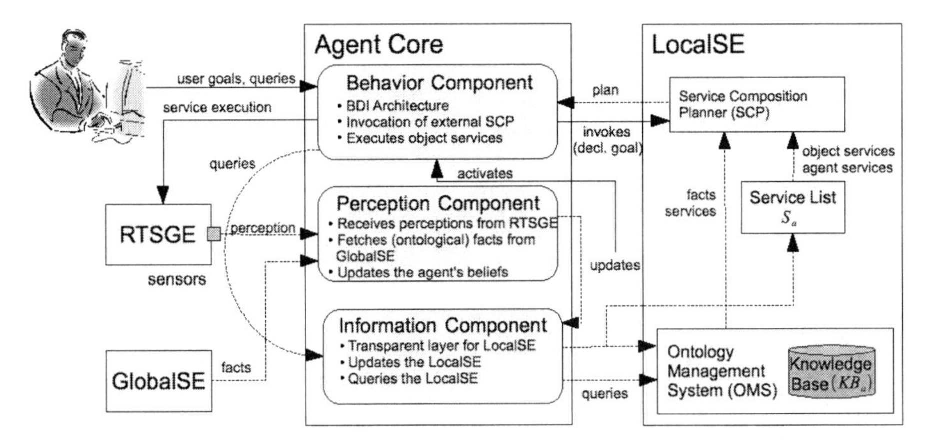

Fig. 3. ISReal agent architecture

and its KB_a. It provides functionality for (i) updating and inserting facts into the KB_a, and (ii) reasoning about information in the KB_a. This functionality is used by the behavior component to access the KB_a. For example, we use SPARQL queries in the context condition of BDI plans.

Behavior Component. Classical first-principles planning starts from a given world state and tries to reach an also given goal state by the application of a set of operators. The whole planning process is done off-line, meaning no changes are incorporated during the planning process. BDI-based planners rely on a plan library that provides the agent with plan templates that guide its execution in certain situations (usually defined by relevance conditions and context condition). Because of the plan library, BDI systems are more efficient than classical planners. Their drawback is that they fail as soon as no plan is applicable in the current situations, even if there exists a combination of their plans that achieves a given goal. A further difference is that BDI agents directly execute the actions of a chosen plan template and incorporate incoming events, while classical planners finish the complete planning process before the plan can be executed. One important distinction is the difference between declarative goals (state descriptions) like they are used in traditional planning problems and procedural goals (goal events) of BDI systems that are used to trigger actions [25].

The ISReal agent architecture combines the efficiency of BDI-planners and the flexibility of classical planners. An agent's core functionality is implemented as BDI plans and (goal) events. We consider two situations in which a BDI agent benefits from the invocation of a classical planner. The first case occurs when there is no applicable plan in the agent's plan library for the current situation. The agent invokes the classical planner to explore new plans that have not been defined at design-time. The second case occurs when the user assigns a declarative goal to the agent to reach a certain state. The agent can pass this goal to its SCP and gets back a plan consisting of a sequence of services. The agent can either map these services to existing BDI plans or invoke the corresponding object services directly.

Of course, the external planner requires a representation of the BDI plans and (declarative) goals in order to explore new solutions. In ISReal, we therefore specify declarative goals g_d for every goal event e_g in the BDI planner describing the facts an agent wants to achieve when e_g triggers. Furthermore, every BDI plan is described as a semantic service and stored in the service list S_a of his LocalSE (see Figure 3). Whenever a goal event g_e triggers and no BDI plan is applicable, the agent invokes the SCP of his LocalSE. The SCP gets (i) the knowledge base KB_a as first, (ii) the declarative goal g_d of the agent transformed to an OWL2-DL ontology as second ontology, and (iii) all known object services S_a as input. As output the SCP returns a sequence of operations (the plan) and a binding that maps the parameters of each operation to facts in the knowledge base. Using this binding, the agent can execute the services. If the SCP fails to find a plan, then the agent fails to achieve the goal. If the SCP finds a plan, the agent has to execute the operations of that plan. For this purpose, the agent checks for all operations o in the plan whether o is (i) a core service implemented by the agent itself or (ii) an object service that is provided by some object. In the case of a core service, the agent maps the service to a BDI plan and executes the applicable plan. In the case of an object service, the agent fetches the grounding information and invokes the appropriate implementation. Using this combined planning approach the agent is able to find new solutions that are not directly encoded in its plan library. As agent systems we use Jack[7] and Jadex[8]. The SCP has been realized with OWLS-XPlan 2.0, a Semantic Web service composition planner for OWL-S 1.1 services [12].

3 Example Scenario

In the following, we show how the ISReal platform can be used. As introduced in Section 2.2, the user can query the GlobalSE and agents in the virtual world with queries a) for object reasoning (e.g. "What is this object?"), b) concept reasoning (e.g. "Are these two objects equivalent?"), c) relational reasoning (e.g. "What is the relation between the red lamp and the machine?"). Furthermore, the user can execute the functionality of the scene (semantic objects), trigger basic actions of the agent, or formulate a goal (e.g. *"Show me what to do in order to get the machine running again."*, *"Open the door!"*) that the agent should achieve.

The scenario considers a new employee whose task is to learn how to use a machine called *Smart Factory* assisted by the ISReal platform. The *Smart Factory* fills pills into cups that are on a transportation belt (see Figure 4). The virtual *Smart Factory* is a simulation of a physical real-world model that is used for demonstrations. In the following the user interacts directly with the machine to turn it on (Section 3.1). After an error occurred he asks the agent for more information about the machine (Section 3.2) and finally let the agent show him how to resolve the error (Section 3.3).

In this simple scenario that we use for demonstration purposes, the user can explore a virtual work room, containing the *Smart Factory* and a virtual assistant controlled by an agent. Please note that the transformation from and to natural language is not in the

[7] `http://www.agent-software.com/index.html`
[8] `http://jadex.informatik.uni-hamburg.de`

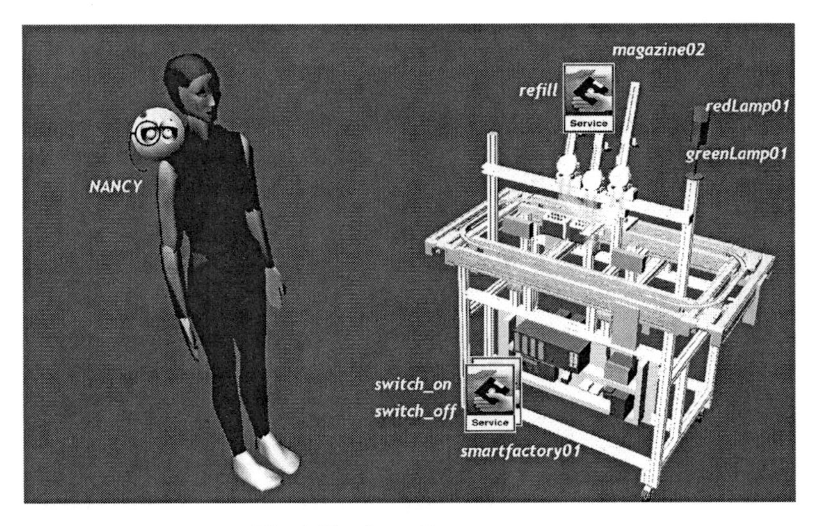

Fig. 4. The Smart Factory use-case

scope of this paper. Therefore, we use only simple template-based transformations. Furthermore, we assume that the user only queries the agent who is familiar with the *Smart Factory* and therefore has an updated local knowledge base that contains all necessary facts.

3.1 Semantic Object Interaction

The user starts exploring the virtual work room as an interactive immersive environment. He recognizes a control panel at the *Smart Factory*. The user's intention is to put the *Smart Factory* into operation by pressing the button on the control panel. Via the UE he can interact with the *Smart Factory* (as a 3D object) and get all object services the *Smart Factory* provides. These services are *switch_on* and *switch_off*. They are triggered through pressing appropriately labeled 3D buttons on the machine. The user chooses the *switch_on* services and invokes it. Figure 5 (left) depicts a sequence diagram of the whole object service execution. The *switch_on* service expects as input parameter an instance of *Machine* ($?self$). Since $?self$ is the instance of the semantic object that provides this service, the parameter is derived automatically. The precondition is defined as follows: *Machine(?self)* $\wedge$ *pluggedIn*($?self, ?x$) $\wedge$ *PowerSupply*($?x$). The effect consists of two conditional effects, (a) one for a failed checkup leading to an error state (represented by an active red lamp) and (b) one for a successful checkup leading to the operation state (represented by an active green lamp). (a) is defined as: *consistsOf*($?self, ?y$) $\wedge$ *Magazine*($?y$) $\wedge$ *Empty*($?y$) $\wedge$ *triggers*($?y, ?z$) $\wedge$ *RedLamp*($?z$), having the effect *Active*($?self$), *Active*($?z$). (b) is defined as: *consistsOf*($?self, ?y$) $\wedge$ *Magazine*($?y$) $\wedge$ *Full*($?y$) $\wedge$ *triggers*($?y, ?z$) $\wedge$ *GreenLamp*($?z$), having the effect *Active*($?self$), *Active*($?z$). Let's assume the *Smart Factory* is provided with a power supply but the magazine is empty (which is a fact the user cannot see). After the user turned on the machine the variable binding after checking the conditions is, e.g. ($?self, smartfactory01$), ($?x, powerSocket01$), ($?y, magazine02$), ($?z, redLamp01$). Based on

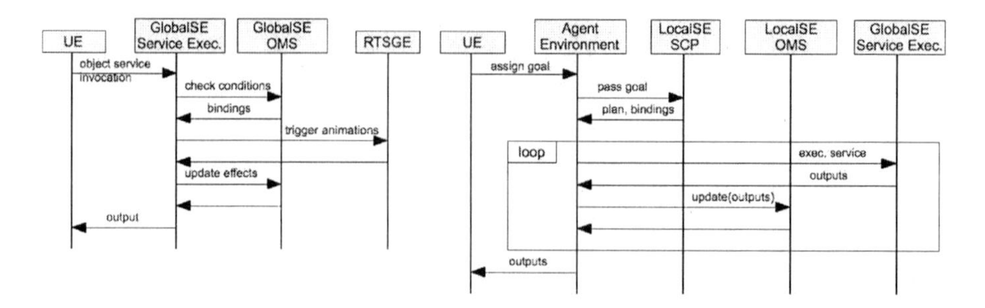

Fig. 5. Sequence diagram for object service execution (left). Sequence diagram for goal assignment (right).

these bindings, the facts *Active*($smartfactory01$), *Active*($redLamp01$) are derived. The appropriate animations are triggered (given by the grounding information of the service that uses endpoints provided by the RTSGE) and the effect is written to the GlobalSE.

3.2 Information Request

Using the ISReal platform the user was able to put the *Smart Factory* into operation but the machine is in an obvious error state (the signal lamp switches to red). To figure out the problem the user gives following query to the agent assistant: *"What is the relation between the red lamp and the Smart Factory?"* The agent transforms the natural language query into a list of entities and passes it to its LocalSE. The LocalSE handles the query as a relational query (cf. type c) in Section 2.2. By computing the Steiner tree between the entities (*'redLamp01'*, *'smartfactory01'*) a graph holding the answer is computed: *"The magazine is a part of the Smart Factory that triggers the red lamp."* This answer is produced from the tree given by: *consistOf(smartfactory01, magazine02), triggers(magazine02, redLamp01)*. In order to get more information about the magazine, the user sends a second query to the agent: *"What can you tell me about this magazine?"* After the agent transformed the question into a simple SPARQL query, the query is passed to the agent's LocalSE. As result, the agent returns a set of statements that can be summarized in natural language: *"The name of this object is magazine02. It is a Magazine and Empty. It is part of the smartfactory01 and triggers redLamp01 and greenLamp01."*

3.3 Goal Assignment

After the user gathered information about the relation of the lamps and the machine, he wants to put the machine in its running state (active green light) and asks the agent: *"Show me what to do in order to get the machine running again."* The agent processes the query and handles it as a planning task. Figure 5 (right) shows how the agent processes the query. The user's query is transformed to a declarative goal description: *Active*($smartfactory01$) $\wedge$ *Active*($greenLamp01$). We already introduced the service *switch_on* above. Additionally the agent knows an object service *refill* of the magazine. This service has the precondition *Magazine*$(?self) \wedge$ *Empty*$(?self) \wedge$ *isPartOf*$(?self, ?x) \wedge$ *Machine*$(?x) \wedge$ *InActive*$(?x)$ and the non-conditional effect *Full*$(?self)$. Please note, that in this example (for the sake of simplicity) we neglect the pills that are actually filled into the magazine. The service *switch_off* checks whether the machine $(?self)$ is *Active*$(?self)$ and sets it to *InActive*$(?self)$.

Using the SCP the agent determines that it has to use the three services in the sequence: (i) *switch_off*, (ii) *refill*, (iii) *switch_on* with the according variable bindings for the input parameters (i) $(?self, smartfactory01)$, (ii) $(?self, magazine02)$, (iii) $(?self, smartfactory01)$ in order to achive his goal. The first service sets the *Smart Factory* to inactive, which is necessary to fulfill the precondition of the service *refill*. Then the second service can be used to trigger the conditional effect (b) at the service *switch_on* that leads to a state where the goal description is fulfilled (see above). Invoking the plan for every service (cf. loop in Figure 5 (right)), the agent either calls the service execution of the GlobalSE, in case of an object service, or triggers the corresponding BDI plan in case of a service describing such a BDI plan. The outputs are used to update the LocalSE of the agent and returned to the user. As visible effect, the agent walks to the machine, unmounts the pill magazine, refills it, mounts it again, and switches the machine back on. In their plans the agents can make use of other services (not discussed here) for navigating and moving in their environment, perform animated actions (like switching a switch on, turning a knob), and the environment will have physical properties and behavior using a physics engine.

4 Related Work

The central idea of the ISReal platform is to use Semantic Web technology to semantically enrich the pure geometric data of the scene to enable intelligent agents to interact with their environment. However, in a different context semantic annotation of 3D environments has been previously discussed, e.g. in [17], [3], and [13]. Kalman et al. [10] proposed the concept of *smart objects* which is a geometrical object enriched with meta information about how to interact with it (e.g. grasp points). Abaci et al. [1] extended smart objects with PDDL data in order to plan with them.

There exist several paper that discuss how classical planning can be incorporated into BDI agent systems [24][5]. Most papers aim on improving the look-ahead behavior of BDI agents and on explicit invokation of a classical planner for sub-goals. We use a SCP to enable the agent to handle (OWL-based) declarative user goals and use the capability of the BDI agent to work in dynamic and partially observable environments to support the SCP.

Lewis et al. [14] motivates the use of computer game engines in scientific research. For example, [15] presents architectural considerations for real-time agents that have been used in the computer game F.E.A.R.. [4] presents an approach how to connect a BDI agent system to the Unreal engine. [16], [22], and [2] present a multi-agent system for general-purpose intelligent virtual environment applications that consists of three types of conceptually discrete components: worlds, agents, and viewers. [23] presents SimHuman consisting of two basic modules: a 3D visualization engine and an embedded physically based modeling engine. Agents can use features such as path finding, inverse kinematics, and planning to achieve their goals. [8] proposes an approach to 3D agent-based virtual communities in which autonomous agents are participants in VRML-based virtual worlds using the VRML *External Authoring Interface* (EAI). The distributed logic programming language DLP has been extended to support 3D agent-based virtual communities. However, what makes ISReal significantly differ from all existing systems is its integration of virtual worlds, Semantic Web and agent technology into one coherent platform for semantically-enabled agent-assisted 3D simulation of realities.

Tutor agents in virtual realities are previously described in [19]. Where agent tutors also show users how to handle and operate devices. However we see ISReal in a wider application field. A tutoring scenario is just one of many possible applications of the ISReal platform and used in this paper as example. ISReal can also be used for virtual prototyping, behaviour and traffic simulations, product demonstrations and many more.

5 Conclusions

Semantically enabled simulated realities as considered by this paper have a great potential for commercial applications. Highly realistic prototypes of buildings, production lines, etc. support companies in the early decision making process and help to avoid expensive error corrections. The possibility of training employees on virtual production lines, long before the actual plant has been built, saves time and money.

The realization of a platform for deploying semantically enabled simulated realities is a cross-discipline endeavor and requires input from various research areas such as computer graphics for realistic animations and rendering of a scene, artificial intelligence for convincing behavior of the entities, and Semantic Web for adding meaning to the purely geometrical objects and describing their functionality. Beside the conceptual integration, requirements such as modularity, scalability, and extensibility are the main drivers for the architecture of the ISReal platform.

In this paper we discussed the basic architecture of the ISReal platform. Key aspects of the future work are on the scalability of core techniques in terms of scene complexity, semantic expressiveness, natural animations, real time performance, and use of many-core hardware.

References

1. Abaci, T., et al.: Planning with Smart Objects. In: WSCG SHORT Papers Proceedings, pp. 25–28. UNION Agency - Science Press (2005)
2. Anastassakis, G., et al.: Multi-agent Systems as Intelligent Virtual Environments. In: Baader, F., Brewka, G., Eiter, T. (eds.) KI 2001. LNCS (LNAI), vol. 2174, pp. 381–395. Springer, Heidelberg (2001)
3. Bilasco, I., et al.: On Indexing of 3D Scenes Using MPEG-7. In: Proc. 13th Annual ACM Intl. Conf. on Multimedia, pp. 471–474. ACM Press, New York (2005)
4. Davies, N.P., et al.: BDI for Intelligent Agents in Computer Games. In: Proc. of the 8th Intl. Conf. on Computer Games: AI and Mobile Systems (CGAMES 2006), University of Wolverhampton (2006)

5. de Silva, L., et al.: First Principles Planning in BDI Systems. In: Proc. of the 8th Intl. Conf. Autonomous Agents and Multiagent Systems (2009)
6. Fensel, D., et al.: Towards LarKC: a Platform for Web-scale Reasoning. IEEE Computer Society Press, Los Alamitos (2008)
7. Ghallab, M., et al.: Automated Planning, Theory and Practice. Morgan Kaufmann Publishers Inc., San Francisco (2004)
8. Huang, Z., et al.: 3D Agent-based Virtual Communities. In: Proc. of the 7th Intl. Conf. on 3D Web Technology, pp. 137–143. ACM, New York (2002)
9. Kasneci, G., et al.: STAR: Steiner Tree Approximation in Relationship-Graphs. In: Proc. of 25th IEEE Intl. Conf. on Data Engineering, ICDE (2009)
10. Kallmann, M.: Object Interaction in Real-Time Virtual Environments. PhD thesis, École Polytechnique Fédérale de Lausanne (2001)
11. Kiryakov, A., et al.: OWLIM - a Pragmatic Semantic Repository for OWL. In: Dean, M., Guo, Y., Jun, W., Kaschek, R., Krishnaswamy, S., Pan, Z., Sheng, Q.Z. (eds.) WISE 2005 Workshops. LNCS, vol. 3807, pp. 182–192. Springer, Heidelberg (2005)
12. Klusch, M., et al.: Semantic Web Service Composition Planning with OWLS-Xplan. In: 1st Intl. AAAI Fall Symposium on Agents and the Semantic Web (2005)
13. Kleinermann, F., et al.: Adding Semantic Annotations, Navigation Paths and Tour Guides for Existing Virtual Environments. In: Wyeld, T.G., Kenderdine, S., Docherty, M. (eds.) VSMM 2007. LNCS, vol. 4820, pp. 100–111. Springer, Heidelberg (2008)
14. Lewis, M., et al.: Game Engines in Scientific Research. Communcation of the ACM 45(1), 27–31 (2002)
15. Orkin, J.: Agent Architecture Considerations for Real-Time Planning in Games. In: Proc. of the Artificial Intelligence and Interactive Digital Entertainment Conf. (AIIDE). AAAI, Menlo Park (2005)
16. Panayiotopoulos, T., et al.: An Intelligent Agent Framework in VRML worlds. In: Advances in Intelligent Systems: Concepts, Tools and Applications, pp. 33–43 (1999)
17. Pittarello, F., et al.: Semantic Description of 3D Environments: A Proposal Based on Web Standarts. In: Proc. of Intl. Web3D Conf. ACM Press, New York (2006)
18. Rao, S., et al.: BDI Agents: From Theory to Practice. In: Proc. of the 1st Intl. Conf. on Multi-Agent Systems, pp. 312–319. AAAI Press, Menlo Park (1995)
19. Rickel, J., Johnson, W.L.: Virtual Humans for Team Training in Virtual Reality. In: Proc. of the 9th Intl. Conf. on AI in Education, pp. 578–585. IOS Press, Amsterdam (1999)
20. Rubinstein, D., et al.: RTSG: Ray Tracing for X3D via a Flexible Rendering Framework. In: Proc. of the 14th Intl. Conf. on 3D Web Technology (Web3D Symposium 2009) (2009) (to appear)
21. Russell, S., et al.: Artificial Intelligence: A Modern Approach, 2nd edn. Prentice-Hall, Englewood Cliffs (2007)
22. Vosinakis, S., et al.: DIVA: Distributed Intelligent Virtual Agents, pp. 131–134. University of Salford (1999)
23. Vosinakis, S., et al.: SimHuman: A platform for real-time virtual agents with planning capabilities. In: de Antonio, A., Aylett, R.S., Ballin, D. (eds.) IVA 2001. LNCS (LNAI), vol. 2190, pp. 210–223. Springer, Heidelberg (2001)
24. Walczak, A., et al.: Augmenting BDI Agents with Deliberative Planning Techniques. In: Proc. of the 5th Intl. Workshop on Programming Multiagent Systems (PROMAS-2006) (2006)
25. Winikoff, M., et al.: Declarative Procedural Goals in Intelligent Agent Systems. In: Proc. of the 8th Intl. Conf. on Principles of Knowledge Representation and Reasoning (KR 2002), pp. 470–481 (2002)

Auction Design and Performance: An Agent-Based Simulation with Endogenous Participation

Atakelty Hailu[1], John Rolfe[2], Jill Windle[2], and Romy Greiner[3]

[1] School of Agricultural and Resource Economics M089
University of Western Australia, 35 Sitrling Highway, Crawley WA 6009, Perth, Australia
[2] Centre for Environmental Management
Central Queensland University, Rockhampton, QLD 4702, Australia
[3] River Consulting, 68 Wellington St, Townsville, QLD 4812, Australia

Abstract. This paper presents results from computational experiments evaluating the impact on performance of different auction design features. The focus of the study is a conservation auction for water quality where auctions are used to allocate contracts for improved land management practices among landholders bidding to provide conservation services. An agent-based model of bidder agents that learn using a combination of direction and reinforcement learning algorithms is used to simulate performance. The auction design features studied include: mix of conservation activities in tendered projects (auction scope effects); auction budget levels relative to bidder population size (auction scale effects); auction pricing rules (uniform versus discriminatory pricing); and endogeneity of bidder participation. Both weak and strong bidder responses to tender failure are explored for the case of endogeneity in participation. The results highlight the importance of a careful consideration of scale and scope issues and that policy-makers need to consider alternatives to currently used pay-as-bid or discriminatory pricing fromats. Averaging over scope variations, the uniform auction can deliver substantially higher budgetary efficiency compared to the discriminatory auction. This advantage is especially higher when bidder participation decisions are more sensitive to auction outcomes.

Keywords: Computational economics, Auction design, Agent-based modelling, Conservation auctions, Procurement auctions.

1 Introduction

Computational experiments are key to testing auction designs. Computational experiments can be used to thoroughly evaluate a variety of design features. Experimental studies are useful for evaluating auction outcomes but without the extreme behavioural assumptions needed to derive results from analytical approaches. In some cases, experiments are also the only means available for evaluating auction outcomes under dynamic settings, e.g. the case of enogeneous participation in auctions.

This paper presents results from an agent-based modelling study undertaken as a component of federally-funded auction trial project undertaken in Queensland, Australia. The trial known as the Lower Burdekin Dry Tropics Water Quality Improvement Tender Project was developed with the aim of exploring issues of scope and scale in tender

J. Filipe, A. Fred, and B. Sharp (Eds.): ICAART 2010, CCIS 129, pp. 214–226, 2011.

design [1]. It involved the conducting of a trial auction, an experimental workshop and this agent-based modelling (or computational experiments) component. The objectives of the project were to assess whether and how increases in scale and scope of a tender may lead to efficiency gains and investigate the extent to which these gains might be offset as a result of higher transaction costs and/or lower participation rates.

Auctions exploit differences in opportunity costs. Therefore, one would expect budgetary efficiency to be enhanced if tenders have wider scale and scope. However, auctions with wider coverage might involve complex design as well as higher implementation costs. Auctions with wider scope might also attract lower participation rates. An evaluation of these trade-offs is essential to the proper design of conservation auctions.

The agent-based modelling study presented here focused on evaluating the impact of auction scale and scope changes in the presence of bidder learning. The study also explored the impact on performance of the use of an alternative auction pricing format, namely, uniform pricing, which pays winners the same amount for the same environmental benefit. These auction design features are evaluated, first, by ignoring the possible ramifications of auction outcomes on the tendency to participate and, second, by allowing bidder participation to be affected by tender experience.In summary, the agent-based modelling study simulated aucton environments and design features that could not be explored through the field trials.

The paper is organized as follows. The next section presents the case for agent-based modelling in the design of auctions. Agent-based computational approaches are being increasingly utilized in the economics literature to complement analytical and human-experimental approaches [2,3]. The distinguishing feature of agent-based modelling is that it is based on experimentation or simulation in a computational environment using an artificial society of agents that emulate the behaviours of the economic agents in the system being studied [3]. These features make the technique a convenient tool in contexts where analytical solutions are intractable and the researcher has to resort to simulation and/or in contexts where modelling outcomes need to be enriched through the incorporation of agent heterogeneity, agent interactions, inductive learning, or other features. Section 3 presents the auction design features explored in the study. These include budget levels, scope of conservation activities, endoegeneity of participation levels, and two alternative pricing formats. Simulated results are presented and discussed in Section 4. In Section 5 we present results from a simulation where bidders are assumed to be more sensitive to bid failures in their participation decisions. The final section summarizes the study and draws conclusions.

2 Agent-Based Auction Model

Auction theory has focused on optimal auction design, but its results are usually valid only under very restrictive assumptions on the auction environment and the rationality of the players. Theoretical analysis rarely incorporates computational limitations, of either the mechanisms or the agents [4]. Experimental results [5,6] demonstrate that the way people play is better captured by learning models rather than by the Nash-Equilibrium predictions of economic theory. So, in practice, what we would observe is people learning over time, not people landing on the Nash equilibrium at the outset of the game. The need to use alternative methods to generate the outcomes of the learning processes has led to an increasing use of human experimental as well as computational approaches such as agent-based modelling.

Our agent-based model has two types of agents representing the players in a procurement auction, namely one buyer (the government) and multiple sellers (landholders) competing to sell conservation services. Each landholder has an opportunity cost that is private knowledge. The procuring agency or government agent has a conservation budget that determines the number of environmental service contracts.

Each simulated auction round involves the following three major steps. First, landholder agents formulate and submit their bids. Second, the government agent ranks the submitted bids based on their environmental benefit score to cost ratios and selects winning bids. The number of successful bids depends on the size of the budget and the auction price format. In the case of discriminatory or pay-as-bid pricing, the government agent allocates the money starting with the highest ranked bidder until the budget is exhausted. In a uniform pricing auction, all winning bidders are paid the same amount per environmental benefit. The cutoff point (marginal winner) for this auction is determined by searching for the bid price that would exhaust the budget if all equally and better ranked bids are awarded contracts. Third, landholder agents apply learning algorithms that take into account auction outcomes to update their bids for the next round. In the very initial rounds, these bids are truthful. In subsequent rounds, these bids might be truthful or involve mark-ups over and above opportunity costs.

Bids are updated through learning. Different learning models have been developed over the last several decades and can inform simulated agent behaviour in the model. A typology of learning models presented by [6] shows the relationship between these learning algorithms. This model combines two types of learning models: a direction learning model [7,8] and a reinforcement learning algorithm [9,10]. These two algorithms are attractive for modelling bid adjustment because they do not require that the bidder know the forgone payoffs for alternative strategies (or bid levels) that they did not utilize in previous bids.

Learning direction theory asserts that ex-post rationality is the strongest influence on adaptive behaviour [11,12]. According to this theory, more frequently than randomly, expected behavioural changes, if they occur, are oriented towards additional payoffs that might have been gained by other actions. For example, a successful bidder, who changes a bid, is likely to increase subsequent bid levels. Reinforcement learning [13,5] does not impose a direction on behaviour but is based on the reinforcement principle that is widely accepted in the psychology literature. An agent's tendency to select a strategy or bid level is strengthened (reinforced) or weakened depending upon whether or not the action results in favourable (profitable) outcomes. This algorithms also allows for experimentation (or generalization) with alternative strategies. For example, a bid level becomes more attractive if similar (or neighbouring) bid levels are found to be attractive.

In our model, we combine the two learning theories because it is reasonable to assume that direction learning is a reasonable model of what a bidder would do in the early stages of participation in auctions. These early rounds can be viewed as discovery rounds where the bidders, through their experience in the auctions, discover their

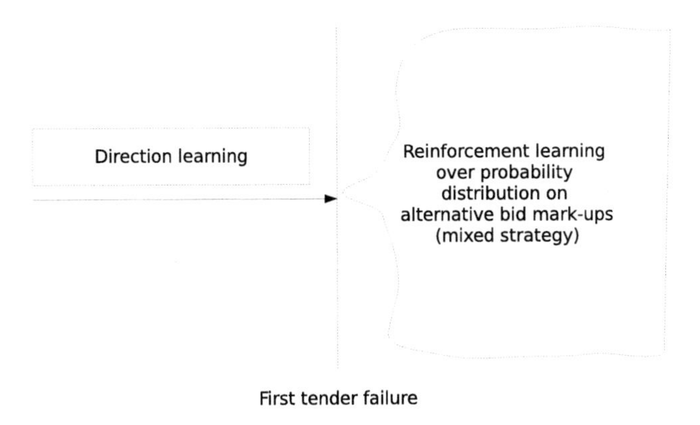

Fig. 1. Combined learning algorithms: direction learning for exploratory phase and reinforcement learning for mixed strategy choices

relative standing in the population of participants. It would thus be reasonable to assume that successful bidders would probabilistically adjust their bids up or leave them unchanged. However, this process of directional adjustment would end once the bidder fails to win in an auction. At this stage, the bid discovery phase can be considered to have finished and the bidder to be in a bid refinement phase where they chose among bid levels through reinforcement algorithm, with the probability or propensity of choice initially concentrated around the last successful bids utilized in the discovery phase (see Figure 1). Further details on the attributes and implementation of the reinforcement algorithm are provided in the paper by Hailu and Thoyer on multi-unit auction pricing formats [10].

3 Design of Experiments

In all experiments reported in this paper, a population of 100 bidding agents is used. This number is chosen to be close to the the number of actual bids (88) submitted in the Burdekin auction trial [14]. The opportunity cost of these bidding agents depend on the mix of water quality enhancing activities that are included in their projects. This dependence of opportunity cost on project activities is determined based on the relationship between costs and activities implied in the actual bids. A data envelopment analysis (DEA) frontier is constructed from the actual bids to provide a mechanism for generating project activities that extrapolate those observed in the actual trial. For a given bundle of conservation activities, this frontier provides the best possible cost estimate. This cost estimate is then adjusted by a random draw from the cost efficiency estimates obtained for the actual bids.

The nature of the bidder opportunity costs, the budget, payment formats and the responses of bidder agents to auction outcomes are varied so that results are generated for experiments that combine these features in different configurations. Further details on these design variations are provided below.

3.1 Scope of Conservation Activities

Changes in scope of the auction are imitated through variations in the coverage of water quality improving activities undertaken by the bidding population. These activities are nitrogen reduction, pesticide reduction and sediment reduction. In the actual bids, the sediment reducing projects came almost entirely from pastoralists while the nitrogen and pesticide reducing activities came from sugar cane growers. The shares of nutrient, pesticide and sediment reduction in the environmental benefits (EBS) score value were varied between 0 and 1 to generate a mix of activities covering a wide range of heterogeneity in projects. For example, to simulate auction performance for a case where the range of allowed activities is on average a 50/50 contribution from nitrogen and pesticide reducing activities, a random population of 100 shares is drawn from a Dirichlet distribution centered at (0.5,0.5). This is then translated into nitrogen, pesticide and sediment quantities using the relationship between environmental benefit scores and reduction activities employed for the actual auction[1].

3.2 Auction Budgets

Two auction budgets are used, $600K and $300K. The first level represents approximately the actual budget used in the field trials, while the second budget indicates a higher level of competition or "degree of rationing" that can be achieved by increasing the scale of participation.

3.3 Endogeneity of Bidder Participation

Auction performance is likely to be influenced by the dynamics of participation. Unless auctions are organized differently (e.g. involving payments that maintain participation levels), one would expect some of the bidders to drop out as a result of failure to win contracts. Therefore, we carry out computational experiments for a case where bidders are assumed to participate even in the case of failure and also for a case where bidders drop out, with some probability. In the second case, if a bidder fails to win contracts, the probability of dropping out increases up to a maximum of 0.5. However, a simulation that allows for a one-way traffic (i.e. exit) would not take into account the fact that the auction can become more attractive as bidders drop out and competition declines. Therefore, we allow for both exit and re-entry into the auctions. Re-entry by inactive bidders occurs with a probability that is increasing with the average net profit participating bidders are making from their contracts.

3.4 Auction Price Format

The choice of payment formats has been an interesting research topic in auction theory. Theory offers guidance on choices in simple cases but has difficulty ranking formats

[1] The method used is an approximation to the actual procedure based on a regression of reduction activity levels and EBS scores. This was done because the actual scoring involved adjustments that credited projects with extra points for other aspects of the project besides nitrogen, pesticide or sediment reduction.

in more complex cases, whether the complexity arises from the nature of the bidder population or the nature of the auction (e.g. multi-unit auction [10]). The Burdekin field trial, like most Market-based instrument (MBI) trials conducted to date, has used a pay-as-bid or discriminatory pricing format. In the agent-based simulations, this payment format is compared to the alternative format of uniform pricing where winning bidders would be paid the same per unit of environmental benefit.

The auction design features discussed above are varied to generate and simulate a range of auction market experiments. Each auction experiment is replicated 50 times to average over stochastic elements involved in the generation of opportunity cost estimates and the probabilistic bid choices that are employed in the learning algorithms. Results reported below are averages over those 50 replications.

4 Auction Performance

The key finding from these experiments is that outcomes vary greatly with the details of the auction and the activities covered in its scope. The results are summarized below.

4.1 Scope Effects

The performance of the auction as measured by benefits per dollar spent is dependent on the scope of conservation activities that are eligible. The benefits per dollar range from a low of 1.91 environmental benefit scores (EBS) per million dollars to a high of 3.62. An increase in the share of sediment reduction reduces benefits obtained; the benefits per million dollars are always less than 3.0 when there is a positive average share of sediment reduction activities. The benefits improve with improvements in the share of pesticide reduction activities.

In Figure 2, the benefits are plotted against the average shares of nitrogen and pesticide contribution to water quality in the projects. The horizontal axis in the figure indicates the relative contribution of nitrogen reduction while the y-axis (diagonal line) indicates the share of pesticide. The residual contribution is the share of sediment reduction. Points at the bottom-right corner have higher shares of nitrogen as opposed to points closer to the top-left corner of the diagram. Points on the top-left to bottom-right diagonal line correspond to a constant combined contribution from nitrogen and pesticide reduction (i.e. they represent a constant sediment contribution). Therefore, points further to the left of the diagonal line stretching from the top-left hand corner to the right-bottom corner include higher shares of sediment reduction. The dots in the figure show the deterioration in benefits per dollar as the scope of the auction covers more and more sediment reduction. Looking at variations other than sediments, the changes in the heights of the lines increasing towards the top-left corner (from the bottom-right) indicate the increase in benefits that occur as the share of pesticide increase at the expense of nitrogen reduction with sediment reduction excluded. See Table 4 in the Appendix for further details on the results.

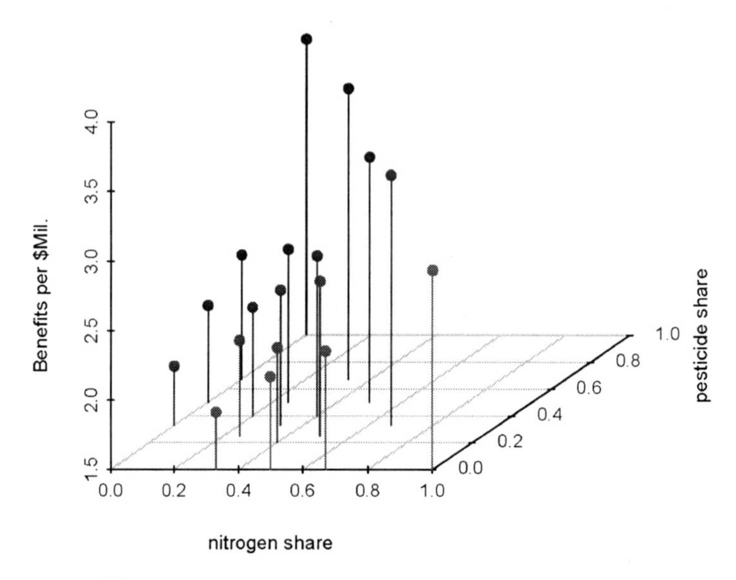

Fig. 2. Conservation activity mix and auction performance

4.2 Budget Levels and Auction Efficiency

For a discriminatory auction with 100 bidders, halving the budget from $600,000 to $300,000 has a large effect on the performance of the auction. The benefit to cost ratios are more than 50% higher for the auction with the $300,000 budget. See Figure 3 where the benefits per million dollars for the auction with the two budgets are shown. The different points represent results for the different mixes in activities discussed above. All the points fall between the two dotted lines which represent ratios of 2 and 1.5 between the values on the y and x axes.

The benefits of the higher degree of rationing are similarly strong for all other experiments where other features of the auction are varied (e.g. pricing formats). Considering all cases together, the benefits per dollar from the auction with a budget of $300,000 can be between 40 and 100% above those where the budget is $600,000. On average, the benefits were 67% higher. The benefits of the higher degree of rationing (or increased competition for a given budget) are notably higher when the scope of the auction is such that it covers higher cost activities.

4.3 Participation and Efficiency

Results for repeated auctions where bidders stay active even after bids are unsuccessful are only marginally higher than for cases where bidders drop out (and re-enter). The weakness of these results seems to be due to the way the participation rules are formulated in the simulation, being biased in favour of bidding. For example, in our experiments, a bidder who just lost in an auction might participate in the next one with a probability of at least 0.5 depending on their net revenue from the contracts in previous rounds. In practice, bidders might be more responsive to bid failures and the results reported here would understate the importance of investments in landholder participation.

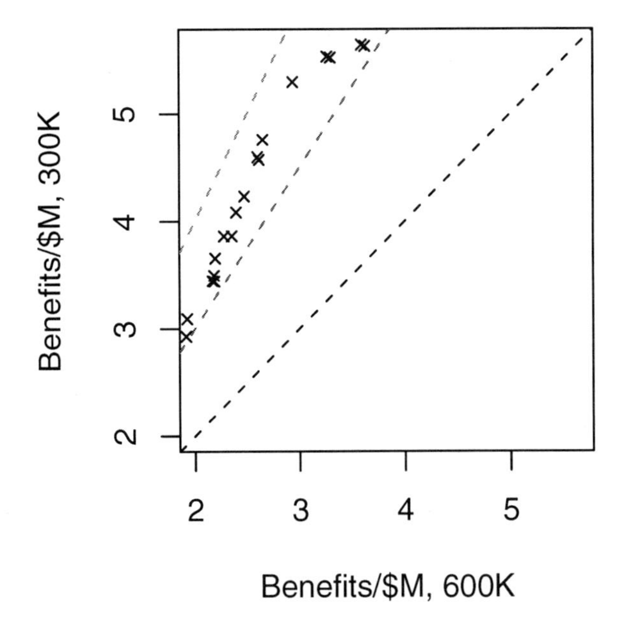

Fig. 3. Benefits per million dollar for auctions with budgets of $300,000 and $600,000

4.4 Uniform versus Discriminatory Pricing

Results for a uniform pricing format where every winning bidder gets paid the same for the same environmental benefit are generated and compared with those obtained under simulation conducted for discriminatory pricing. The key results are summarized in Table 1 where we report the ratios of values from the uniform price auction to those from the discriminatory price auction for both budget levels and activity threshold specifications. A "yes" value for activity threshold or endogenous participation indicates that the results in the row are for simulation where bidders drop out as a result of failure to win. Each row reports the results for the corresponding budget and activity threshold averaged over all the activity scope variations covered in the experiments.

In terms of performance, the results reported in the third column indicate that a uniform auction delivers benefits that are at least 25% higher than those obtained under a discriminatory auction when performance measures are averaged over all scope variations. The relative benefits of the uniform auction are highest when competition is tight (budget of $300,000) and for the case where bidders tend to drop out when they fail to win contracts.

The higher benefits obtained under the uniform price auction are the result of the fact that it encourages less overbidding than the discriminatory pricing auction. This can be seen in the last two columns of Table 1 where the figures indicate the ratios of the bid prices from uniform to those from discriminatory prices for the highest (column 4) and for the lowest ranked (column 5) bidders in the auction. The first row results show that the benefit price (bid to benefit ratio) of the highest ranked and the lowest ranked bidders under the uniform auction were only 52 and 45 percent of the corresponding figures for the discriminatory auction. In other words, the bids under the uniform auction are

Table 1. Ratio of results from uniform auction to those from discriminatory auction

Budget (000s)	Endogeneous participation	Ratio of benefits per dollar	Ratio of first ranked bid price	Ratio of last ranked bid price
300	no	1.30	0.52	0.45
300	yes	1.36	0.49	0.37
600	no	1.25	0.29	0.51
600	yes	1.28	0.28	0.43

Table 2. Results from a regression of environmental benefits per dollar on auction design features

Variable	Coef. estimate	t-stat
(Intercept)	0.00811	60.987
Share of Pesticide	0.00057	4.309
Share of Sediment	-0.00322	-20.727
Budget (dummy, with 600K equal to 1)	-0.00001	-31.243
Discriminatory pricing (dummy)	-0.00101	-16.253
Activity threshold (dummy, 0 if bidders do not drop out)	-0.00007	-1.169
R-squared	0.93	

more truthful or involve less overbidding. This is because the bidder does not have the incentive (unless they are the marginal bidder) to misrepresent their bid under the uniform auction; subject to winning, a bidder's payoff does not depend on their own bid but on that of the most expensive winner. The disparity between the bids of the best ranked bidders are highest for the higher budget auction.

In summary, the performance of the auction is dependent on the variations considered in the computation experiments. To investigate the individual contributions of the different auction features, the benefits per dollar are regressed against variables representing the design features. These results are presented in Table 2 which highlights the importance of the auction design features, with 93% of the benefit per dollar variations explained by the design features. Increases in budgets (for a given pool of bidders), share of sediment reduction contribution, discriminatory pricing and the presence of activity threshold in a bidders decision to participate all contribute towards lower benefit value per dollar spent. Higher shares for pesticide reduction activities, on the other hand, increase efficiency.

5 Effect of Higher Sensitivity to Bid Failure

The results on the impact of participation heterogeneity discussed above indicate that outcomes for auctions where bidders drop out due to bid failure were only marginally better than outcomes for auctions where bidders continue to participate regardless of bid outcomes. As indicated there, the weakeness of the result relating to endogeneous participation was observed probably because bidder response to bid failure was weak. In particular, in the experiment above, the probability of a bidder dropping out of the tender increases upto a maximum of 0.5. We test the effect of a stronger response by allowing the probability of participation to be even lower. Specifailly, we assume that

Table 3. Results from a regression of environmental benefits per dollar on auction design features: The case of more sensitive bidder participation

Variable	Coef. estimate	t-stat
(Intercept)	0.00304	64.541
Share of Pesticide	0.00061	14.071
Share of Sediment	-0.00124	-22.565
Budget (dummy, with 600K equal to 1)	-0.00001	-25.285
Discriminatory pricing (dummy)	-0.00048	-21.269
Activity threshold (dummy, 0 if bidders do not drop out)	-0.00015	-6.462
R-squared	0.92	

the probability of participation, p, for a bidder as a function of average auction income in the previous three rounds, x, takes the following form:

$$p = 1 - \frac{1}{e^x} \tag{1}$$

Unlike in the previous formulation, the probability of participation (in the next round) by a bidder who was not selected in all three previous rounds (and, therefore, earned zero incomes) is zero, rather than 0.5.[2] This formulation takes effect only if the bidder lost in the previous round, i.e. the probility of participaiton by a winning bidder is always unity.

We similarly adjust the probability of re-entry by a bidder so that the value has a lower bound of zero (if average auction incomes are zero) rather than 0.5. The probability of an inactive agent participating in the next auction round as a function of the average income earned by participating (and successful) bidders, y, formulated as follows:

$$p = 1 - e^{-y} \tag{2}$$

In summary, in this experiment, bidders are more likely to leave the auction if they become unsuccessful and they require that the average income earned by successful bidders be higher for them to be drawn back into participation.

Regression model results explaining benfits per dollar as a function of auction design features under the new participation formulations are reported in Table 3. All features have statistically significant effects on auction outcomes, at the 95 or 99% level of significance. Compared to the results reported in Table2 above, the activity threshold variable indicating whether bidders drop out or not is not statistically significant; and, it coefficient sign indicates that aucton outcomes deteriorate if bidder drop out (and re-enter).

The effects of these increased bidder sensitivities affect the relative performance of the two auction pricing format. The benefits per dollar obtained from the uniform pricing auction are higher. This pricing format yields about 40% more benefits compared to a discriminatory auction for both the low and high budget levels.

[2] Previously, the probability was formulated as $p = 1/(1 + e^{-x})$.

6 Conclusions

This study conducted computational experiments to evaluate the impact on auction performance of several design features, including: the scope of water quality improving activities allowed in projects; the scale of the auction as measured by the budget size relative to participating bidder numbers; and the choice of pricing format. These design features were conducted for two cases of bidder responses to failures in auctions. In the first case, bidder numbers were assumed to be constant regardless of auction outcomes. In the second case, bidders were assumed to drop out with a probability if they lose in tenders and to re-enter in with a probability that increases with the net revenue from contracting that is obtained by active bidders.

The results consistently indicate that auction performance as measured by environmental benefits per dollar is highly dependent on the mix of conservation activities allowed in the projects. In particular, increases in the average share of sediment reduction activities are detrimental to the performance of the auction. The environmental benefits generated per dollar of funding fall consistently as the average share of sediment reduction activities in projects rises. This outcome is a reflection of the more costly nature of sediment based activities and highlights the need for the identification of scope/efficiency trade-offs based on the nature of conservation activities that prevail in different industries. It demonstrates that narrowly scoped auction focused on activities with high opportunity costs can perform very poorly compared to more broadly scoped auctions.

Improvements in the scale of participation are highly beneficial for auction performance. The benefits of a higher degree of rationing obtained through higher participation numbers relative to budgets are very strong. In this case study, the benefits per dollar from the auction with a budget of $300,000 can be between 40 and 100% above those where the budget is $600,000. The benefits of the higher degree of rationing or increased competition for a given budget are notably higher when the scope of the auction is such that it covers higher cost activities.

Results for repeated auctions where bidders stay active even after bids are unsuccessful are higher than for cases where bidders drop out and re-enter, especially if bidders are more sensitive to bid failures in their deicision to participate or drop out.

Finally, the use of uniform pricing rather than discriminatory pricing in repeated auctions would lead to higher benefits per conservation dollar. With uniform pricing, bidders get paid the price of the marginal winner. Their own bids influence whether they win or not but not how much they get paid (unless they are the most expensive winner). This auction leads to more truthful bidding or to less overbidding. The simulations indicate that with uniform pricing in repeated auctions, one could increase the benefits per dollar by between 15 and 55%. The benefits from uniform pricing are especially higher if bidders tend to drop out following bid failures.

Acknowledgements. The Lower Burdekin Dry Tropics Water Quality Improvement Tender Project was funded by the Australian Government through the National Market Based Instruments Program, with additional support provided by the Burdekin Dry Tropics Natural Resource Management Group. The project was conducted as a partnership between Central Queensland University, River Consulting, the University of

Western Australia and the Burdekin Dry Tropics Natural Resource Management Group. The views and interpretations expressed in these reports are those of the authors and should not be attributed to the organisations associated with the project.

References

1. Rolfe, J., Greiner, R., Windle, J., Hailu, A.: Identifying scale and scope issues in establishing conservation tenders: Using Conservation Tenders for Water Quality Improvements in the Burdekin. Research Report 1, Central Queensland University, Rockhampton (2007)
2. Epstein, J.M., Axtell, R.: Growing artificial societies: Social sciences from the bottom up. Brookings Institution Press, Washington (1996)
3. Tesfatsion, L.: Agent-Based Computational Economics: Growing Economies from the Bottom Up. Artificial Life 8, 55–82 (2002)
4. Arifovic, J., Ledyard, J.: Computer testbeds: the dynamics of Groves-Ledyard mechanisms. Technical Report, Simon Fraser University and California Institute of Technology (2002)
5. Erev, I., Roth, A.E.: Predicting how people play games with unique, mixed strategy equilibria. American Economic Review 88, 848–881 (1998)
6. Camerer, C.F.: Behavioral Game Theory. Princeton University Press, New York (2003)
7. Hailu, A., Schilizzi, S.: Are auctions more efficient than fixed price schemes when bidders learn? Australian Journal of Management 29, 147–168 (2004)
8. Hailu, A., Schilizzi, S.: Learning in a 'Basket of Crabs': An Agent-Based Computational Model of Repeated Conservation Auctions. In: Lux, T., Reitz, S., Samanidou, E. (eds.) Nonlinear Dynamics and Heterogeneous Interacting Agents, pp. 27–39. Springer, Berlin (2005)
9. Hailu, A., Thoyer, S.: Multi-unit auction format design. Journal of Economic Interaction and Co-ordination 1, 129–146 (2006)
10. Hailu, A., Thoyer, S.: Designing multi-unit multiple bid auctions: An agent-based computational model of uniform, discriminatory and generalized Vickrey auctions. Economic Record 83, S57–S72 (2007)
11. Selten, R., Stoecker, R.: End behaviour in sequance finite prisoners' dilemma supergames: A learning theory approach. J. of Economic Behaviour and Organization 7, 47–70 (1986)
12. Selten, R., Abbink, K., Cox, R.: Learning direciton theory and winners curse. Discussion Paper 10, Department of Economics, University of Bonn (2001)
13. Roth, A.E., Erev, I.: Learning in extensive form games: experimental data and simple dynamic models in the intermediate term. Games and Economic Behavior 8, 164–212 (1995)
14. Greiner, R., Rolfe, J., Windle, J., Gregg, D.: Tender results and feedback from ex-post participant survey. Research Report 5, Central Queensland University, Rockhampton (2008)

Appendix

Table 4. Benefits per million dollars for alternative activity scopes

Nitrogen share	Pesticide share	Sediment share	Benefits per Mill. $
0.00	1.00	0.00	3.62
0.00	0.50	0.50	2.20
0.00	0.33	0.67	1.93
0.00	0.67	0.33	2.40
1.00	0.00	0.00	2.94
0.50	0.00	0.50	2.17
0.33	0.00	0.67	1.92
0.50	0.50	0.00	3.26
0.33	0.33	0.33	2.47
0.25	0.25	0.50	2.19
0.33	0.67	0.00	3.59
0.25	0.50	0.25	2.60
0.20	0.40	0.40	2.28
0.67	0.00	0.33	2.35
0.67	0.33	0.00	3.30
0.50	0.25	0.25	2.62
0.40	0.20	0.40	2.18
0.40	0.40	0.20	2.65

A Framework for the Development and Maintenance of Adaptive, Dynamic, Context-Aware Information Services

Manel Palau[1], Ignasi Gómez-Sebastià[2], Luigi Ceccaroni[3], Javier Vázquez-Salceda[2], and Juan Carlos Nieves[2]

[1] TMT Factory, C/ Moià 1, planta 1, E08006, Barcelona, Spain
[2] Knowledge Engineering & Machine Learning Group, UPC
C/Jordi Girona 1-3, E08034, Barcelona, Spain
[3] Barcelona Digital Centre Tecnològic, C/Roc Boronat 117, 5ª planta, E08018 Barcelona, Spain
manel.palau@tmtfactory.com,
{igomez,jvazquez,jcnieves}@lsi.upc.edu,
lceccaroni@bdigital.com

Abstract. This paper presents an agent-based methodological approach to design distributed service-oriented systems which can adapt their behaviour according to changes in the environment and in the user needs, even taking the initiative to make suggestions and proactive choices. The highly dynamic, regulated, complex nature of the distributed, interconnected services is tackled through a methodological framework composed of three interconnected levels. The framework relies on coordination and organisational techniques, as well as on semantically annotated Web services to design, deploy and maintain a distributed system, using both a top-down and bottom-up approach. We present results based on a real use case: interactive community displays with tourist information and services, dynamically personalised according to user context and preferences.

Keywords: Multi-agent systems, Coordination and organisational theory, Context awareness, Personalised recommendation, Semantic Web services.

1 Introduction

Urban information services are often provided in ways which have not changed much in a century. This scenario brings up the opportunity to improve services provided to people living in or visiting a city, with the novel possibility of ubiquitously accessing personalised, multimedia content [12]. On one hand, there are numerous, dynamic services that have to be composed and coordinated in order to provide higher-value services. For instance, an advanced entertainment service can be provided by combining information coming from cinema, restaurant and museum services, along with transport and mapping services. These services are not static, as existing services can leave the system and new ones can enter it; and the service used in a given moment for a given task might not be available later or it might happen that a more suitable service becomes available. On the other hand, the incoming content has to be filtered and adapted, to make it compatible with user's context (e.g. location, time and date),

J. Filipe, A. Fred, and B. Sharp (Eds.): ICAART 2010, CCIS 129, pp. 227–239, 2011.

preferences and requirements, and with existing regulations. For example, the system should not suggest a pub to an underage user if local laws do not allow it.

An additional challenge for systems in highly dynamic environments, where unexpected events can arise at any time (e.g. transport not in time due to a traffic jam), is to be able to react and adapt to these events.

We consider that this complex scenario can benefit from the combination of multi-agent techniques and semantic Web services [6] to enable dynamic, context-aware service composition [21], thus providing users with relevant high-level services depending on their current context. Moreover, technologies concerning organisational and coordination theories applied to (intelligent) Web services [14] are also important in order to effectively maintain a system operating in such a constrained (due to user's preferences and local laws) and dynamic environment.

Additionally, the scenario presents the need of integrating new functionalities, new services or new actors (humans or artificial intelligence systems) into an existing running system. This integration is especially difficult taking into account that the scenario presents a system formed by active, distributed and interdependent processes.

In the ALIVE European project [1] a new software-engineering methodology is being explored [22]. The approach aims to bring together leading methods from coordination technology, organisation theory and *model driven design* (MDD) to create a framework for software engineering to address a reality composed of live, open systems of active services. Apart from the ones described in this paper, other examples of these services can be found in Quillinan et al. [19]. The ALIVE's framework is a multi-level architecture composed of three levels:

- the organisational level, which provides context for the other levels, supporting an explicit representation of the organisational structure of the system (composed by roles, objectives and the dependencies among them), and effectively allowing a structural adaptation of distributed systems over time;
- the coordination level, which provides the means to specify, at a high level, the patterns of interaction among services, transforming the organisational representation coming from Organisational Level into coordination plans (including information flows, constraints, tasks and agents);
- the service level, which allows the selection of the most appropriate services for a given task based on the semantic description of services and, effectively supporting high-level, dynamic service composition.

In this paper, results of the application of the ALIVE's approach (introduced in section 3) for designing dynamic, adaptive systems are presented. In particular, the design for each level is described (see section 4). First, a complete organisational level design of an *interactive community display* (ICD) scenario (introduced in section 2) is presented. This design shows that the correct identification of roles in the organisational level allows a dynamic adaptation in the coordination level. Then, the complete design of the coordination level of the ICD scenario is outlined. This design shows how, given a set of landmarks (which are states of special interest), a system can be dynamically adaptable. Finally, the specifications of the Web services involved are presented. In Sections 5 and 6, a discussion of related and future work is outlined.

2 The Use Case

We will use a personalised recommendation tool for entertainment and cultural activities as the basis for exemplifying the scenario. The personalisation is offered via ICDs, which are multimedia information points offering interactive services in public areas [4] [10]. The aim is to bring city services closer to people living in or visiting a city by interconnecting people, service providers and locations. In the scenario, it is taken into account that services and information provided, and how user information is stored, processed and distributed, are all subject of various municipal, national and European regulations.

The scenario considered for this purpose starts when a user interacts with the system's interface (the ICD). The system accesses the user profile (if available) or a group profile from a remote repository. Then, the system adapts the interface format and the interaction mode presenting the initial interface composed by several services, such as cinema or monument. If the user requests one of these services, the system manages the user request considering (if available) user personal data, preferences, requirements and, above all, time, date and location (i.e. user context). Furthermore, environmental context and legislations including components such as weather, traffic reports and legal adult age can be considered.

Finally, the system presents an ordered list of activities located on a map together with basic information, such as a brief description, address and pictures. Moreover, it informs on transportation (e.g., bus and metro) to reach the venue and, if time is appropriate, it suggests a restaurant along the way, thus composing information from different services such as cinemas, restaurants, maps and transport (see Figure 1).

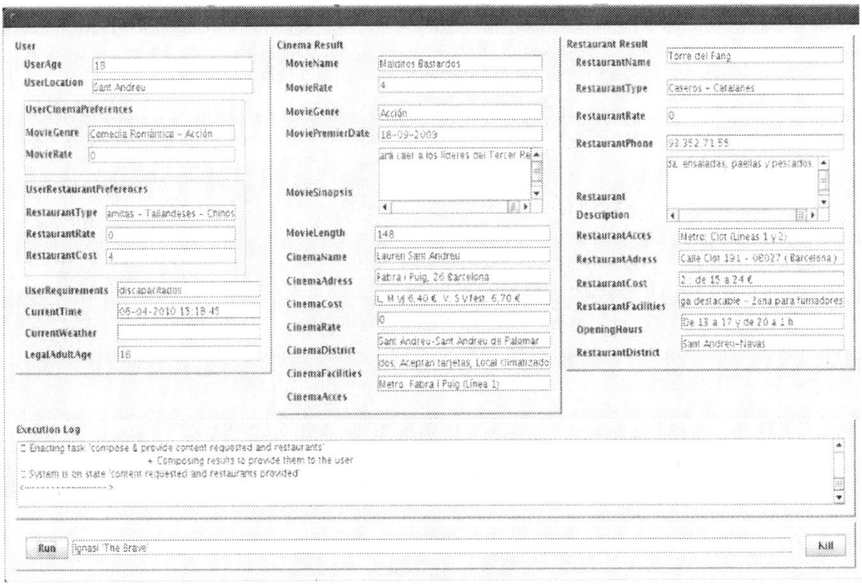

Fig. 1. Application interface showing food and movies suggestions

3 The ALIVE Framework

The ALIVE framework is being developed in collaboration by several universities and enterprises within the frame of the European project ALIVE. It combines MDD and agent-based system engineering with coordination and organisational mechanisms, providing support for "live" (that is, highly dynamic) and open systems of services. ALIVE's multi-level approach (see Figure 2 and following sections) helps to design, deploy and maintain distributed systems by combining, reorganising and adapting services. As shown in Section 4, this framework is suitable for scenarios with new services entering the system and existing services leaving it at run-time.

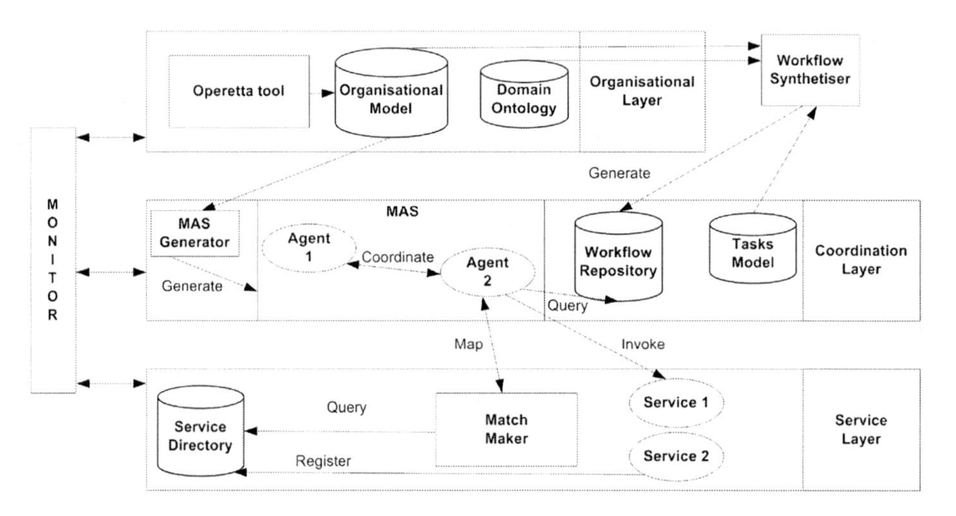

Fig. 2. Main components of the ALIVE architecture

3.1 Organisational Level

The organisational level provides an explicit representation of the organisational structure of the system. The organisational model is the main component of the organisational level, representing the organisation as a social system created by autonomous actors (i.e. they have their own interests) to achieve common goals. Stakeholders and their relations are represented, together with formal goals and restrictions. The model is formalised according to the Opera methodology [8] including goals (e.g. gather petition type, obtain personalised content or gather and compose content); roles (e.g. user, interaction task manager or content provider) that are groups of activity types played by actors (i.e. the agents or human users); and landmarks (e.g. user data collected, external data collected and content to be provided personalised).

Objectives are assigned to roles, among which three kinds of relations exist: the hierarchical relation, where a parent role can delegate an objective to a child role; the market relation where a child role can request the assignment of an objective to the parent role; and the network relation, where both parent and child roles can request an objective to the other one. Each relation is assigned according to what interaction type the designer expects to happen. For instance, in the presented scenario, there is a market relation when the content adaptor role asks the content provider to obtain food suggestion. The set of all roles and the relations among them constitutes the social structure.

Landmarks are important states in the achievement of a goal, and landmark patterns impose an ordering over landmarks to be reached. A set of landmarks and their relations is known as *scene* (see section 4). Scene transitions can be modelled by organising them in an *interaction structure* (see section 4). The organisational level supports the definition of norms, rights and obligations (suitable for highly regulated scenarios) of the actors, effectively forming a normative structure. The social structure, the interaction structure and the normative structure are the three components of the organisational model.

The OperettA tool [17] supports system designers in specifying and visually analysing an organisational model. The domain ontology represents the shared understanding of the domain, providing a common vocabulary about all concepts and their properties, definitions, relations and constraints, and can be defined using existing ontology-editors [5].

The organisational model is used by the multi-agent system (MAS) generator in the coordination level to create the agents that populate the system. For each role defined in the organisational model one or more agents are generated.

3.2 Coordination Level

The coordination level provides the patterns of interaction among actors, transforming the organisational model into coordination plans, or workflows. Workflows are defined using *generalised partial global planning* (GPGP), a framework for coordinating multiple AI systems that are cooperating in a distributed network [13]. Workflows bring the system from a landmark state to the next one (see Figure 3) and are formed by chains of tasks.

Tasks, defined on the coordination level's task model, contain both pre- and postconditions that describe the state of the system before and after the task is performed. Tasks are the bridge between the Coordination and Service levels, containing information that binds them to abstract services on the service level (e.g. inputs and outputs) and to elements of the organisational model (e.g. roles assigned to the task).

The workflow synthesiser uses information from the organisational model, domain ontology and tasks model in order to generate the workflows the agents will enact. These workflows are stored in the workflow repository where they are retrieved when required.

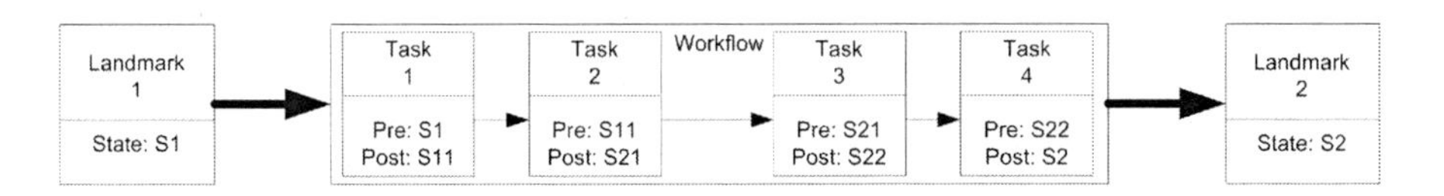

Fig. 3. Workflow example connecting two landmarks

A set of intelligent agents (MAS) deployed on the AgentScape platform [18] enacts the workflows in a coordinated and distributed fashion. Agents analyse and monitor workflow execution, reacting to unexpected events, either by enacting other workflows or by communicating the incident to other levels.

Each agent includes the following components (see Figure 4): the brain module, which provides reasoning and decision-making capabilities; the normative plan analyser, which scans the workflows in order to determine if enacting them will violate any of the norms defined in the organisational model; the Agent Communication Language (ACL) module, which provides agents with the capability of communicating with other agents in the system by sending messages; the GPGP scheduler, which provides an interface for the agents to coordinate and distribute tasks; and the enactment component, which facilitates the invocation of services.

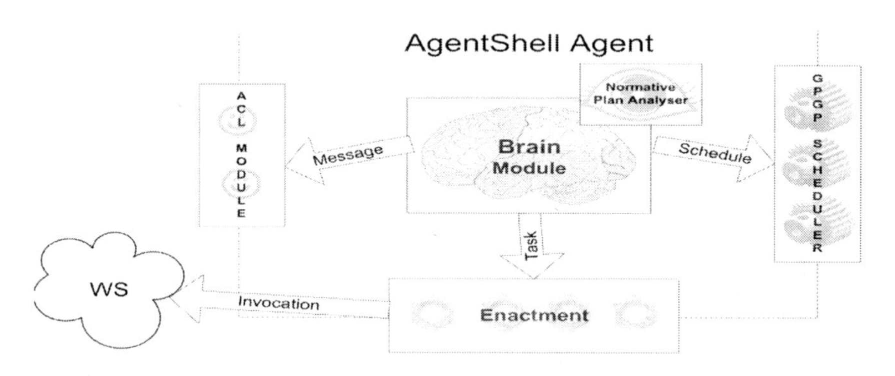

Fig. 4. Agent architecture diagram

3.3 Service Level

Appropriate services are selected for each abstract task in the workflows, using the information included in the service description and in the task description. These descriptions are defined in terms of OWL-S service profiles [16], facilitating the process of composing services [20] and finding alternative services. The reassignment of services to tasks, when a given service is not available, is carried out on the fly.

The match maker component receives an abstract task description from an agent and looks for services that can fulfil this task. It queries the service directory and selects the most appropriate one (if several ones are available), based on the task's semantic description and on quality of service parameters (such as average response time). The service chosen is returned to the agent, and the task is executed and monitored.

At the service level, service composition within the scope of a task is carried out, too. For instance, if a given task requires providing information of a venue on a map, and there are two available services, one to obtain venue information and another to show information on a map, then the task can be bound to the composition of these services.

3.4 The Monitor Tool

The monitor tool is the back-bone of the ALIVE framework [2], connecting all the three levels allowing the exchange of events among them, from a service invocation that fails to an update on the Organisational design (e.g. a new role or objective is introduced) that affects the agents in the coordination level.

As seen on sections 3.2 and 3.3, agents enact their roles by interacting either via direct communication (coordinating among themselves) or via service invocation.

The monitor tool observes these interactions and matches them with the normative and organisational states (e.g. obligations, permissions, roles) effectively allowing agents to reason about the effects (in a normative sense) of their actions.

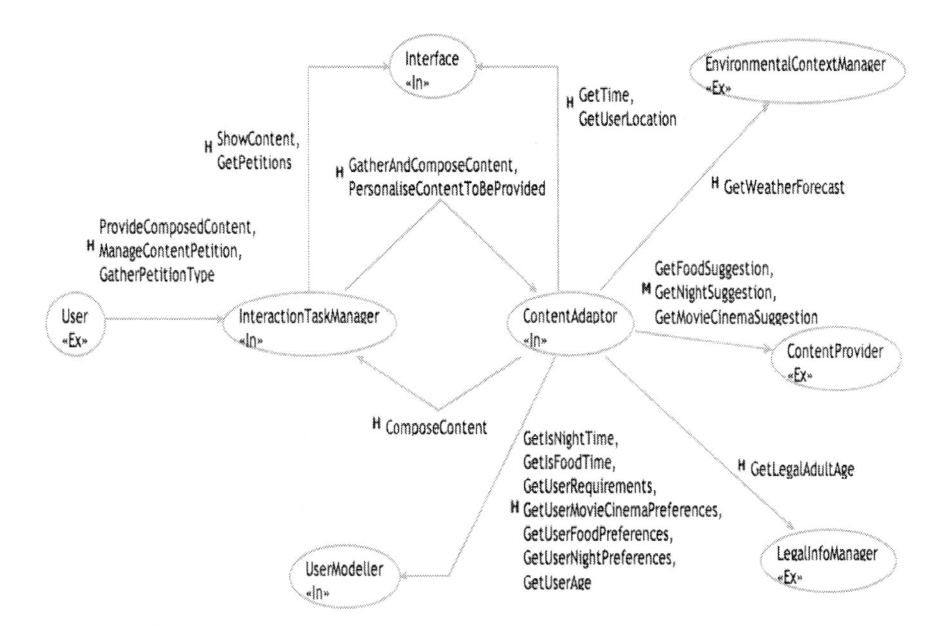

Fig. 5. Operetta model of the social structure

4 Scenario Modelling

This section describes how the ALIVE framework has been applied to the scenario presented in Section 2, and how the system manages user's petitions by means of the defined roles, objectives, scenes, landmarks, workflows, plans, tasks and services.

An organisational social structure is defined using OperettA (see Figure 5). It includes several (external/internal) roles (e.g. content adaptor, user modeller and content provider) in the domain, and their (hierarchical/market) dependencies in terms of delegated sub-goals (e.g. get user food preferences and get food suggestion).

Roles are represented as nodes and sub-objective dependencies as directional arrows. The objectives or goals considered for the role user (sign up; sign in; obtain personalised content; change profile, preferences or requirements; and change the interface format or the interaction mode) are related with the petitions to be managed by the system. Each user's goal is subdivided into sub-goals and delegated to other roles. These roles can delegate sub-goals to other roles, too.

From here on, we will focus on the user objective to obtain personalised content. Figure 5 shows how this objective is delegated to the interaction task manager that identifies the type of petition (sub-objective gather petition type) once the interface gets the user petitions (sub-objective get petitions) and then manages the petition. The interaction task manager demands the content adaptor to gather, personalise and compose the content to be provided. To this end, this role relies on information provided by other roles: the interface (providing information about the user context, such as localisation, time and date), the user modeller (providing the user's personal data, preferences and requirements gathered from the user model) and external information-providers (providing content, and information related to the environmental context and the legislation). The connection with these external information-providers is performed by the match maker component.

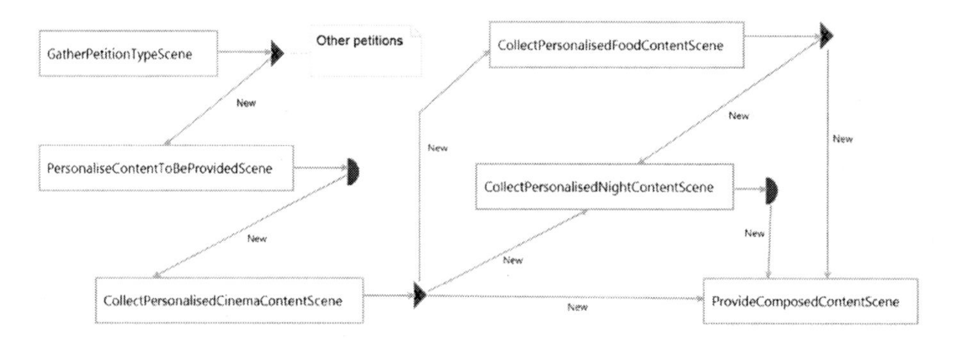

Fig. 6. Interaction structure

The interaction structure for the chosen scenario (see Figure 6) is composed by several scenes representing the petition requested by the user. The transitions among these scenes, starting with the *"gather petition type"* scene and finalising with the *"provide composed content"* scene, show several sequences of interactions. Once the petition type is gathered, there are several possible scenes (one per petition type); however we consider here only a movie cinema petition for simplicity. Hence, next scenes are *"personalise content to be provided"* and *"collect personalised movie cinema content"*. There are two possible scenes before providing the composed content, which are *"collect personalised food content"* and *"collect personalised night content"*, and are performed depending on user preferences and context.

Each scene is defined by the roles playing within it and a landmark pattern imposing an ordering over the important states (landmarks) that should be reached in the achievement of the goals in the scene. For instance, the personalise content to be provided scene contains four players (interaction task manager, interface, content adaptor and user modeller) and several landmarks (user data collected, external data collected and content to be provided personalised), which follow a partial order among them (see Figure 7).

The plan repository within the coordination level contains the workflows the agents must follow in order to accomplish the landmarks defined in the organisational model. Focusing in the scene personalise content to be provided, there is a distributed plan where several agents coordinate performing tasks in parallel (e.g. the agent enacting the interface role fetches user's context while the agent enacting the user modeller role looks for user's personal data, preferences and requirements). The plan personalise content to be provided is adapted depending on several variables, such as night time, food time, user preferences, requirements and context. For instance, the plan can also provide food suggestion taking into account user food time preferences.

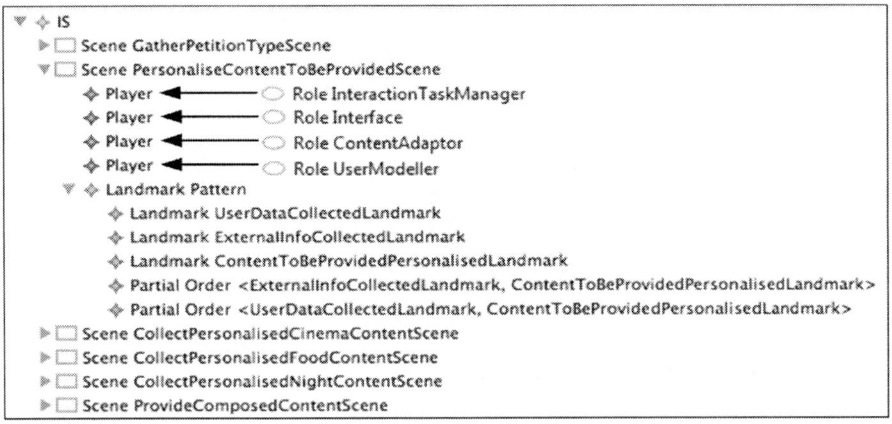

IS
Scene GatherPetitionTypeScene
Scene PersonaliseContentToBeProvidedScene
Player ← Role InteractionTaskManager
Player ← Role Interface
Player ← Role ContentAdaptor
Player ← Role UserModeller
Landmark Pattern
Landmark UserDataCollectedLandmark
Landmark ExternalInfoCollectedLandmark
Landmark ContentToBeProvidedPersonalisedLandmark
Partial Order <ExternalInfoCollectedLandmark, ContentToBeProvidedPersonalisedLandmark>
Partial Order <UserDataCollectedLandmark, ContentToBeProvidedPersonalisedLandmark>
Scene CollectPersonalisedCinemaContentScene
Scene CollectPersonalisedFoodContentScene
Scene CollectPersonalisedNightContentScene
Scene ProvideComposedContentScene

Fig. 7. Landmark pattern for the personalised content to be provided scene

The Coordination level describes a sequence of composite and atomic tasks for each plan. Figure 8 shows the tasks required to provide content to a user considering age and preferences regarding food and night time.

There are several norms applied to the interaction among agents. For instance, the content provider has the obligation of performing his task before a deadline (10 slots time). This norm prevents the user from having to wait too long before its petition is processed.

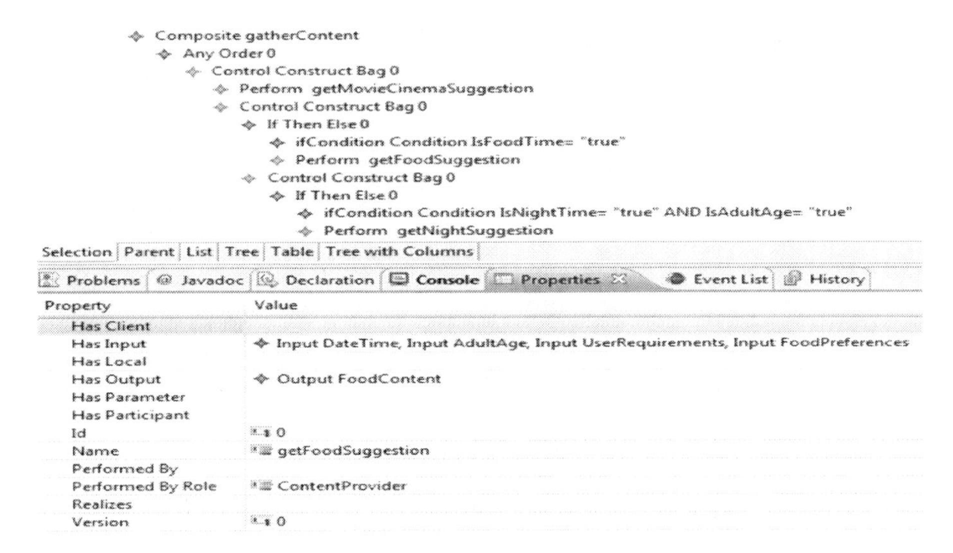

Fig. 8. Composite and atomic tasks to gather content

Tasks are implemented as Web services, which are semantically annotated and described in terms of OWL-S service profiles (i.e. inputs, outputs, preconditions and effects) as shown in Figure 9. In order to perform the task *"get food suggestion"* the match maker component maps *abstract content providers* to concrete ones (e.g. *lanetro* [http://www.lanetro.com] restaurants). Doing it this way allows the system to dynamically readapt to failures of a *concrete content provider* (e.g. remapping to *Atrapalo* [http://www.atrapalo.com] restaurants if *lanetro* restaurants are not available).

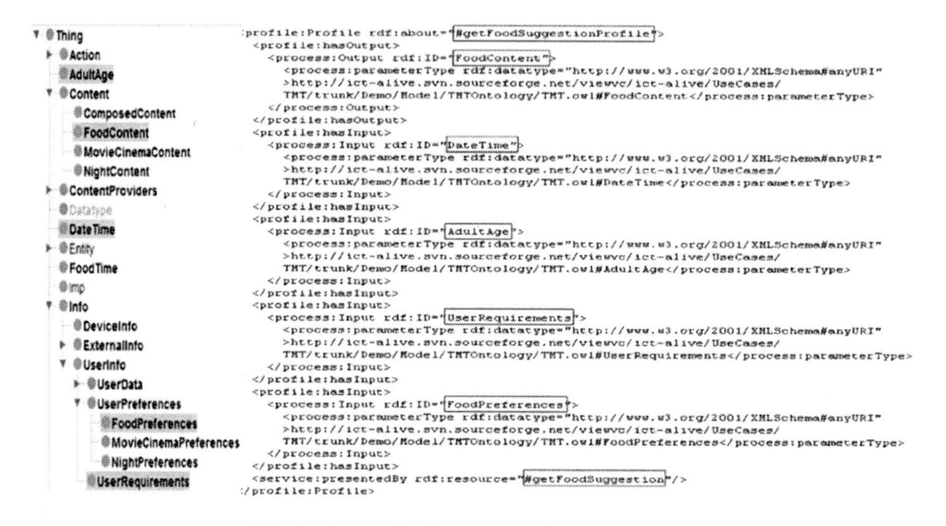

Fig. 9. Ontological concepts related to the task

The modelling presented in this section covers the levels presented in Section 3: organisational level, coordination level, and service level.

5 Conclusions

Users' presence and context can be exploited to provide personalised, dynamic and composed services fulfilling their expectations, needs and functional diversity. As seen on Section 3, this paper presents the design of a multi-agent system that adapts its behaviour according to the environment and the user, and takes the initiative to make suggestions and proactive choices.

Orchestration and the Re-planning. Several tourism-related projects, such as E-travel [11] and Deep Map [15], take advantage of agent technology integrated with the semantic Web. The ALIVE project is also close to the Interactive Collaborative Information Systems (ICIS) project [9], in which a MAS takes into consideration unexpected events that happen in the real world in order to obtain a steady and reliable system in dynamic and changing environments, and a higher level view is used to take advantage of the service orchestration and the re-planning.

Dynamic service-composition is an issue that has been tackled via pre-defined workflow models where nodes are not bound to concrete services, but to abstract tasks at runtime. This work presents a similar approach (through the mapping performed by the match maker component) with the difference that workflows used are not predefined, but dynamically generated from the information provided by an organisational level, and thus, workflows evolve as the organisational information evolves. Due to the connection among levels, a change in the organisational level can trigger changes both in the coordination level (via plan and agent generators) and in the service level (new plans will result in the execution of new tasks and, possibly, the invocation of new services).

As outlined in Section 4, intelligent agents at the coordination level present an option for providing both exception handling and organisational-normative awareness capabilities to the system. Exception handling is common in other SOA architectures, however, most approaches tend to focus on low-level (i.e. service) exception handling. The ALIVE approach enables managing of exceptions at multiple levels either substituting services (service level) looking for alternative workflows to connect two landmarks (coordination level) or even looking to achieve alternative landmarks among the same scene (organisational level). Agents at coordination level enable this medium and high-level exception handling, which are not commonly seen in other SOA approaches. Regarding organisational-normative awareness, to the best of authors' knowledge, no attempts have been made to include normative information in workflows. However, normative agents are common in the literature [3]. Making normative agents reason about the workflows (and the tasks included in them) before performing them, and discarding the ones that do not comply with organisational norms, adds organisational awareness to the execution of the workflows.

6 Future Work

As future work we plan to extend the system integrating further services, such as booking, payment or planning routes. Considering the unexpected events of the real world, for instance, it might happen that user does not arrive on time to a booked cinema session because of a traffic jam. In this case alternatives must be provided to the user, for instance, booking a ticket in a session that starts a couple of hours later, and suggesting some nearby shops to spend some time while the new session starts. Furthermore, work on the integration of on-time reorganisation mechanisms and Model Driven Design will be performed to be able to promote reliability and stability for services, enabling to keep slowly changing elements separate from dynamic aspects of the environment.

Another interesting aspect that requires additional studies is the improvement of the handling of the user profile and her preferences in the system by means of a logic framework based on Answer Set Programming and possibilistic logic [7]. The framework, called logic programs with possibilistic ordered disjunction (or LPPOD) proposes a flexible formalism which supports: 1) the creation and adaptation of user profile by means of preference rules and necessity values; 2) the reasoning about user preferences to obtain an order among them. In this way it is possible to add dynamically user preferences and to query the user profile against her most preferred interests whenever a service has to be selected. The implementation and integration of the LPPOD framework in the system presented in this paper is currently being investigated. Once the framework is implemented we aim to encapsulate the framework functionality in a dedicated service, which can be queried against user profile and preference at any time.

Acknowledgements. This work has been partially supported by the FP7-215890 ALIVE project. Javier Vázquez-Salceda's work has been also partially funded by the *Ramón y Cajal* program of the Spanish Ministry of Education and Science. The authors would like to acknowledge the contributions of their colleagues from the ALIVE consortium. The views expressed in this paper are not necessarily those of the ALIVE consortium.

References

1. ALIVE European project, `http://www.ist-alive.eu`
2. Aldewereld, H., Álvarez-Napagao, S., Dignum, F., Vázquez-Salceda, J.: Making Norms Concrete. Accepted in: 9th International Conference on Autonomous Agents and Multiagent Systems (2010)
3. Castelfranchi, C., Frank, D., Jonker, C., Treur, J.: Deliberative Normative Agents: Principles and Architecture. In: Jennings, N.R. (ed.) ATAL 1999. LNCS, vol. 1757, pp. 364–378. Springer, Heidelberg (2000)
4. Ceccaroni, L., Codina, V., Palau, M., Pous, M.: PaTac - Urban, ubiquitous, personalized services for citizens and tourists. In: 3rd International Conference on Digital Society, pp. 7–12. IEEE Computer Society, Los Alamitos (2009)
5. Ceccaroni, L., Kendall, E.: A graphical environment for ontology development. In: 2nd International Joint Conference on Autonomous Agents and Multiagent Systems, pp. 958–959. ACM, New York (2003)
6. Comas, Q., R-Roda, I., Ceccaroni, L., Sànchez-Marrè, M.: Semiautomatic learning with quantitative and qualitative features. In: VIII Conference of the Spanish Association for Artificial Intelligence, pp. 17–25. Asociación Española para la Inteligencia Artificial, Spain (1999)
7. Confalonieri, R., Nieves, J.C., Osorio, M., Vázquez-Salceda, J.: Possibilistic Semantics for Logic Programs with Ordered Disjunction. In: Link, S., Prade, H. (eds.) FoIKS 2010. LNCS, vol. 5956, pp. 133–152. Springer, Heidelberg (2010)
8. Dignum, V.: A Model for Organizational Interaction: Based on Agents, Founded in Logic. SIKS Dissertation Series 2004-1. Utrecht University. PhD Thesis (2004)
9. Ghijsen, M., Jansweijer, W., Wielinga, B.B.: Towards a Framework for Agent Coordination and Reorganization, AgentCoRe. In: Sichman, J.S., Padget, J., Ossowski, S., Noriega, P. (eds.) COIN 2007. LNCS (LNAI), vol. 4870, pp. 1–14. Springer, Heidelberg (2008)
10. Gómez-Sebastià, I., Palau, M., Nieves, J.C., Vázquez-Salceda, J., Ceccaroni, L.: Dynamic orchestration of distributed services on interactive community displays: The ALIVE approach. In: 7th International Conference on Practical Applications of Agents and Multi-Agent Systems, ASC, vol. 55, pp. 450–459. Springer, Heidelberg (2009)
11. Gordon, M., Paprzycki, M.: Designing Agent Based Travel Support System. In: 4th International Symposium on Parallel and Distributed Computing, pp. 207–216. IEEE Computer Society, Los Alamitos (2005)
12. Greenfield, A.: Everyware: The Dawning Age of Ubiquitous Computing, 1st edn. New riders publishing in association with AIGA (2006)

13. Lesser, V., Decker, K., Wagner, T., Carver, N., Garvey, A.: Evolution of the GPGP/TÆMS Domain-Independent Coordination Framework. In: Autonomous Agents and Multi-Agent Systems, vol. 9(1), pp. 87–143. Kluwer Academic Publishers, Dordrecht (2004)
14. Luck, M., McBurney, P., Shehory, O., Willmott, S.: Agent Technology: Computing as Interaction. A Roadmap for Agent Based Computing. AgentLink (2005)
15. Malaka, R., Zipf, A.: Deep Map challenging IT research in the framework of a tourist information system. In: 7th International Congress on Tourism and Communication, pp. 15–27. Springer, Heidelberg (2000)
16. Martin, D., Burstein, M., Hobbs, J., Lassila, O., McDermott, D., McIlraith, S., Narayanan, S., Paolucci, M., Parsia, B., Payne, T.R., Sirin, E., Srinivasan, N., Sycara, K.: OWL-S: Semantic Markup for Web Services. Technical Report, Member Submission, W3C (2004)
17. Okouya, D.M., Dignum, V.: OperettA: A prototype tool for the design, analysis and development of multi-agent organizations. In: 7th International Conference on Autonomous Agents and Multiagent Systems, pp. 1677–1678. ACM, New York (2008)
18. Overeinder, B.J., Brazier, F.: Scalable middleware environment for agent-based internet applications. In: Dongarra, J., Madsen, K., Waśniewski, J. (eds.) PARA 2004. LNCS, vol. 3732, pp. 675–679. Springer, Heidelberg (2006)
19. Quillinan, T.B., Brazier, F., Aldewereld, H., Dignum, F., Dignum, V., Penserini, L., Wijngaards, N.: Developing Agent-based Organizational Models for Crisis Management. In: 8th International Conference on Autonomous Agents and Multiagent Systems (2009)
20. Sirin, E., Parsia, B., Hendler, J.: Filtering and selecting semantic web services with interactive composition techniques. IEEE Intelligent Systems 19(4), 42–49 (2004)
21. Vallée, M., Ramparany, F., Vercouter, L.: A Multi-Agent System for Dynamic Service Composition in Ambient Intelligence Environments. In: 3rd International Conference on Pervasive Computing. Doctoral Colloquium (2005)
22. Vázquez-Salceda, J., Ceccaroni, L., Dignum, F., Vasconcelos, W., Padget, J., Clarke, S., Sergeant, P., Nieuwenhuis, K.: Combining Organisational and Coordination Theory with Model Driven Approaches to develop Dynamic, Flexible, Distributed Business Systems. In: 1st International Conference on Digital Businesses, Springer, Heidelberg (2009)

Decision Making in Complex Systems with an Interdisciplinary Approach

Marina V. Sokolova[1,2,3], Antonio Fernández-Caballero[2,3], and Francisco J. Gómez[2]

[1] Kursk State Technical University, Kursk, 305007, Russia
[2] University of Castilla-La Mancha, Polytechnical Superior School of Albacete 02071-Albacete, Spain
[3] Albacete Research Institute of Informatics, 02071-Albacete, Spain
marina.v.sokolova@gmail.com, {caballer,fgomez}@dsi.uclm.es

Abstract. This article presents a framework for the creation of decision support and expert systems for complex natural domains. The framework uses the numerous advantages of intelligent methods of data manipulation and use agents to make decentralized decisions. The qualitative improvement in decision making is obtained by using interdisciplinary approach. The frameworks combines, on the one hand, the numerous advantages of intelligent methods for data treatment and, on the other hand, supports software systems life cycle. The approach contributes to decentralization and local decision making within the standard workflow.

1 Introduction

When the idea of creating a symbiotic human-computer system to increment accessible knowledge for decision making in complex problems appeared, it was firstly applied for managerial and business domains. Later, the initial domains of decision support systems (DSS) application were widened, and the concept of DSS was spread out to manifold spheres and fields of human activities, extending not only to technical but also to complex ill-determined domains (environmental, medical, social issues, etc). With time diverse DSS have appeared, which enhanced a number of models for use such as preprocessing, optimization, hybrid and simulation models.

The use of DSS is of great importance now for complex natural phenomena studies, because they allow specialists to quickly gather information and analyze it in order to understand the real nature of the processes, their internal and external dependencies, and the possible outcomes while making actions and correcting decisions. The technical areas where these systems could help vary from the storing and retrieving of necessary records and key factors, examination of real-time data gathered from sensors, analysis of tendencies of complex natural processes, retrospective time series, making short and long-term forecasting, and in many other cases [9], [1], [12].

In this article we will present our framework for creation of agent-based decision support and expert systems, focusing firstly on its principal formalisms and phases of decision making process for a complex natural domain (part 2) and revise related works (part 3), then presenting our approach for creation a framework for complex system design in complex domains (part 4) and demonstrating an application of the framework for the case study and discuss obtained results (part 5) and, lastly, we will comment directions of future work (part 6).

J. Filipe, A. Fred, and B. Sharp (Eds.): ICAART 2010, CCIS 129, pp. 240–250, 2011.

2 Decision Making for Complex Systems

Human activity increases constantly, and both the scale and speed of the human influence upon the natural, social, economical, and other processes grows significantly, so, it is impossible now not to take it into account as one of the motive forces in "human - nature - technology" arena. The science of today has reached significant results in modeling and control over man-made technical systems. Notwithstanding, effective managing natural complex phenomena often lies beyond of our possibilities.

Generally speaking, a complex systems (CS) is a composite object, which consists of many heterogeneous (and on many occasions complex as well) subsystems, and has emergent features, which arise from interactions within the different levels. Such systems behave in non-trivial ways, originated in composite functional internal flows and structures of the CS. As a rule, researchers encounter difficulties when trying to model, simulate and control complex systems. Due to this, it would be correct to say that to puzzle out the natural complex systems paradigm is one of the crucial issues of modern science. Because of high complexity of CS, traditional approaches fail in developing theories and formalisms for their analysis. Such study can only be realized by a cross-sectoral approach, which uses knowledge and theoretical backgrounds from various disciplines as well as collaborative efforts of research groups and interested institutions.

The term "system" has a number of definitions and one of them sounds like "a set of interdependent or temporally interacting parts", where parts are generally systems themselves, and are composed of other parts in their turn [5]. If any part is being extracted from the system, it looses its particular characteristics (emergency), and converts into an array of components or units. Thus, an effective approach to CS study is to follow the principles of the system analysis [8], when we have to switch over to the abstract view of the system and perform the following flow of tasks:

- Description of a system. Identification of its main properties and parameters.
- Study of interconnections amongst parts of the system.
- Study of external system interactions with the environment and with other systems, etc.
- System decomposition and partitioning.
- Study of each subsystem or system part.
- Integration of results obtained on the previous stages.

However, it is obvious that it is impossible to create an overall tool, which would play a role of a "magic wand" to study any complex adaptive problem. The process of DSS and expert system (ES) design is laborious, and has many requirements, the crucial one are: collaboration of the specialists from various domains (experts, developers, analysts, mathematics, programmers, etc.). The discussed peculiarities and difficulties of CS study constrain multiple requirements on experts and developers, those have to do not with the problem in question, but with theories and formalisms from different disciplines. Thus, the process of decision making and expert systems design should be facilitated, and a portfolio of methods and tools have to be offered to specialists working with complex adaptive systems.

3 Contemporary Approaches to Complex System Study

Recognition that a complex natural phenomenon has to be studied using the principles of the system approach, induced some theoretical insights and their practical realization. But, it should be said, that many researchers still tend to use mono-discipline approaches and do not zoom in the complex emergent behavior of CS and their causal-effect relations with the environment.

Having revised a number of research works and publications, we can mark out some general tendencies for complex systems study. Some research group offer "island solutions", which, as a rule, are oriented to evaluation or assessment of a few parameters or indicators, in other words, they are dedicated to resolve specific goals. For example, a system which provides indicator assessment for a particular case study or for a limited area. Domain ontologies for such systems are limited though may suffer from possible heterogeneity. As a rule, such systems are effective when working within the application domain and very sensitive to any unforeseen changes [4],[15]. Multi-functional systems provide multiple analysis of input information, can be based upon hybrid techniques, possess tools and methods of data pre- and post-processing, modelling, simulation. They are less sensitive to changes in application domain, as they have tools to deal with uncertainty and heterogeneity [7], [14].

Methodologies for software applications development may support the mayor part of the system life cycle phases, starting with the initial system planning, which include system analysis and domain (problem) analysis phases, and then assist and provide EIS design, coding, testing, implementation, deployment and maintenance. In this case, consolidate cooperation of specialists from various domains and with various backgrounds is necessary [2], [10]. Basically, in spite of the diversity of existing approaches of DSS creation, obtained and reviewed outcomes, it is not possible to elaborate a uniform overall tool, capable of dealing with various domains and create adequate solutions. However, the quality of solutions and multi-function tools upsurges, because they perform better results when these tools are oriented to limited and specific domains. Even then, there exists a necessity to elaborate a general methodology for DSS design oriented to distributed heterogeneous domains, and powerful enough to facilitate work not only for small, but also for distributed and numerous research groups.

4 Our Approach to the Framework Creation

4.1 Interdisciplinary Approach

In general terms, a framework facilitates development of an informational system, as it offers a consequence of goals and works to do. In our case the informational system in the concept definition given above (see part 2), represents a CS or a natural phenomenon. Following the stages, determined in a framework, developers and programmers will have more possibilities to specialize in the domain of interest, meet software requirements for the problem in term, thereby reducing overall development time.

An effective framework should, on the one hand, be general enough to be applicable for various domains and, on the other hand, be adaptable for specific situations and problem areas. Moreover, the framework should be based of the interdisciplinary approach, which results from the melding of two or more disciplines. Being applied to a complex system, the approach includes the principal works: (1) the complex system is decomposed into components if necessary, the process of decomposition is repeated; (2) then, each of the subsystems is studied by means of the "proper" techniques, belonging to the respected discipline, or/and of hybrid methods; (3) managerial process of decision making is realized, with feedback and possible solutions generation.

As we have accentuated in the previous part, the most important principles of CS is that they can not be studies from a mono-discipline viewpoint, and it is necessary to provide a complex hybrid application of methods and techniques from various disciplines. Using intelligent agents seems to be an optimal solution in this case [16]. Actually, MAS helps to create cross-disciplinary approaches for data processing, and, hence, for CS study. An agent may include nontraditional instruments to bear from different domains. Roles played by an agent depend on the system (or subsystem) functions and aims. There are no restrictions or limitations put on knowledge and rule base, used by each agent [13].

4.2 The Main Phases of the Framework

The purpose of our framework is to provide and facilitate complex systems analysis, simulation, and, hence, their understanding and managing. From this standpoint, and taking into account results and insights, given in the previous parts, we implement the principles of system approach. The overall approach we use is straightforward: we decompose the system into subsystems, and apply intelligent agents to study them, then we pool together obtained fragments of knowledge, and model general patterns of the system behavioral tendencies [11].

The framework consists of the three principal phases:

Preliminary Domain and System Analysis. This is the initial and preparation phase when an analyst in collaboration with experts study the domain of interest, extract entities and discover their properties and relations. Then, they state main and additional goal of the research, possible scenarios and functions of the system. During this exploration analysis, they search answer to the questions: *what* the system has to do and *how* it has to do it. As a result of this collaboration it appears meta-ontology and knowledge base. This phase is supported by the Protege Knowledge editor, that implements the meta-ontology, and by the Prometheus Design Kit, which we use to design multi-agent system and generate skeleton code for further implementation of the DSS.

System Design and Codification. The active "element" of this phase is a developer. As supporting software at this phase, we used the Jack^{TM} Intelligent Agents and Jack^{TM} Development Environment. Once the codification is finished and the system is tested, the second phase is completed.

Simulation and Decision Making. This is the last phase of the framework and it has a special mission. During this phase the final users - decision makers - can interact with the system, constructing solutions and policies, estimating consequences of possible actions on the basis of simulation models.

5 Case Study: Environmental Impact Assessment on Human Health

5.1 Intelligent Agents

The general system organization is coherent to the principal phases,described above.

Data search, retrieval, fusion and pre-processing are realized by two intelligent agents: the *Data Aggregation agent* (DAA) and the *Data Preprocessing agent* (DPA), which do a number of tasks, following the workflow:

Information Search - Data Source Classification - Data Fusion - Data Preprocessing - Believes Creation

Data mining procedures are based on the *Function Approximation agent* (FAA) and its team of agents. The principal tasks to be solved here are:

- to state the environmental pollutants that impact on every age and gender group and determine if they are associated with examined diseases groups;
- to create the models which explain dependencies between diseases, pollutants and groups of pollutants.

Here we are aimed to discover the knowledge in form of models, dependencies and associations from the pre-processed information which comes from the previous logical layer. The workflow of this level includes the following tasks:

State Input and Output Information Flows - Create models - Assess Impact - Evaluate models - Select Models - Display The Results

Decision generation, simulation and human-computer interaction are realized by the *Computer Simulation agent* (CSA). The system works in a cooperative mode, and it allows the decision maker to modify, refine or complete decision suggestions, provided by the system and validate them. This process of decision improvement is repeated until the consolidated solution is generated. The workflow is represented below:

State Factors for Simulation - State the Values of Factors - Simulate - Evaluate Results - Check Possible Risk - Display The Results - Receive Decision Maker Response - Simulate - Evaluate Results - Check Possible Risk - Display The Results

Agents communicate to each other and are triggered by events and messages that they send. Agents also share common data. The preliminary system design was realized in the Prometheus Development Kit.

5.2 Methods Used by Agents

The DAA has a number of subordinate agents under its control; these are the *Domain Ontology agent* and the fusion agents: the *Water Data Fusion agent*, the *Petroleum Data Fusion agent*, the *Mining Data Fusion agent*, the *Traffic Pollution Fusion agent*, the *Waste Data Fusion agent* and the *Morbidity Data Fusion agent*. At the beginning of the execution the DAA firstly send a message to the Domain Ontology agent, which

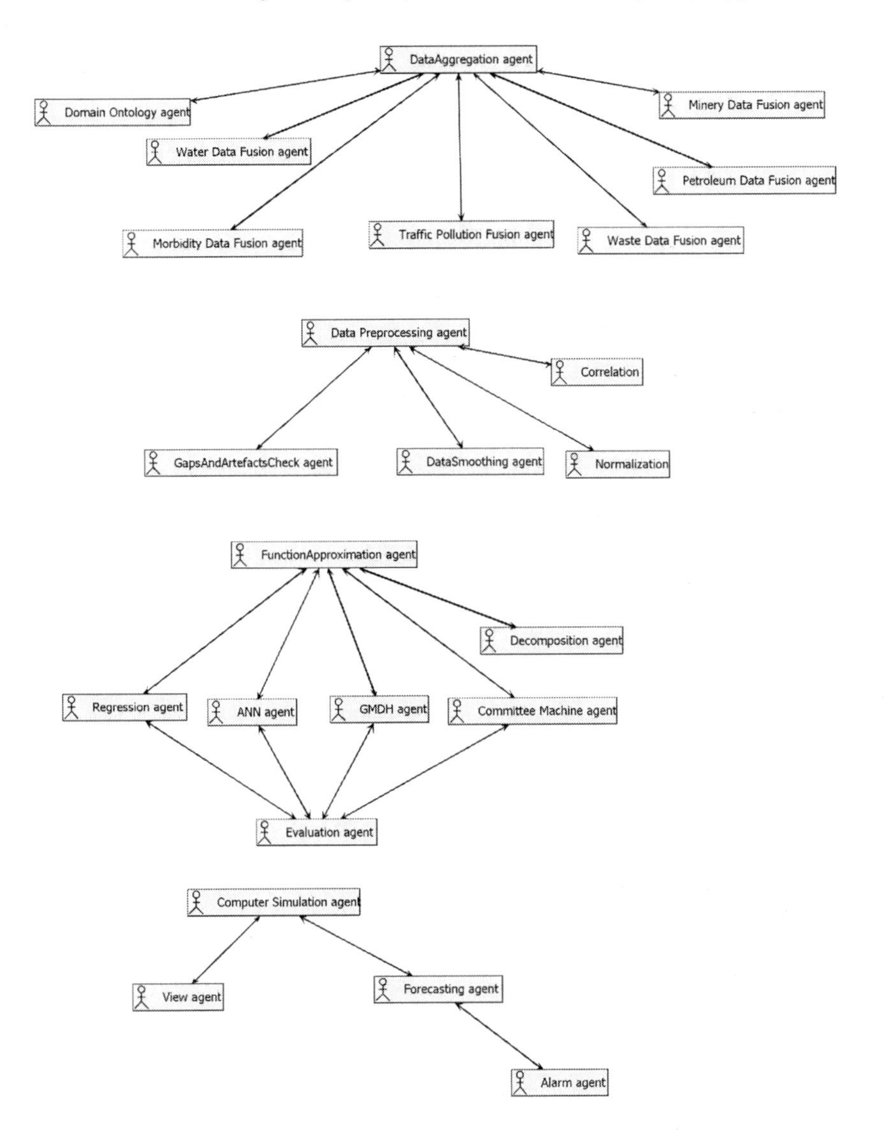

Fig. 1. The agent teams created for the case study

reads information from metaontology, and returns in to the DAA. Then, the DAA starts to searches for information sources and reviews them trying to find if there was a key ontological concept there. If the file contains the concept, the Data Aggregation agent sends an internal event to start data retrieval to the specialized fusion agent. The plan responsible for execution with the identified concept starts reading the information file and searching for terms of interest.

The DPA provides data preprocessing and has a number of subordinate agents which specialize in different data clearing techniques: the *Normalization agent*,the *Correlation agent*, the *Data Smoothing agent*,the *Gaps and Artifacts Check agent*. They perform all data preprocessing procedures, including outliers and anomalies detection, dealing with missing values, smoothing, normalization, etc.

The FAA has a hierarchical team of subordinate agents, which serve to support the roles: "Impact Assessment", "Decomposition" and "Function Approximation". FAA has under its control a number of data mining agents: the *Regression agent*, the *ANN agent*, and the *GMDH agent*, which work in a concurrent mode, reading input information and creating models. Then, if any agent from this group finishes modeling, it calls for the *Evaluation agent*, which evaluates obtained models, and return the list of the accepted ones, while the others are banned and deleted. The FAA pools the output of the agents work, creates the list with the accepted models and then, once the *Regression agent*, the *ANN agent*, and the *GMDH agent* finished their execution, calls for the *Committee Machine agent*, which creates the final models in form of committees for each of the dependent variables, and validates them.

The methods which execute agents of the FAA team are the following:

- Statistical modelling - include linear, non-linear simple and multiple models and their evaluation.
- Artificial Networks modelling - include feed-forward neural networks, trained by back-propagation algorithm, genetic algorithms, radial-based function networks.
- Group Method of Data Handling (GMDH), which is one of the powerful methods of self-organization. We used the combinatorial algorithm of GMDH [6].
- Committee Machines, which provide creation of weighted hybrid models with the condition that the best models selected for the committee have more impact on the final result [3].

The CSA interacts with user and performs a set of tasks within "Computer Simulation", "Decision Making" and "Data Distribution" roles. It has the agent team, which includes *Forecasting agent*, *Alarm agent* and *View agent*.

The process of simulation and generation of possible solutions is interactive. The human-computer interaction sessions are organized by the *View agent*, which offer window -forms, graphical and textual documentation as supporting tools. The *Forecasting agent* calculated predicting values for the CSA, and the *Alarm agent* checks if these values satisfy permitted standards. Figure 1 gives a look at agent teams.

5.3 Simulation Results

The MAS has an open agent-based architecture, which would allow us an easy incorporation of additional modules and tools, enlarging a number of functions of the system. The system belongs to the organizational type, where every agent obtains a class of tools and knows *how* and *when* to use them. Actually, such types of systems have a planning agent, which plans the orders of the agents' executions. In our case, the main module of the JACK^{TM}program carries out these functions.

The *View agent* displays the outputs of the system functionality and realize interaction with the system user. As the system is autonomous and all the calculations are executed by it, the user has only access to the result outputs and the simulation window.

To evaluate the impact of environmental parameters upon human health in Castilla-La Mancha, in general, and in the city of Albacete in particular, we have collected

Table 1. Part of the Table with the outputs of impact assessment

N	Disease	Pollutant
1	Neoplasm	Nitrites in water; Miner products; DBO5; Asphalts; Fuel-oil; Water: solids in suspension; Petroleum liquid gases; Dangerous chemical wastes; Non-dangerous chemical wastes
2	Diseases of the blood and blood- forming organs, the immune mechanism	Miner products; Fuel-oil; Nitrites in water; Solids in suspension; Dangerous metallic wastes Miner products; DBO5
3	Pregnancy, childbirth and the puerperium	Kerosene; Petroleum liquid gases; Solids in suspension; Petroleum; DQO; Fuel-oil; Gasohol; DBO5; Water: Nitrites; Petroleum; Asphalts; Petroleum autos

retrospective data since year 1989, using open information resources offered by the Spanish Institute of Statistics and by the Institute of Statistics of Castilla-La Mancha. As indicators of human health and the influencing factors of the a environment, which can cause negative effect upon the noted above indicators of human health, the levels of contamination of potable water, air, soil, and indicators of traffic activity, hazardous wastes, energy and used annually, etc.

The DSS has recovered data from plain files, which contained the information about the factors of interest and pollutants, and fused in agreement with the ontology of the problem area. It has supposed some necessary changes of data properties (scalability, etc.) and their pre-processing. After these procedures, the number of pollutants valid for further processing has decreased from 65 to 52. This significant change was caused by many blanks related to several time series, as some factors have started to be registered recently. After considering this as an important drawback, it was not possible to include them into the analysis. The human health indicators, being more homogeneous, have been fused and cleared successfully. The impact assessment has shown the dependencies between water characteristics and neoplasm, complications of pregnancy, childbirth and congenital malformations, deformations and chromosomal abnormalities.

The MAS has a wide range of methods and tools for modeling, including regression, neural networks, GMDH, and hybrid models, based on committee machines. The function approximation agent selected the best models, which were: simple regression - 4381 models; multiple regression - 24 models; neural networks - 1329 models; GMDH - 2435 models. The selected models were included into the committee machines. We have foretasted diseases and pollutants values for the period of four years, with a six month step, and visualized their tendencies, which are going to overcome the critical levels. Control under the "significant" factors, which cause impact upon health indicators, could lead to decrease of some types of diseases.

Committee machines provide universal approximation, as the responses of several predictors (experts) are combined by means of a mechanism that does not involve the

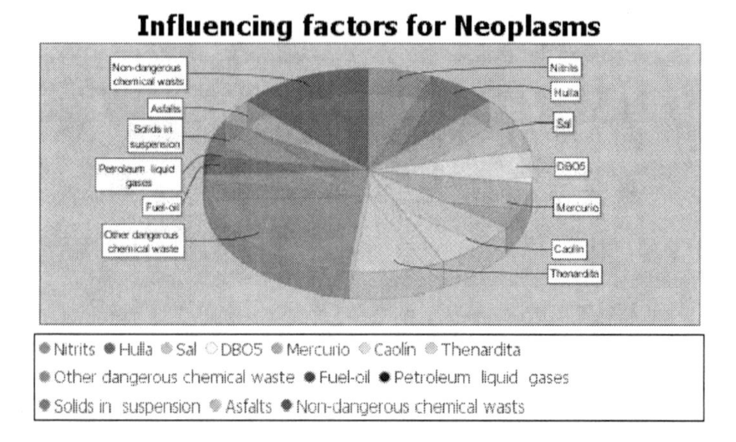

Fig. 2. The outcomes of the impact assessment for one of the diseases of the case study

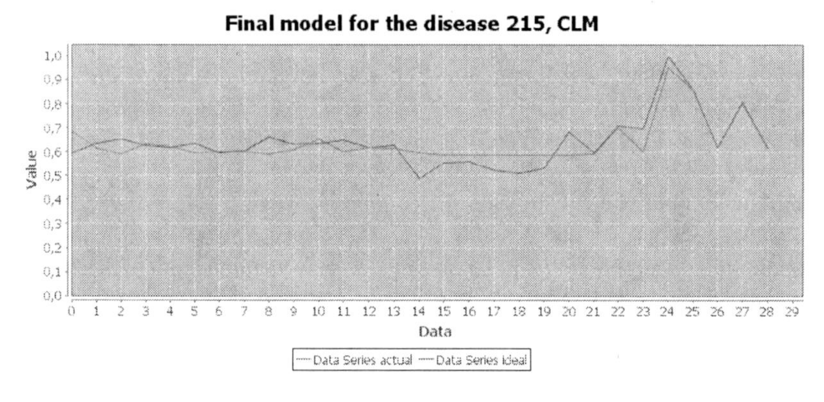

Fig. 3. The final model for one of the diseases of the case study

input signal, and the ensemble average value is received. As predictors we used regression and neural network based models. An example of the final model, received by committee machine for the "'Neoplasms"' for the region of Castilla La Mancha, Spain, is given on Figure 3, and the results of the impact assessment are shown on Figure 2.

6 Conclusions

Complex system analysis and modelling is still a complicated problem, and an efficient framework which supports decision making and expert systems creation facilitates research efforts for various research groups. The framework we have demonstrated in the article, is one of the approximations to solve this complicated task. And, to conclude with, we should note some essential advantages we have reached, and some directions for future research.

First, the framework is interdisciplinary and is flexible for changes: it can be adapted to comply with specific features of the domain of interest. The prototype of the DSS, created in accordance with the framework, supports decision makers in choosing the behaviour line (set of actions) in such a general case, which is potentially difficult to analyze and foresee. As for any complex system,the human choice is decisive. Second, in spite of our time consuming modeling work, we are looking forward to both revising and improving the system and deepening our research. Third, we consider making more experiments varying as with data structure, trying to apply the system to the other.

The framework supports all the necessary steps for standard decision making procedure by utilizing intelligent agents. The teams of intelligent agents, that are logically and functionally connected, have been presented. Real-time interaction with the user provides a range of possibilities in choosing one course of action from among several alternatives, which are generated by the system through guided data mining and computer simulation. The system is aimed to regular usage for adequate and effective management by responsible municipal and state government authorities.

We used both traditional data mining techniques, hybrid and specific methods, with respect to data nature (incomplete data, short data sets, etc.). Combination of different tools let us gain in quality and precision of the reached models, and, hence, in recommendations, which are based on these models. Received dependencies of interconnections and associations between the factors and dependent variables help to correct recommendations and avoid errors.

Acknowledgements

This work was partially supported by Spanish Ministerio de Ciencia e Innovacin TIN2007-67586-C02-02 and TIN2010-20845-C03-01 grants, and Junta de Comunidades de Castilla-La Mancha PII2I09-0069-0994 and PEII09-0054-9581 grants.

References

1. Athanasiadis, I.N., Mitkas, P.A.: An agent-based intelligent environmental monitoring system. CoRR, cs.MA/0407024 (2004)
2. Gorodetski, V.I., Karsaev, O., Samoilov, V., Konushy, V., Mankov, E., Malyshev, A.: Multiagent system development kit. In: Intelligent Information Processing (2004)
3. Haykin, S.: Neural Networks: A Comprehensive Foundation. Macmillan, New York (1998)
4. Karaca, F., Anil, I., Alagha, O., Camci, F.: Traffic related pm predictor for besiktas, turkey. In: Athanasiadis, I.N., Mitkas, P.A., Rizzoli, A.E., Gómez, J.M. (eds.) ITEE, pp. 317–330. Springer, Heidelberg (2009)
5. Levin, M.S.: Composite Systems Decisions (Decision Engineering). Springer, New York (2006)
6. Madala, H.R., Ivakhnenko, A. (eds.): Inductive Learning Algorithms for Complex Systems Modelling. CRC Press Inc., Boca Raton (1994)
7. Nastar, M., Wallman, P.: An interdisciplinary approach to resolving conflict in the water domain. In: Information Technologies in Environmental Engineering Proceedings of the 4th International ICSC Symposium Thessaloniki, Greece (2009)
8. Rechtin, E.: Systems architecting of organizations: why eagles can't swim. CRC Press, Boca Raton (1999)

9. Riaño, D., Sànchez-Marrè, M., Roda, I.R.: Autonomous agents architecture to supervise and control a wastewater treatment plant. In: Monostori, L., Váncza, J., Ali, M. (eds.) IEA/AIE 2001. LNCS (LNAI), vol. 2070, p. 804. Springer, Heidelberg (2001)
10. Rotmans, J.: Tools for integrated sustainability assessment: A two-track approach. Integrated Assessment 6(4) (2006)
11. Sokolova, M.V., Fernández-Caballero, A.: A multi-agent architecture for environmental impact assessment: Information fusion, data mining and decision making. In: ICEIS 2007 - Proceedings of the Ninth International Conference on Enterprise Information Systems, Funchal, Portugal, vol. AIDSS (2007)
12. Sokolova, M.V., Fernández-Caballero, A.: Facilitating mas complete life cycle through the protégé-prometheus approach. In: Agent and Multi-Agent Systems: Technologies and Applications, KES-AMSTA, Incheon, Korea (2008)
13. Sokolova, M.V., Fernández-Caballero, A.: Data mining driven decision making. In: Proceedings of the International Conference on Agents and Artificial Intelligence, ICAART 2009, Porto, Portugal (2009)
14. Bossomaier, T., Jarratt, D., Anver, M.M., Scott, T., Thompson, J.: Data integration in agent based modelling. Complexity International, 11 (2005)
15. Urbani, D., Delhom, M.: Water management policy selection using a decision support system based on a multi-agent system. In: AI*IA (2005)
16. Weiss, G. (ed.): Multiagent systems: a modern approach to distributed artificial intelligence. MIT Press, Cambridge (1999)

Author Index